W0107264

Cover design

Pictures on the cover were created by Jordi Vives Nebot (Dept. de Física Fonamental, Universitat de Barcelona, Diagonal 647, E-08028, Barcelona, Spain). They represent sections $x + y\mathbf{e}_{12} + z\mathbf{e}_1$ of two Julia-type 3-dimensional Clifford fractals viewed from an isometric perspective. Each image is a result of 10 iterations of the map $c \rightarrow c^2 + b$, where c and b are elements of the Clifford algebra $C\ell_{p,q}$, $p + q = 2$. Convergence of the resulting sequences of Clifford elements was determined by using a spinorial norm (the scalar part of the Clifford number times its Clifford-conjugate). The following metrics and base points b have been used to create these images:

- Plane Fractal (larger image): $b = -0.152815 + 0.656528\mathbf{e}_{12} \in C\ell_{2,0}$;

- Quaternionic Fractal (smaller image): $b = -0.152815 + 0.5\mathbf{e}_1 + 0.656528\mathbf{e}_{12} \in C\ell_{0,2}$.

The graphics have been produced on a 486 DX-100 at a resolution of 1024 x 768 pixels. It took 24 hours of computing time to create the Plane Fractal and 15 hours to create the Quaternionic Fractal.

Clifford Algebras with Numeric and Symbolic Computations

Rafał Abłamowicz
Pertti Lounesto
Josep M. Parra

Editors

1996

Birkhäuser
Boston • Basel • Berlin

Rafał Abłamowicz
Department of Mathematics
Gannon University
Erie, PA 16541
USA

Pertti Louesto
Helsinki University of Technology
02150 Espoo 15, Finland

Josep M. Parra
Department Física Fonamental
Facultat de Física
Universitat de Barcelona
E-08028 Barcelona, Spain

Printed on acid-free paper

Birkhäuser ⑬ ®

© 1996 Birkhäuser Boston
Softcover reprint of the hardcover 1st edition 1996

Copyright is not claimed for works of U.S. Government employees.
All rights reserved. No part of this publication may be reproduced, stored in a retrieval system, or transmitted, in any form or by any means, electronic, mechanical, photocopying, recording, or otherwise, without prior permission of the copyright owner.

Permission to photocopy for internal or personal use of specific clients is granted by Birkhäuser Boston for libraries and other users registered with the Copyright Clearance Center (CCC), provided that the base fee of $6.00 per copy, plus $0.20 per page is paid directly to CCC, 222 Rosewood Drive, Danvers, MA 01923, U.S.A. Special requests should be addressed directly to Birkhäuser Boston, 675 Massachusetts Avenue, Cambridge, MA 02139, U.S.A.

ISBN 978-1-4615-8159-8   ISBN 978-1-4615-8157-4 (eBook)
DOI 10.1007/978-1-4615-8157-4

Typeset by the Editors in TeX.

9 8 7 6 5 4 3 2 1

TABLE OF CONTENTS

Preface
Rafał Abłamowicz, Pertti Lounesto, and Josep M. Parra vii

List of Contributors xvi

1. VERIFYING AND FALSIFYING CONJECTURES 1
Counterexamples in Clifford algebras with CLICAL
Pertti Lounesto 3

2. DIFFERENTIAL GEOMETRY, QUANTUM MECHANICS, SPINORS
   AND CONFORMAL GROUP 31
The use of computer algebra and Clifford algebra in teaching
mathematical physics
Jayme Vaz, Jr. 33

General Clifford algebra and related differential geometry calculations
with MATHEMATICA
Josep M. Parra and Llorenç Roselló 57

Pauli-algebra calculations in MAPLE V
W. E. Baylis 69

The generative process of space-time and strong interaction quantum
numbers of orientation
Bernd Schmeikal 83

On a new basis for a generalized Clifford algebra and its application
to quantum mechanics
A. Granik and M. Ross 101

Vector continued fraction algorithms
D. E. Roberts 111

LUCY: A Clifford algebra approach to spinor calculus
Jörg Schray, Robin W. Tucker and Charles H.-T. Wang 121

Computer algebra in spinor calculations
Franco Piazzese 145

Vahlen matrices for non-definite metrics
J. Cnops 155

3. GENERALIZED CLIFFORD ALGEBRAS AND NUMBER SYSTEMS,
   PROJECTIVE GEOMETRY AND CRYSTALLOGRAPHY 165
On Clifford algebras of a bilinear form with an antisymmetric part
Rafał Abłamowicz and Pertti Lounesto 167

A unipodal algebra package for MATHEMATICA
Garret Sobczyk 189

Octonion X-product orbits
Geoffrey Dixon 201

A commutative hypercomplex algebra with associated function theory
Clyde M. Davenport 213

On generalized Clifford algebras – recent applications
W. Bajguz and A. K. Kwaśniewski 229

Oriented projective geometry with Clifford algebra
Richard C. Pappas 233

The applications of Clifford algebras to crystallography using
MATHEMATICA
A. Gómez, J. L. Aragón, O. Caballero, and F. Dávila 251

4. NUMERICAL METHODS IN CLIFFORD ALGEBRAS 267
Orthonormal basis sets in Clifford algebras
G. Bergdolt 269

Complex conjugation - relative to what?
Alexander Soiguine 285

Object-oriented implementations of Clifford algebras in C++: a prototype
Arvind Raja 295

INDEX 317

PREFACE

> The human mind is seldom satisfied, and is certainly never exercising its highest functions, when it is doing the work of a calculating machine. What the man of science, whether he is a mathematician or a physical inquirer, aims at is, to acquire and develope clear ideas of the things he deals with. For this purpose he is willing to enter on long calculations, and to be for a season a calculating machine, if he can only at last make his ideas clearer. — James Clerk Maxwell, 1870. [1]

These words of James Clerk Maxwell, who in September of 1870 gave a Presidential Address at a meeting of Mathematical and Physical Sections of the British Association, fully express and justify our goals in planning and assembling this contributed volume. Although we often indeed become "for a season" (or more) these "calculating machines", we look more and more into those marvelous time-saving machines we call computers as our new allies capable of symbolic, geometric, and algebraic methods. It is the intelligent use of computers that has given scientists more freedom, more time and more insight in their highest task of developing a scientific knowledge of Nature.

Computer algebra systems such as AXIOM, CAYLEY, CLICAL, DERIVE, MAC-SYMA, MAPLE, MATHEMATICA, MATLAB, and REDUCE have contributed considerably to mathematical research in the past few years. In addition to encouraging experimentation, their greatest significance has been in providing a common language and approach to a variety of mathematical problems. This way, they have contributed enormously to mutual understanding of people with different scientific backgrounds.

We need to realize that we live in an "exceptional era" in which the new computer tools not previously known to mathematicians and physicists are now available. Cartan's classification of real simple Lie algebras derived at the beginning of this century required an enormous amount of computation and time. Today, computations of that complexity would only take a few minutes of the computer time. Thus, it is important to develop confidence with computers doing numeric and symbolic computations in lieu of our becoming "calculating machines." Computers can be used for the following tasks:

- interactive and creative experimentation in creating new knowledge;
- verification of one's own and of others' hypotheses;
- falsification of conjectures, theorems, theories, paradigms, etc., by counter-examples;

---

[1] James Clerk Maxwell, 'Address to the Mathematical and Physical Sections of the British Association,' Liverpool, September 15, 1870, in: *The Scientific Papers of James Clerk Maxwell*, Vol. 2, ed. W. D. Niven, Dover Publications, Inc., New York, 1965, p. 219, lines 5 – 10.

- verification and selection of right alternatives by elimination of implausible alternatives.

However, it is very important to know the limitations of the given computer program being used. Programs that have been used, some successfully and some unsuccessfully, to perform computations with Clifford algebras can be generally divided into three groups:

- numeric, such as FORTRAN, C++, PASCAL (internal language of CLICAL), muSIMP (a function oriented language derived from LISP, an internal language of muMATH);

- semi-symbolic, such as CLICAL;

- symbolic, such as MACSYMA, MAPLE, MATHEMATICA, DERIVE (previously known as muMATH), AXIOM (previously known as SCRATCHPAD), MATLAB, REDUCE.

For example, while MATLAB (Marcus, 1993) is used by physicists in quantum mechanics for matrix computations, the multivector approach offered by CLICAL (until recently the only computer program capable of such an approach) is much better precisely because instead of matrices it uses the much more efficient formalism of Clifford algebras. CLICAL has been used successfully by several of our contributors, namely Lounesto, Schmeikal, Pappas, and Soguine. Anthony Hearn's REDUCE (Hearn, 1968) was created in an unsuccessful attempt to find symbolic solutions to the Dirac equation with different potentials and for scattering computations with contracted Dirac matrices (Hearn, 1971). Later REDUCE was used in Gent, Belgium, for symbolic computations with Clifford algebras. Use of REDUCE in teaching mathematical physics is discussed in our volume by Jayme Vaz. For historical accuracy, we mention that the first computer algebra system which could actually run on a PC was muMATH created by Albert Rich and David Stoutemyer (Wooff and Hodgkinson, 1987; Freese et al., 1986), a precursor of DERIVE (Rich et al., 1989). Another group in Namur, Belgium, has used MACSYMA for symbolic computations with Clifford algebras. None of these programs, muMATH, DERIVE and MACSYMA, is featured in this volume.

A Gröbner basis is a basis for an ideal generated by a given set of so called "distributed multivariate polynomials" (Geddes et al., 1993). The basis is generally not unique since it may depend on the order in which variables are specified. Gröbner bases facilitate computations with multivariate polynomials, such as deciding if the given algebraic system of polynomial equations has a solution and, if so, solving it. Their introduction was very important for the development of symbolic programs such as MACSYMA (see first chapter of Davenport et al., 1988), AXIOM (with its precursor SCRATCHPAD), MATHEMATICA (with its precursor SMP computer algebra system), and MAPLE, capable of handling polynomials, solving systems of linear and non-linear equations, solving differential equations, etc. SCRATCHPAD, created by IBM researchers in the mid-70's, has developed into AXIOM, which offers a categorical approach to computer mathematics and contains a Clifford algebra domain for symbolic computations with these algebras (Jenks and Sutor, 1992). For a review of computer algorithms used in MACSYMA, muMATH, REDUCE, and SCRATCHPAD see (Davenport *et al.*, 1988). For a brief history of computer algebra systems see (Geddes *et al.*, 1993, pages 1 – 10).

The two symbolic computer algebra systems used by most of our contributors are MAPLE and MATHEMATICA. The development of MAPLE started in 1980. An important property of the system is that most of its algebraic facilities are implemented in its high-level language which can also be used as a programming language by the user (Char *et al.*, 1992; Geddes *et al.*, 1993). Several of our contributors, namely Baylis, Schray *et al.*, and Abłamowicz, have benefited from that feature by creating extensive packages for Clifford algebra symbolic computations with MAPLE. MATHEMATICA, announced by Stephen Wolfram in 1988, also offers a high-level programming language (Wolfram, 1991). Parra and Roselló, Sobczyk, Gómez *et al.* have used MATHEMATICA in such diverse areas as mathematical physics and its teaching, unipodal number systems, and crystallography. Finally, two of our contributors, Bergdolt and Raja, opted for the numerical languages FOR-TRAN and C++, respectively.

Clifford algebras are at a crossing point in a variety of research areas, including abstract algebra, crystallography, projective geometry, quantum mechanics, differential geometry and complex analysis. For many researchers working with these algebras, the computer algebra systems have become an indispensable tool in a discovery of new knowledge, in gaining a better understanding of the existing theory and its applications, and in the classroom teaching mathematical physics in the formidable language of Clifford algebras.

In organizing our volume we have given priority to:

— manuscripts describing results on Clifford algebras which have been obtained with one or more of the above computer systems, including original packages written by our contributors;

— manuscripts summarizing contributors' own experiences in using one or more of these systems in Clifford algebra teaching;

— articles of high scientific quality which would be of interest to Clifford algebra researchers and those wishing to learn about Clifford algebras and their applications in various fields.

In the following section, we briefly review each contribution in an effort to guide the reader through the theory of Clifford algebras and some of its applications, and some new developments in both areas afforded by the use of computers.

Chapter 1 contains a contribution from Pertti Lounesto, and it accomplishes several goals. First, it teaches us the need for critical reading of scientific literature not only because authors and referees make mistakes but mainly as a path to deeper understanding and creative thinking. Second, it proves beyond any doubt that well-designed computer programs are most useful – we dare to say *vital*, due to the present-day complexity of mathematical science – for that critical learning task. They make possible a continuous exchange between general abstract theory and specific models or examples which is so essential in keeping science alive. James Clerk Maxwell publicly said it in an unmistakable way:

There are men who, when any relation or law, however complex, is put before them in a symbolical form, can grasp its full meaning as a relation among abstract quantities. Such men sometimes treat with indifference the further statement that quantities actually exist in nature which fulfil this relation. The mental image of the concrete reality seems rather to disturb than to assist their contemplations.

But the great majority of mankind are utterly unable, without long training, to retain in their minds the unembodied symbols of the pure mathematician, so that, if science is ever to become popular, and yet remain scientific, it must be by a profound study and copious application of those principles of the mathematical classification of quantities which, as we have seen, lie at the root of every truly scientific illustration.
— James Clerk Maxwell, 1870. [2]

Third, it vindicates the falsifying strategy whenever there is any doubt about the validity of any proposition (and, most interestingly, when there seems to be no doubt at all in the minds of most). As a large sample of counter-examples shows, this is not at all a destructive task. Exactly the opposite is true: by recognizing where our present knowledge fails us, we are given an opportunity to correct it, extend it, and to select a better strategy for advancing it. Fourth, it will qualify the tenacious if not advanced reader as a "Clifford algebra critical expert" thanks to its wide spectrum of topics, such as spinor norm, intricacies of the covering groups and general scalar spinor products to the Dirac theory of electron, a relationship between Clifford and exterior algebras, as well as the method of exposition that requires a direct confrontation with well-documented sources. This accomplished, the following chapters offer a wide arena for critical learning, and, due to the open nature of this contributed monograph, also for a public debate. A newcomer to the field should not mistake Chapter 1 for an elementary introduction to Clifford algebras and should be prepared to postpone reading until tools offered in the remaining chapters or in Crumeyrolle (1990) are mastered.

Chapter 2 presents those aspects of Clifford algebras that have established them as an essential part of what has been called *physical mathematics*. The use of Clifford geometric algebra facilitates formulation of fundamental physical laws and brings better understanding to physics, at least for those of us who prefer geometrical images or "illustrations" when connecting mathematical abstraction with physical reality.

Jayme Vaz presents rotations and Lorentz transformations, electromagnetism, the Dirac equation, etc., in the language of Clifford algebras which has become a well-established practice in physics. Through a use of REDUCE packages, we are also led to interesting applications of Clifford algebras in differential geometry and gauge theories. We note here that REDUCE allows an extremely simple definition of (and manipulation with) Clifford algebras.

Parra and Roselló contribute a MATHEMATICA package for algebraic and differential geometry. Although specially devised for physicists' use, the package allows for arbitrary dimension, signature, and orthogonal curvilinear coordinates. Flexibility of its input/output facilities and simplicity of expression have made it an ideal

---

[2] James Clerk Maxwell, 'Address to the Mathematical and Physical Sections of the British Association,' Liverpool, September 15, 1870, *ibid.*, pp. 219–220, lines -4 – 8.

teaching aid. To this 'core program', the authors have added, in separate files, a set of selected applications in three and four dimensions as well as a numerical implementation in MATLAB.

Baylis gives us a MAPLE V implementation of the most basic geometric algebra, the one that corresponds to Euclidean three-space. His approach puts an emphasis on an isomorphism between the algebra and Pauli matrix algebra widely used in physics. Complex paravectors, viewed as complexified quaternions, are introduced in order to represent general elements of the algebra. Surprising relativistic applications of this algebra are most naturally explained by the algebra isomorphisms $Cl_{3,0} \cong Cl_{3,1}^{+} \cong Cl_{1,3}^{+}$.

Schmeikal contributes a theoretical paper on fundamentals of elementary particle theory. Much in the same way as in the Spin Gauge theory, a full set of inner quantum numbers with orientation properties is introduced and analyzed in the language of a geometric real Clifford algebra. In this formulation, quarks are viewed as quantized states of orientation and the Gell-Mann–Nishijima relation is derived from geometric calculus. All computations in this paper have been done with the numeric-symbolic calculator CLICAL.

Granik gives us an abstract operator equivalent of all the usual matrix calculus based on the SU(2) group that plays an essential role in quantum theory of angular momentum and in the now fashionable field of quantum computing. Since this operational form is explicitly derived from a *generalized Clifford algebra* for the Euclidean plane, it may lead to new geometrical interpretations as well as to a similar analysis for general SU(n) groups.

Roberts contributes an unusual application for Clifford algebras: Padé approximants for vector-valued complex functions defined by power-series with real vector coefficients. His use of Clifford algebra in arbitrary dimension improves the theory in providing a rationale for vector division as well as new algorithms for an effective calculus of the approximants. The paper opens a path to further work on computer implementation of the algorithms and on their relationship to Clifford analysis.

Schray, Tucker and Wang present LUCY, a complete package written in MAPLE, for Clifford algebra differential calculus even with a non-orthogonal set of coordinates. A well-known connection between Clifford algebras and their matrix representation is carefully elaborated. Spinors are discussed in depth, including the covariant spinor derivative fundamental to any discussion of the wave equation in curved space-times. LUCY provides a good help to those beginning their study of spinor fields since it is effective with spinor calculus due to its well-chosen notation.

Piazzese's contribution can easily be put into correspondence with Baylis', including his use of MAPLE. The Clifford algebra $Cl_{3,0}$ is relied on to explain the appearance of spinor quantities in a description of physical systems. A basic geometrical and 'non-quantum' character of spinors is made evident through a detailed treatment of rotations and the Kepler problem.

Cnops' contribution begins with a clear and concise presentation, in Clifford algebra terms, of a relationship among Möbius transformations, orthogonal groups, and Lipschitz group. Then, he states and proves a theorem on a form of a general element in Lipschitz group. It turns out to be a Vahlen-type matrix, originally discovered by Vahlen in 1902 in his study of special conformal transformations. Its relevance to physics follows naturally from the importance of the transformation groups considered in the paper.

In Chapter 3 we have collected contributions on generalized Clifford algebras, commutative Clifford algebras, octonions, and very promising new applications of the Clifford algebra formalism to crystallography of quasicrystals and projective geometry.

Abłamowicz and Lounesto investigate properties of a Clifford algebra $C\ell(B)$ of a bilinear form B with an antisymmetric part A. They explicitly show how the operations of reversion and contraction in $C\ell(B)$ depend on A. In particular, they show that if $A \neq 0$, then reversion mixes grades of $\bigwedge V$ and introduces a new kind of A-dependent gradation in $C\ell(B)$. The authors clarify the nature of the isomorphism $C\ell(Q) \simeq C\ell(B)$ where Q is the quadratic form uniquely determined by the symmetric part of B (in characteristic $\neq 2$.) As an example, it is shown how this isomorphism may be constructed in dimension 3 and then extended to all dimensions ≤ 9. All computations in $C\ell(B)$ are performed with an original MAPLE V package 'Clifford' developed by Abłamowicz specifically for the symbolic computations with Clifford algebras of the arbitrary bilinear form.

Sobczyk gives us his MATHEMATICA package for computations with a unipodal number system defined as an algebraic extension of the complex field $\mathbb{C}$ by a "unipotent" element u such that $u^2 = 1$ but $u \neq \pm 1$. The resulting algebra is an example of a *commutative* Clifford algebra. Entire functions of a complex variable are extended to entire functions of a unipodal variable. As an application, Sobczyk shows how to solve a classical cubic equation with unipodal numbers and how to solve a structure equation of an arbitrary Clifford number once its minimal polynomial is known.

Dixon presents an analysis of various bases and the resulting multiplication tables in the real octonion division algebra $\mathbb{O}$. The new bases are obtained by permutation of the indices in any orthogonal basis representing pure octonions. There are only 480 distinct multiplication tables for $\mathbb{O}$, of which 4 are singled out "for their elegance and symmetry". These distinct products can be defined on $\mathbb{O}$ via a certain deformation of the standard octonion product called by the author an "X-product" Each X-product is parameterized by a unit octonion $X \in \mathbb{O}$ and gives rise to an isomorphic copy of $\mathbb{O}$. A connection is made between the X-products and permissible quaternionic index triples. Orbits of various copies of $\mathbb{O}$ are explicitly calculated with a spreadsheet.

Davenport develops a function theory for the double algebra of the complex field, ($^2\mathbb{C}$ in Porteous' notation), in such a way that it formally looks like an ordinary vector algebra. The commutativity of the structure allows straightforward generalization of the analysis in one complex variable, but its applicability in Physics may be rather limited. However, due to an algebraic isomorphism with certain subalgebras of the geometric algebras, these applications cannot be excluded.

Bajguz and Kwaśniewski provide us with a concise review of current applications of Clifford algebras and their generalizations to k-ubic forms, quasi-number systems, spin lattice systems, $\mathbb{Z}_k$-Potts models, finite-dimensional quantum mechanics and its proposed use in digital image processing.

Pappas discusses formulation of projective geometry $\mathbf{P}^{n-1}$ in the language of Geometric Algebra developed by Hestenes and Sobczyk. In that language, concepts of a point, a line, a plane, etc., acquire clear algebraic interpretation in terms of equivalence classes of vectors, bivectors, trivectors, etc. While the operations of duality, meet, and join are expressed in terms of algebraic operations in the algebra. An argument is made for usefulness of "oriented" projective geometry which allows one to solve some betweenness problems related to the non-orientability of classical projective geometry. Projective transformations on oriented spaces are defined and discussed. Examples of numerical computations done with CLICAL are provided.

The next paper by Gómez *et al.* is related to Pappas' in that it makes an extensive use of projective geometry concepts like that of a flat, which is an oriented subspace of a real oriented projective space, and Plücker coordinates. It gives us a new and fascinating application of Clifford algebras to the theory of quasicrystals defined as "... metallic alloys whose diffraction patterns exhibit sharp spots but non-crystallographic symmetry". This means that the underlying structure is non-periodic: it is "quasiperiodic" while the underlying lattice is then referred to as "quasilattice". Since a d-dimensional quasilattice may be regarded as a projection of a certain subset of a lattice in an n-dimensional space $\mathbb{R}^n$, $(n > d)$, there is an immediate need to do geometry in spaces of dimension higher than 3, including non-Euclidean spaces. For example, structures of "metallic glasses" and "fullerenes" require that geometry of hyperbolic spaces be considered. In order to describe morphology of quasicrystals, an algebra capable of expressing geometric relationships in such spaces is needed: the choice is of course to use Clifford algebras. A MATHEMATICA package is presented for solving, with these algebras, problems of faceting and phason degrees of freedom in quasicrystals.

In Chapter 4 we have collected papers relying on numerical computations with FORTRAN and C++.

Bergdolt presents a discussion of the so called "orthonormal basis sets" in Clifford algebras which are defined as sets of multivectors satisfying "scalar product relations". For example, since the fundamental identity $e_i e_j + e_j e_i = 2g_{ij}1$ in $Cl_{p,q}$ is satisfied by any orthonormal set of basis vectors $\{e_1, e_2, \ldots, e_n\}$ in $\mathbb{R}^{p,q}$, $p + q = n$, it is a "product relation" which defines that set as one "orthonormal basis set". However, there might be other sets of multivectors in $Cl_{p,q}$ which satisfy this relation and if so, they define a Clifford map from $\mathbb{R}^{p,q}$ to some associative algebra A. If A happens to be another Clifford algebra $Cl_{p',q'}$, then this Clifford map may be extended to an isomorphism $Cl_{p,q} \simeq Cl_{p',q'}$ of Clifford algebras. In his paper, Bergdolt uses FORTRAN to compute all possible orthonormal basis sets when $n = 4$.

Soiguine discusses complex conjugation as an operation in the Geometric Algebra of a Euclidean plane which preserves its even subalgebra. Then, he extends the discussion to the Geometric Algebra of three-dimensional space. In Section 4, he reviews geometric features of CLICAL and proposes development of a "geometrical/graphical" interface that would allow the user to visualize corresponding objects being manipulated with CLICAL. It seems that C++, with its object-oriented features, is most suitable for that task. Some C++ code related to computations with complex numbers is shown.

Raja presents an attempt to reformulate the CLICAL library written in PASCAL into C++. The purpose of this effort is to lay a foundation for a new interactive program which is presumably going to include visualization of geometric aspects of Clifford algebras (similar idea to Soiguine's). At present, fundamental operations with elements of Clifford algebras such as Clifford and exterior products, outer powers, octonionic product, and outer exponentiation of bivectors can be performed. Emphasis is placed on object-oriented features of the language.

We close this introduction as we opened it. In his Presidential Address, Maxwell had the insight to predict the following:

> Time would fail me if I were to attempt to illustrate by examples the scientific value of the classification of the quantities. I shall only mention the name of that important class of magnitudes having direction in space which Hamilton has called vectors, and which form the subject-matter of the Calculus of Quaternions, a branch of mathematics which, when it shall have been thoroughly understood by men of the illustrative type, and clothed by them with physical imagery, will become, perhaps under some new name, a most powerful method of communicating truly scientific knowledge to persons apparently devoid of the calculating spirit. — James Clerk Maxwell, 1870.[3]

The year before Maxwell's death in 1879, William Kingdon Clifford put into its proper place the Calculus of Quaternions, the even subalgebra of the Euclidean Geometric Algebra. This momentous synthesis of the works of Hamilton and Grassmann established a new language for a classification of the physico-geometrical magnitudes mentioned by Maxwell. When later extended to other signatures, dimensions, and fields, it has become to be known as the theory of Clifford algebras. The importance and applicability of the theory to various branches of science has been understood only in very recent times by a still small minority of scientists, some of whom have contributed to this volume. Without any hesitation, we may declare that computer-assisted Clifford algebra geometric tools provide one of the "most powerful method of communicating truly scientific knowledge to persons apparently devoid of the calculating spirit".

Note to the Reader

Selected electronic materials submitted by our contributors can be found at a Web site:

http://www.birkhauser.com/books/ISBN/0-8176-3907-1.

These materials contain original packages, worksheets, notebooks, computer programs, etc., that were used in deriving results presented in this book. They have been carefully organized into groups following the Table of Contents. Each group of files contains a `readme.txt` file with a brief description of each file in the group and additional instructions how to use them. The reader is encouraged to browse through the Web site, download files of interest, and then check computations presented in the articles.

---

[3] James Clerk Maxwell, 'Address to the Mathematical and Physical Sections of the British Association,' Liverpool, September 15, 1870, *ibid.*, p. 220, lines -11 – -3.

References

B. W. Char, K. O. Geddes, G. H. Gonnet, B. L. Leong, M. B. Monagan, S. M. Watt: 1992, *MAPLE V: Library Reference Manual*, 2nd printing, Springer Verlag, New York.

A. Crumeyrolle: 1990, *Orthogonal and Symplectic Clifford Algebras: Spinor Structures*, Kluwer, Dordrecht.

J. H. Davenport, Y. Siret, and E. Tournier: 1988, *Computer Algebra: Systems and Algorithms for Algebraic Computation*, Academic Press, New York (see Chapter 1 devoted primarily to MACSYMA).

R. Freese, P. Lounesto, and D. A. Stegenga: 1986, 'The use of muMATH in the calculus classroom,' *J. Computers in Math. and Sci. Teaching* 6 (1), pp. 52 – 55.

K. O. Geddes, S. R. Czapor, and G. Labahn: 1993, *Algorithms for Computer Algebra*, Kluwer Academic Publishers, Boston.

A. C. Hearn: 1968, 'REDUCE: a user-oriented interactive system for algebraic simplification', in *Interactive Systems for Experimental Applied Mathematics*, eds. M. Klerer and J. Reinfelds, Academic Press, New York, pp. 79 – 90.

A. C. Hearn: 1971, 'Applications of symbol manipulation in theoretical physics,' *Communication of ACM* **14**, pp. 511 – 516. 'Calculation of traces of products of gamma matrices,' in "Proc. of the Second Colloquium on Advanced Computing Methods in Theoretical Physics, C.N.R.S., Marseilles", pp. I-30 – I-44.

R. D. Jenks and R. S. Sutor: 1992, *AXIOM: The Scientific Computation System*, Springer-Verlag, New York.

P. Lounesto, R. Mikkola, and V. Vierros: 1987, 'CLICAL User Manual', Helsinki University of Technology, Institute of Mathematics, *Research Reports* **A248**, Helsinki.

M. Marcus: 1993, *Matrices and MATLAB: A Tutorial*, Prentice Hall, Englewood Cliffs, New Jersey.

A. Rich, J. Rich, and D. Stoutemyer: 1989, *DERIVE – A Mathematical Assistant*, Soft Warehouse, Inc., Honolulu.

S. Wolfram: 1991, *MATHEMATICA: A System for Doing Mathematics by Computer*, 2nd ed., Addison-Wesley Publishing Company, Redwood City, California.

C. Wooff and D. Hogkinson: 1987, *muMATH: A Microcomputer Algebra System*, Academic Press, London.

It is our pleasure to thank Lauren E. Lavery and Wayne Yuhasz from Birkhäuser for their support in this project and for preparing the Web site.

March 1996   *Rafał Abłamowicz*   *Pertti Lounesto*   *Josep M. Parra*

LIST OF CONTRIBUTORS

Rafał Abłamowicz
Department of Mathematics
Gannon University
Erie, PA 16541, U.S.A.
e-mail: ablamowicz@cluster.gannon.edu

J. L. Aragón
Instituto de Física
Universidad Nacional Autónoma de México
Apartado Postal 20-364
01000 México, Distrito Federal, México
e-mail: aragon@ifunam.ifisicacu.unam.mx

W. Bajguz
Institute of Physics
Warsaw University Campus Białystok
ul. Przytorowa 2 A
15–104 Białystok, Poland

W. E. Baylis
Department of Physics
University of Windsor
Windsor, Ontario, Canada N9B 3P4
e-mail: baylis@uwindsor.ca

G. Bergdolt
Centre de Recherches Nucleaires, C.N.R.S.
B.P.28
F-67037 Strasbourg Cedex-2, France
e-mail: BERGDOLT@frcpn11.in2p3.fr

O. Caballero
Instituto de Física
Universidad Nacional Autónoma de México
Apartado Postal 20-364
01000 México, Distrito Federal, México

J. Cnops
Universiteit Gent
Galglaan 2, B-9000 Gent, Belgium
e-mail: jc@cage.rug.ac.be

Clyde M. Davenport
4124 Guinn Road
Knoxville, TN 37931
e-mail: cmdaven@use.usit.net

F. Dávila
Departamento de Matematicas
Escuela Superior de Física y
Matemáticas-I.P.N.
U.P. Adolfo López Mateos, Edificio 9
07300 México, Distrito Federal, México

Geoffrey Dixon
Department of Mathematics or Physics
Brandeis University
Waltham, MA 02254
e-mail: dixon@binah.cc.brandeis.edu

A. Gómez
Instituto de Física
Universidad Nacional Autónoma de México
Apartado Postal 20-364
01000 México, Distrito Federal, México

A. Granik
Physics Department
University of the Pacific
Stockton, CA 95211
e-mail: galois@ix.netcom.com

A. K. Kwaśniewski
Institute of Physics
Warsaw University Campus Białystok
ul. Przytorowa 2 A
15–104 Białystok, Poland
e-mail: Kwandr@cksr.ac.bialystok.pl

Pertti Lounesto
Institute of Mathematics
Helsinki University of Technology
FIN-02150 Espoo, Finland
e-mail: lounesto@dopey.hut.fi

Richard C. Pappas
Department of Mathematics
Widener University
Chester, PA 19013
e-mail: pappas@kuratowski.math.widener.edu

Josep M. Parra
Laboratori de Física Matemàtica, I.E.C.
Dept. de Física Fonamental, U. de Barcelona
Diagonal 647, E-08028, Barcelona, Spain
e-mail: jmparra@hermes.ffn.ub.es

Franco Piazzese
Department of Physics, Politecnico
Corso Duca degli Abruzzi 24
10129 Torino, Italy
e-mail: PIAZZESE@polfis.polito.it

Arvind Raja
Institute of Mathematics
Helsinki University of Technology (HUT)
Otakaari 1, FIN-02150 Espoo, Finland
e-mail: Arvind.Raja@hut.fi

D. E. Roberts
Department of Mathematics
Napier University
219 Colinton Road
Edinburgh, EH14 1DJ
e-mail: davidr@maths.napier.ac.uk

Llorenç Roselló
Dept. L.S.I.
U. Politècnica de Catalunya
Pau Gargallo 5, E-08028, Barcelona, Spain
e-mail: lrosello@goliat.upc.es

M. Ross
Physics Department
University of the Pacific
Stockton, CA 95211
e-mail: galois@ix.netcom.com

Bernd Schmeikal
Biofield Laboratory – ICBHR
Kundmanngasse 26/8
A-1030 Vienna, Austria
e-mail: schmeika@isis.wu-vien.ac.at

Jörg Schray
School of Physics and Chemistry
Lancaster University
Lancaster LA1 4YB, UK
e-mail: j.schray@lancaster.ac.uk

Garret Sobczyk
Universidad de las Americas
Departamento de Fisico-Matematicas
Apartado Postal #100, Santa Catarina Mártir
72820 Puebla, Pue., México
e-mail: sobczyk@udlapvms.pue.udlap.mx

Alexander M. Soiguine
191011 St. Petersburg
Fontanka 53 #37
Russia

Robin W. Tucker
School of Physics and Chemistry
Lancaster University
Lancaster LA1 4YB, UK
e-mail: r.tucker@lancaster.ac.uk

Jayme Vaz, Jr.
Department of Applied Mathematics – IMECC
State University at Campinas (UNICAMP)
CP 6065, 13081-970 Campinas, S.P., Brazil
e-mail: vaz@ime.unicamp.br

Charles H.-T. Wang
School of Physics and Chemistry
Lancaster University
Lancaster LA1 4YB, UK
e-mail: c.wang@lancaster.ac.uk

1

Verifying and Falsifying Conjectures

COUNTEREXAMPLES IN CLIFFORD ALGEBRAS WITH CLICAL

PERTTI LOUNESTO
Institute of Mathematics
Helsinki University of Technology
FIN-02150 Espoo, Finland
e-mail: Pertti.Lounesto@hut.fi

Abstract. This article presents a collection of counter-examples to conjectures about Clifford algebras, spinors, spin groups and the exterior algebra. In research, counter-examples serve as indicators of a wrong way. They tell one where not to go in a new domain to be explored. The modern era is exceptional in the history of mathematics: for the first time we can use computers to test the validity of our conjectures. The counter-examples presented in this article have been mostly discovered with CLICAL, a computer program specifically designed for Clifford algebra calculations.

Key words: Clifford algebra – spin groups – spinors – counter-examples

1. Progress in science via counter-examples

Ideally, scientists publish papers for the purpose of testing and evaluating their ideas in public scrutiny. This ideal has been obscured by the peer review/refereeing system, which pretends to guarantee correctness of ideas – prior to a public scrutiny, and the tendency to publish in order to get a position in academia. Traditionally, science has progressed through public debates about new ideas: statements, counter-examples, refined statements and new counter-examples, etc.

In mathematics, proving theorems, finding gaps and errors in the proofs, correcting the theorems, detecting errors in the corrected theorems, etc. is a normal activity. This is even more so in advanced mathematics because our cognitive charts are less accurate in new frontiers of knowledge. See Lakatos 1976.

In evaluating the validity of a mathematical theorem, one should either check every detail of its proof or point out a flaw in the chain of deductions or line of thoughts. After a counter-example has been presented, it is often easier to settle whether it fulfills all the assumptions than to check all the details of the proof. As in science, also in mathematics we are faced with the fact that a single counter-example can falsify a theorem or a whole theory. See Popper 1972.

What follows is a collection of counter-examples to some published statements about Clifford algebras. I have paid special attention to interpreting the texts in the way the authors have intended so that my counter-examples reveal interior inconsistencies.

Informing colleagues about their own errors is more subtle. It offers them an opportunity to learn more mathematics. It might also result in a feeling of insufficiency, a cognitive conflict, and instigate a learning process, and thus indirectly lead into the same final situation, namely cognitive growth. See Ginsburg & Opper 1988.

In order to benefit from mathematical arguments presented in this paper, the reader should have references at his possession, and he should follow the reasoning of the counter-examples line by line.

2. Description of CLICAL

The counter-examples presented in this article have been mostly discovered with CLICAL, a computer program specifically designed for Clifford algebra calculations. CLICAL deals with non-degenerate real Clifford algebras. It uses the multivector structure, and elements are keyed in by means of an orthonormal basis. CLICAL can evaluate elementary functions with variables in Clifford algebras, like the exponential, square root and the inverse function. The user can call in his own functions, and write external files to be executed step by step.

Quite a few of the counter-examples stem from the failure of the authors to check their statements in low dimensions, or for low values of indices.

3. First counter-examples

Consider the Clifford algebra $Cl_{3,1} \simeq \mathrm{Mat}(4, \mathbb{R})$ of the Minkowski space-time $\mathbb{R}^{3,1}$. Take an element

$$a = (1 + e_1)(1 + e_{234}) = 1 + e_1 + e_{234} + e_{1234},$$

which has the Clifford-conjugate

$$\bar{a} = (1 + e_{234})(1 - e_1) = 1 - e_1 + e_{234} + e_{1234}.$$

Compute the products of a and $\bar{a}$ in different orders to find

$$a\bar{a} = 4(e_{234} + e_{1234}),$$
$$\bar{a}a = 0.$$

This serves as a counter-example to Harvey 1990, Lemma 10.45, (c,d), p. 202, ll. 4-5, since $\bar{a}a = 0 \in \mathbb{R}$ but $a\bar{a} = 4(e_{234} + e_{1234}) \notin \mathbb{R}$. [1] Gilbert & Murray 1991 denote $\Delta(x) = \bar{x}x$ and prove in Theorem 5.16 that for x such that $\Delta(x) \in \mathbb{R}$ it necessarily follows that $\Delta(\bar{x}) = \Delta(x)$ [p. 41, l. 19] and in particular that $\Delta(x) = 0$ forces $\Delta(\bar{x}) = 0$ [p. 42, ll. 2-3]. Choose $x = a$ to find $\Delta(a) = 0 \in \mathbb{R}$ although $\Delta(\bar{a}) = \bar{\bar{a}}\bar{a} = 4(e_{234} + e_{1234}) \neq 0$, and you have a counter-example to Gilbert & Murray. [2] The element $\bar{a}$ also serves as a counter-example to Knus 1991, p. 228, l. 13, since $\bar{\bar{a}}\bar{a} = 4(e_{234} + e_{1234}) \notin Cl_{3,1}^+$, although a simpler counter-example is $x = e_1 + e_{23} \in Cl_3 \simeq \mathrm{Mat}(2, \mathbb{C})$ for which $\bar{x}x = -2e_{123} \notin Cl_3^+$. [3] In particular, $\bar{x}x = -2e_{123} \notin \mathbb{R}$ for $x = e_1 + e_{23} \in Cl_3$, and we have a counter-example to Dabrowski 1988, p. 7, l. 12, who observed his error [see the errata sheet distributed along with his monograph].

---

[1] Harvey introduces the Clifford-conjugation $\bar{a}$ on p. 183; he calls it a hat involution and denotes by $\hat{a}$.

[2] Gilbert & Murray's conjugation is the Clifford-conjugation, see p. 17.

[3] Knus introduces the Clifford-conjugation $\bar{x}$ on p. 195; he calls it standard involution $\sigma(x)$.

In the Clifford algebra $C\ell_3$ of the Euclidean space $\mathbb{R}^3$ there are elements whose exponentials are vectors, like $\mathbf{e}_3 = \exp[\pm\frac{\pi}{2}(\mathbf{e}_{12} - \mathbf{e}_{123})]$. Therefore, the multivalued inverse of the exponential satisfies

$$\log \mathbf{e}_3 = \pm\frac{\pi}{2}(\mathbf{e}_{12} - \mathbf{e}_{123}).$$

This shows that also vectors can have logarithms in a Clifford algebra, and serves as a counter-example to Hestenes 1986, p. 75 [I showed this counter-example to the author, who corrected his error in the next edition of 1987]. Two pages earlier Hestenes 1986/87, p. 73, is mistaken, when he claims that

$$e^A e^B = e^{A+B}$$

holds if and only if $AB = BA$. Clearly, $AB = BA$ implies $e^A e^B = e^{A+B}$, but the converse is not true, as can be seen by the counter-example

$$A = 3\pi i, \quad B = 4\pi j,$$

where i, j are unit quaternions in $\mathbb{H}$.

4. Counter-examples about spin groups

The Lipschitz group $\boldsymbol{\Gamma}_{p,q}$, also called the Clifford group although invented by Lipschitz 1880/86, could be defined as the subgroup in $C\ell_{p,q}$ generated by invertible vectors $\mathbf{x} \in \mathbb{R}^{p,q}$, or equivalently by either of the following ways

$$\boldsymbol{\Gamma}_{p,q} = \{s \in C\ell_{p,q} \mid \forall \mathbf{x} \in \mathbb{R}^{p,q}, \; s\mathbf{x}\hat{s}^{-1} \in \mathbb{R}^{p,q}\},$$
$$\boldsymbol{\Gamma}_{p,q} = \{s \in C\ell_{p,q}^+ \cup C\ell_{p,q}^- \mid \forall \mathbf{x} \in \mathbb{R}^{p,q}, \; s\mathbf{x}s^{-1} \in \mathbb{R}^{p,q}\}.$$

Note the presence of the grade involution $s \to \hat{s}$, and/or restriction to the even/odd parts $C\ell_{p,q}^{\pm}$. The Lipschitz group $\boldsymbol{\Gamma}_{p,q}$ has a subgroup, normalized by the reversion $s \to \tilde{s}$,

$$\mathbf{Pin}(p, q) = \{s \in \boldsymbol{\Gamma}_{p,q} \mid s\tilde{s} = \pm 1\},$$

with an even subgroup

$$\mathbf{Spin}(p, q) = \mathbf{Pin}(p, q) \cap C\ell_{p,q}^+,$$

which contains as a subgroup the two-fold cover

$$\mathbf{Spin}_+(p, q) = \{s \in \mathbf{Spin}(p, q) \mid s\tilde{s} = 1\}$$

of the connected component $SO_+(p, q)$ of $SO(p, q) \subset O(p, q)$.

Although $SO_+(p, q)$ is connected, its two-fold cover $\mathbf{Spin}_+(p, q)$ need not be connected. In particular,

$$\mathbf{Spin}_+(1, 1) = \{x + y\mathbf{e}_{12} \mid x, y \in \mathbb{R}; \; x^2 - y^2 = 1\}$$

has two components, two branches of a hyperbola [and so the group

$$\mathbf{Spin}(1, 1) = \{x + y\mathbf{e}_{12} \mid x, y \in \mathbb{R}; \; x^2 - y^2 = \pm 1\}$$

has four components]. This serves as a counter-example to Choquet-Bruhat et al. 1989, p. 37, ll. 2-3, p. 38, ll. 22-23 [see also p. 27, ll. 4-5].

Although the two-fold covers $\mathbf{Spin}(n) = \mathbf{Spin}(n,0) \simeq \mathbf{Spin}(0,n)$, $n \geq 3$, and $\mathbf{Spin}_+(n-1,1) \simeq \mathbf{Spin}_+(1,n-1)$, $n \geq 4$, are simply connected, $\mathbf{Spin}_+(3,3)$ is not simply connected, an therefore not a universal cover of $SO_+(3,3)$, since the maximal compact subgroup $SO(3) \times SO(3)$ of $SO_+(3,3)$ has a four-fold universal cover $\mathbf{Spin}(3) \times \mathbf{Spin}(3)$. The two-fold cover $\mathbf{Spin}_+(3,3)$ of $SO_+(3,3)$ is doubly connected, contrary to the claims of Lawson & Michelsohn 1989, p. 57, l. 22, and Göckeler & Schücker 1987, p. 190, l. 17. [4] As a consequence, $\mathbf{Spin}_+(3,3) \simeq SL(4,\mathbb{R})$, and so $\mathbf{Spin}_+(3,3)/\{\pm 1\} \not\simeq SL(4,\mathbb{R})$ contrary to the claims of Harvey 1990, p. 272, l. 24, and Lawson & Michelsohn 1989, p. 56, l. 21.

Moreover, the element $e_1 e_2 \ldots e_6 \in \mathbf{Spin}(3,3)\backslash\mathbf{Spin}_+(3,3)$ is not in $\mathbf{Spin}_+(3,3)$, since it is a preimage of $-I \in SO(3,3) \setminus SO_+(3,3)$, contrary to the claims of Lawson & Michelsohn 1989, p. 57, ll. 29-30.

Comment on Bourbaki 1959. The groups $\mathbf{Pin}(p,q)$ and $\mathbf{Spin}(p,q)$, obtained by normalizing the Lipschitz group $\Gamma_{p,q}$, are two-fold coverings of the orthogonal and special orthogonal groups, $O(p,q)$ and $SO(p,q)$, respectively. If one defines, instead of the Lipschitz group, a slightly different group

$$\mathbf{G}_{p,q} = \{s \in C\ell_{p,q} \mid \forall \mathbf{x} \in \mathbb{R}^{p,q}, \; s\mathbf{x}s^{-1} \in \mathbb{R}^{p,q}\},$$

one obtains, only in even dimensions, a cover of $O(p,q)$. Furthermore, for odd $n = p+q$, an element of $\mathbf{G}_{p,q}$ need not be even or odd, but might have an inhomogeneous central factor $x + y e_{12\ldots n} \in \mathbb{R} \oplus \bigwedge^n \mathbb{R}^{p,q}$. Thus Bourbaki 1959, p. 151, Lemme 5, does not hold, as has been observed by Deheuvels 1981, p. 355, Moresi 1988, p. 621, and by Bourbaki himself [see Feuille d'Errata No. 10 distributed with Chapters 3, 4 of Algèbre Commutative 1961]. ∎

The confusion about proper covering of $O(p,q)$ in $C\ell_{p,q}$ pops up frequently.

In the Lipschitz group every element $s \in \Gamma_{p,q}$ is of the form $s = \rho g$, where $\rho \in \mathbb{R} \setminus \{0\}$, $g \in \mathbf{Pin}(p,q)$. The group $\mathbf{G}_{p,q}$ does not have this property in odd dimensions. For instance, the central element $z = x + y e_{123} \in C\ell_3$, with non-zero $x, y \in \mathbb{R}$, satisfies $z \in \mathbf{G}_3$, but $z \neq \rho g$, $g \in \mathbf{Pin}(3)$. This serves as a counter-example to Baum 1981, p. 57, l. -1. [Baum's $C_{n,k}$ means $C\ell_{k,n-k}$, see p. 51, and her $\mathbf{Pin}(n,k)$ means $\mathbf{Pin}(k,n-k)$, see p. 53. Note that the two-fold cover of $O(3)$,

$$\mathbf{Pin}(3) = \mathbf{Spin}(3) \cup e_{123}\mathbf{Spin}(3) \simeq SU(2) \cup iSU(2),$$

is a subgroup of $\mathbf{G}_3$, but since the actions are defined differently, $\mathbf{G}_3$ does not cover $O(3)$.]

For all $s \in \mathbf{G}_3$, $s\tilde{s} > 0$. Therefore, if we normalize $\mathbf{G}_3$ by the reversion, the central factor is not eliminated, but instead we get the group $\{s \in \mathbf{G}_3 \mid s\tilde{s} = 1\} \simeq U(2)$, which does not cover $O(3)$ but covers $SO(3)$ with kernel $\{x + y e_{123} \mid x, y \in \mathbb{R}; \; x^2 + y^2 = 1\} \simeq U(1) \not\simeq \{\pm 1\}$. Compare this to Figueiredo 1994, p. 230, ll. -4.

---

[4] Lawson & Michelsohn 1989 give also correct information about the connectivity properties of the rotation groups $SO_+(p,q)$, see p. 20, ll. 6-8.

Exponentials of bivectors. There are two possibilities to exponentiate a bivector $\mathbf{B} \in \bigwedge^2 \mathbb{R}^{p,q}$: the ordinary/Clifford exponential $e^{\mathbf{B}}$, and the exterior exponential $e^{\wedge \mathbf{B}}$, where the product is the exterior product. If the exterior exponential $e^{\wedge \mathbf{B}}$ is invertible with respect to the Clifford product, then it is in the Lipschitz group $\boldsymbol{\Gamma}_{p,q}$. For the ordinary exponential we always have $e^{\mathbf{B}} \in \mathbf{Spin}_+(p,q)$.

All the elements of the compact spin groups $\mathbf{Spin}(n,0) \simeq \mathbf{Spin}(0,n)$ are exponentials of bivectors [when $n \geq 2$]. Among the other spin groups the same holds only for $\mathbf{Spin}_+(n-1,1) \simeq \mathbf{Spin}_+(1,n-1)$, $n \geq 5$, see M. Riesz 1958/1993 pp. 160, 172. In particular, the two-fold cover $\mathbf{Spin}_+(1,3) \simeq SL(2,\mathbb{C})$ of the Lorentz group $SO_+(1,3)$ contains elements which are not exponentials of bivectors: take $(\gamma_0 + \gamma_1)\gamma_2 \in \bigwedge^2 \mathbb{R}^{1,3}$, $[(\gamma_0 + \gamma_1)\gamma_2]^2 = 0$, then $-e^{(\gamma_0 + \gamma_1)\gamma_2} = -1 - (\gamma_0 + \gamma_1)\gamma_2 \neq e^{\mathbf{B}}$ for any $\mathbf{B} \in \bigwedge^2 \mathbb{R}^{1,3}$. [5] However, all the elements of $\mathbf{Spin}_+(1,3)$ are of the form $\pm e^{\mathbf{B}}$, $\mathbf{B} \in \bigwedge^2 \mathbb{R}^{1,3}$. Therefore, the exponentials of bivectors do not form a group, contrary to a statement of Dixon 1994, p. 13, ll. 8-10.

Every element L of the Lorentz group $SO_+(1,3)$ is an exponential of an antisymmetric matrix, $L = e^A$, $gA^{\top}g^{-1} = -A$; a similar property is not shared by $SO_+(2,2)$, see M. Riesz 1958/1993, pp. 150-152, 170-171. There are elements in $\mathbf{Spin}_+(2,2)$ which cannot be written in the form $\pm e^{\mathbf{B}}$, $\mathbf{B} \in \bigwedge^2 \mathbb{R}^{2,2}$; for instance $\pm e_{1234}e^{\beta \mathbf{B}}$, $\mathbf{B} = e_{12} + 2e_{14} + e_{34}$, $\beta > 0$. This serves as a counter-example to Doran 1994, p. 41, l. 26, formula (3.16). [6]

Counter-example to Delanghe & Sommen & Souček 1992. On page 126 the authors define the spin groups in the complex case

$$\mathrm{Pin}_{\pm}(n,\mathbb{C}) = \{g \in C\ell(\mathbb{C}^n) \,|\, N(g) = \pm 1; \, \forall \mathbf{x} \in \mathbb{C}^n, \, g\mathbf{x}\hat{g}^{-1} \in \mathbb{C}^n\},$$
$$\mathrm{Spin}_{\pm}(n,\mathbb{C}) = \{g \in C\ell^+(\mathbb{C}^n) \,|\, N(g) = \pm 1; \, \forall \mathbf{x} \in \mathbb{C}^n, \, g\mathbf{x}g^{-1} \in \mathbb{C}^n\}$$

and claim that the subgroups $\mathrm{Pin}_+(n,\mathbb{C})$ and $\mathrm{Spin}_+(n,\mathbb{C})$ are connected. These groups are not connected, for instance, $\mathrm{Pin}_+(1,\mathbb{C}) = \{\pm 1, \pm ie_1\}$, $\mathrm{Spin}_+(1,\mathbb{C}) = \{\pm 1\}$ and each $\mathrm{Pin}_+(n,\mathbb{C})$, $n \geq 2$, has two components.

This can be seen as follows. For $n = 1$, take a basis vector $\mathbf{e}_1 \in \mathbb{C}^1$ such that $\mathbf{e}_1^2 = 1$ and an arbitrary element $\alpha + \beta\mathbf{e}_1 \in C\ell(\mathbb{C}^1)$ with $\alpha, \beta \in \mathbb{C}$. Then

$$N(\alpha + \beta\mathbf{e}_1) = (\alpha + \beta\mathbf{e}_1)(\alpha - \beta\mathbf{e}_1) = \alpha^2 - \beta^2.$$

For an element $\alpha + \beta\mathbf{e}_1 \in \mathrm{Pin}_+(1,\mathbb{C})$ we have $\alpha^2 - \beta^2 = 1$, so that

$$(\alpha + \beta\mathbf{e}_1)^{-1} = \frac{\alpha - \beta\mathbf{e}_1}{\alpha^2 - \beta^2}$$

---

[5] Exercise: Locate an error about this in the literature. Note, that in $\mathbf{Spin}_+(4,1) \simeq Sp(2,2)$ we have $-e^{(e_1 + e_5)e_2} = -1 - (e_1 + e_5)e_2 = e^{(e_1 + e_5)e_2 + \pi e_{34}}$.

[6] Riesz also showed, by the same construction on pp. 170-171, that there are bivectors which cannot be written as sums of simple and completely orthogonal bivectors; for instance $\mathbf{B} = e_{12} + 2e_{14} + e_{34} \in \bigwedge^2 \mathbb{R}^{2,2}$. Exercise: Locate an error about this in the literature.

and

$$(\alpha + \beta e_1)e_1(\alpha - \beta e_1)^{-1}$$

$$= (\alpha + \beta e_1)e_1 \frac{\alpha + \beta e_1}{\alpha^2 - \beta^2}$$

$$= \frac{(\alpha + \beta e_1)^2}{\alpha^2 - \beta^2} e_1$$

$$= \frac{\alpha^2 + \beta^2 + 2\alpha\beta e_1}{\alpha^2 - \beta^2} e_1 \in \mathbb{C}^1$$

which implies that either $\alpha = 0$ or $\beta = 0$. If $\alpha = 0$ then $\beta^2 = -1$ and so $\beta = \pm i$, and if $\beta = 0$ then $\alpha^2 = 1$ and so $\alpha = \pm 1$. Thereby $\text{Pin}_+(1, \mathbb{C}) = \{\pm 1, \pm i e_1\}$, which is not connected. The even subgroup $\text{Spin}_+(1, \mathbb{C}) = \{\pm 1\}$ is also disconnected.

In general, $\text{Pin}_+(n, \mathbb{C})$ covers continuously a 2-component group $O(n, \mathbb{C})$, and contains thereby at least two components, and consequently cannot be connected.

The group of non-zero real numbers $\dot{\mathbb{R}} = \mathbb{R}\backslash\{0\}$ has two square classes, namely $\dot{\mathbb{R}}^2 = \{x^2 \,|\, x \in \dot{\mathbb{R}}\}$ and $-\dot{\mathbb{R}}^2 = \{-x^2 \,|\, x \in \dot{\mathbb{R}}\}$. Thereby, in defining the groups

$$\mathbf{Pin}(p, q) = \{g \in C\ell_{p,q} \,|\, N(g) = \pm 1; \, \forall \mathbf{x} \in \mathbb{R}^{p,q}, \, g\mathbf{x}\hat{g}^{-1} \in \mathbb{R}^{p,q}\}$$

we set $N(g) = \pm 1$, that is, we take a representative out of each square class $\pm\dot{\mathbb{R}}^2$ of $\dot{\mathbb{R}}$ [instead of $N(g) = g\bar{g}$ we could also use $Q(g) = g\tilde{g}$ as before]. In contrast, since $\dot{\mathbb{C}} = \dot{\mathbb{C}}^2$ has only one square class, it is customary to define

$$\mathbf{Pin}(n, \mathbb{C}) = \{g \in C\ell(\mathbb{C}^n) \,|\, N(g) = 1; \, \forall \mathbf{x} \in \mathbb{C}^n, \, g\mathbf{x}\hat{g}^{-1} \in \mathbb{C}^n\}$$

in order to get a two-fold cover of $O(n, \mathbb{C})$. Delanghe & Sommen & Souček overlooked the square-class structure of fields. ∎

Counter-example to Hestenes & Sobczyk 1984, 1987. There is an error on page 106 line -2 word -2. The mistake could be rectified by deleting the erroneous word *two*.

Let us first recall some nomenclature of Hestenes & Sobczyk. On p. 103, l. -7, the authors define a *versor* as a 'products of vectors' whereby they mean the Clifford products of vectors in $\mathbb{R}^{p,q} \subset C\ell_{p,q}$ [they denote the orthogonal space $\mathbb{R}^{p,q}$ by $\mathcal{A}_{p,q}$ and the Clifford algebra $C\ell_{p,q}$ by $\mathcal{G}(\mathcal{A}_{p,q})$, see p. 42]. On p. 106, l. 18, the authors define their *spinor* to be such an element $\psi \in C\ell_{p,q}$ that for any vector $\mathbf{x} \in \mathbb{R}^{p,q}$ also $\psi\mathbf{x}\tilde{\psi} \in \mathbb{R}^{p,q}$, where $\psi \to \tilde{\psi}$ is the reversion [they denote the reversion by $\psi \to \psi^\dagger$, see p. 5].

Then on p. 106, l. -2, the authors say that *a spinor can always be expressed as the sum of two even versors*. This statement is incorrect, as the following counter-example shows. Consider the Euclidean space $\mathbb{R}^8 = \mathbb{R}^{8,0}$ and its Clifford algebra $C\ell_8 = C\ell_{8,0} \simeq \text{Mat}(16, \mathbb{R})$. Let $e_1, e_2, \ldots, e_8$ be an orthonormal basis of $\mathbb{R}^8$. Consider the element

$$\psi = (1 - e_{12\ldots8})(1 + \mathbf{w}) \in C\ell_8$$

where $e_{12\ldots8} = e_1 e_2 \ldots e_8$ and

$$\mathbf{w} = e_{1236} - e_{1257} - e_{1345} + e_{1467} + e_{2347} - e_{2456} - e_{3567}$$

[$\mathbf{w} = -\mathbf{v}\mathbf{e}_{12...7}$ where $\mathbf{v} = \mathbf{e}_{124} + \mathbf{e}_{235} + \mathbf{e}_{346} + \mathbf{e}_{457} + \mathbf{e}_{561} + \mathbf{e}_{672} + \mathbf{e}_{713}$]. This element ψ satisfies $\psi \mathbf{x} \tilde{\psi} = 0$ for all $\mathbf{x} \in \mathbb{R}^8$, and is thereby a spinor [in the sense of Hestenes & Sobczyk]. I will show that ψ cannot be a sum of *two* even versors. Recall that all even versors are known to be of the form αa where $\alpha \in \mathbb{R}$ and $a \in \mathbf{Spin}(8)$ [so that $\mathbf{x} \to a\mathbf{x}\tilde{a}$ is a rotation of $\mathbb{R}^8$].

The spinor ψ is a multiple of an idempotent, since $\psi^2 = 16\psi$. Note that ψ was written as a product $\psi = (1 - \mathbf{e}_{12...8})(1 + \mathbf{w})$ in such a way that the first factor was of the same form as the first factor of the formula (8.11) on p. 106, l. -5, and that the first factor of ψ was not invertible. The second factor in the formula (8.11) is by definition invertible whereas the second factor $1 + \mathbf{w}$ of the above ψ is again non-invertible [since $(1 + \mathbf{w})^2 = 8(1 + \mathbf{w})$]. This shows that there is something wrong in the justification of Hestenes & Sobczyk, since ψ cannot be of the form (8.11). However, although the argument of Hestenes & Sobczyk is flawed, their end-result might still hold, and to invalidate it we should show that $\psi \neq \alpha a + \beta b$ for any $\alpha, \beta \in \mathbb{R}$ and $a, b \in \mathbf{Spin}(8)$. To do this, compute

$$(\psi - \beta b)\mathbf{x}(\psi - \beta b)^{\tilde{}} = \psi\mathbf{x}\tilde{\psi} - \beta(\psi\mathbf{x}\tilde{b} + b\mathbf{x}\tilde{\psi}) + \beta^2 b\mathbf{x}\tilde{b}$$

for $\mathbf{x} \in \mathbb{R}^8$. The first term on the right is 0, as already mentioned, and the last term on the right is a vector, since $b \in \mathbf{Spin}(8)$. The middle term is odd and equals its own reverse; thereby it is a sum of a vector and a 5-vector, in other words,

$$\psi\mathbf{x}\tilde{b} + b\mathbf{x}\tilde{\psi} \in \mathbb{R}^8 + \overset{5}{\bigwedge} \mathbb{R}^8.$$

Denote the k-vector part of $u \in C\ell_8$ by $\langle u \rangle_k$ [$\in \bigwedge^k \mathbb{R}^8$]. Then

$$\langle \psi\mathbf{x}\tilde{b} + b\mathbf{x}\tilde{\psi} \rangle_1 = 2\langle \psi\mathbf{x}\tilde{b} \rangle_1 \quad \text{and} \quad \langle \psi\mathbf{x}\tilde{b} + b\mathbf{x}\tilde{\psi} \rangle_5 = 2\langle \psi\mathbf{x}\tilde{b} \rangle_5$$

since $(\psi\mathbf{x}\tilde{b})^{\tilde{}} = b\mathbf{x}\tilde{\psi}$. To see that the statement of Hestenes & Sobczyk is incorrect it is sufficient to show that the 5-vector part of $\psi\mathbf{x}\tilde{b}$ never vanishes for $\mathbf{x} \neq 0$. Since $\mathbf{x} \to \langle \psi\mathbf{x}\tilde{b} \rangle_1$ is a rotation [related to the triality of b, or more precisely induced by $\mathbf{e}_8 \operatorname{trial}(\operatorname{trial}(b)) \mathbf{e}_8^{-1}$],

$$|\langle \psi\mathbf{x}\tilde{b} \rangle_1| = |\mathbf{x}|.$$

The norm on $\mathbb{R}^8$ can be naturally extended to all of $C\ell_8$ by defining $|u|^2 = \langle u\tilde{u} \rangle_0$. A direct, but quite tedious, computation shows that

$$|\langle \psi\mathbf{x}\tilde{b} \rangle_5|^2 = 7|\mathbf{x}|^2$$

[the factor 7 comes from $\mathbf{w}$ having 7 terms; the element $\mathbf{w}$ is related to the octonion multiplication]. Thereby the 5-vector part of $\psi\mathbf{x}\tilde{b}$ never vanishes for a non-zero vector $\mathbf{x}$, and consequently, $\psi\mathbf{x}\tilde{b} + b\mathbf{x}\tilde{\psi}$ never reduces to a vector. We conclude that $\psi - \beta b$ cannot be a multiple of any $a \in \mathbf{Spin}(8)$.

This error of Hestenes & Sobczyk was most fascinating: it offered a true challenge in constructing a counter-example [after a flaw in the deduction had been detected]. The above counter-example was constructed with CLICAL. In this exploration, the design goals of CLICAL proved useful: the ease of use allowed iterative experimentation, and the user interface adapted to the multivector structure was essential in finding the relation between the norms of vectors and 5-vectors. It seems that general purpose computer programs are not so efficient research tools as programs with special features targeted to some active research field, like in CLICAL. ∎

5. Counter-examples about spinors

There is a natural way to introduce scalar products of spinors, generalizing the scalar product of Dirac spinors

$$\langle \psi | \varphi \rangle = \psi_1^* \varphi_1 + \psi_2^* \varphi_2 - \psi_3^* \varphi_3 - \psi_4^* \varphi_4.$$

This leads to 32 different scalar products of spinors for the real Clifford algebras $Cl_{p,q}$, 4 different scalar products for the complexified Clifford algebras $\mathbb{C} \otimes Cl_{p,q}$, with a sesquilinear product on $\mathbb{C} \otimes \mathbb{R}^{p,q}$, and 8 different scalar products for the complex Clifford algebras $Cl(\mathbb{C}^n)$, with a bilinear product on $\mathbb{C}^n$.

  Without loss of generality, we can regard spinor spaces as minimal left ideals $S = Cl_{p,q}f$ in $Cl_{p,q}$ [in the real case], where f is a primitive idempotent in $Cl_{p,q}$. The spinor spaces are right linear spaces over a division ring $\mathbb{R}, \mathbb{C}, \mathbb{H} \simeq \mathbb{F} = fCl_{p,q}f$. [For semisimple Clifford algebras $Cl_{p,q}$, $p - q = 1 \mod 4$, we could also consider double spinor spaces $S \oplus \hat{S}$ over double rings $^2\mathbb{F} = \mathbb{F} \times \mathbb{F} \simeq \mathbb{F} \oplus \hat{\mathbb{F}}$.] In a fixed $Cl_{p,q}$, there are up to equivalence two scalar products of spinors $S \times S \to \mathbb{F}$ with images

$$s\tilde{\psi}\varphi \quad \text{and} \quad s\bar{\psi}\varphi,$$

where $s \in Cl_{p,q}$ is an invertible element chosen so that the scalar product has values in $\mathbb{F}$. These scalar products are intrinsic in the sense that they are defined within one algebra by its anti-involutions. Both anti-involutions induce 32 different classes of scalar products of spinors, for the collection of Clifford algebras $Cl_{p,q}$.

  More information about scalar products of spinors can be found in Porteous 1969/1981, p. 271, Lounesto 1981, and Harvey 1990.

Benn & Tucker 1987. On page 76 in the Table 2.14, the authors present, for some lower dimensional cases, the automorphism groups of the scalar products of spinors, induced by the anti-involutions of Clifford algebras. Below the Table 2.14 is reproduced in a corrected form [for the number-codes see pp. 67, 75, and for the index see p. 66]:

		reversion ξ		Clifford-conjugation $\xi\eta$
$Cl_{1,0} = Cl_1$	z	$^2O(2)$	a,v	$GL(1,\mathbb{R})$
$Cl_{2,0} = Cl_2$	z	$O(2)$		$Sp(2,\mathbb{R})$
$Cl_{3,0} = Cl_3$	z	$U(2)$		$Sp(2,\mathbb{C})$
$Cl_{4,0} = Cl_4$	z	$Sp(4)$		$Sp(2,2)$
$Cl_{0,1}$	a	$O(1,\mathbb{C})$	z	$U(1)$
$Cl_{0,2}$	a,w	$SO^*(2)$	z	$Sp(2)$
$Cl_{0,3}$	a,w	$GL(1,\mathbb{H})$	z	$^2Sp(2)$
$Cl_{0,4}$		$Sp(2,2)$	z	$Sp(4)$
$Cl_{1,1}$		$O(1,1)$		$Sp(2,\mathbb{R})$

In this table the small letters mean

a = not neutral (since 1-dimensional over $\mathbb{R}, \mathbb{C}, \mathbb{H} = fC\ell_{p,q}f$)

w = would be neutral if considered over $\mathbb{C}$ rather than $\mathbb{H}$

v = vanishes identically ($\Rightarrow$ not maximal)

z = zero index

and the symplectic groups are denoted similarly as in Lounesto 1981.

There are more cases with zero index than those which appear in the authors' list, namely $\xi : C\ell_{1,0}$ and $\xi\eta : C\ell_{0,1}, C\ell_{0,2}, C\ell_{0,3}$. The cases $\xi : C\ell_{0,3}$ and $\xi\eta : C\ell_{1,0}$ could hardly be called maximal, since the scalar product of spinors vanishes identically. The cases $\xi : C\ell_{0,1}, C\ell_{0,2}, C\ell_{0,3}$ and $\xi\eta : C\ell_{1,0}$ are not neutral over $\mathbb{C}, \mathbb{H}, {}^2\mathbb{H}$ (or $\mathbb{H}$) and ${}^2\mathbb{R}$ (or $\mathbb{R}$), respectively; although the case $\xi : C\ell_{0,2} \simeq \mathbb{H}$ could be regarded as neutral over $\mathbb{C}$ [as could the non-vanishing scalar product of double-spinors in the case of $\xi : C\ell_{0,3} \simeq \mathbb{H} \oplus \mathbb{H}$].

Regarding exercise 10.3 on pages 285-286, I have explored $\mathbb{R}^{1,7}$ with the Clifford algebra $C\ell_{1,7} \simeq \text{Mat}(16, \mathbb{R})$, but I have not managed to find a spinor $\psi \in (\mathbb{C} \otimes C\ell_{1,7})f$ such that $\mathbf{J} \simeq \langle \psi\tilde{\psi}^* \rangle_1$ would be space-like, $\mathbf{J}^2 < 0$ [as claimed on p. 286, l. 2]. I have asked the authors to provide me with such a spinor, preferably in the form $\psi = uf$, where $u \in \mathbb{C} \otimes C\ell_{1,7}$ and

$$f = \frac{1}{2}(1 + \gamma_0)\frac{1}{2}(1 + i\gamma_{12})\frac{1}{2}(1 + i\gamma_{34})\frac{1}{2}(1 + i\gamma_{56}).$$

While the authors have not provided such a spinor, or shown that my interpretation is not the one they intended, I remain confident that no such spinor exists. ∎

Coquereaux 1988. The author considers real Clifford algebras $C\ell_{p,q}$. On p. 178, l. -3 he claims that the scalar product of spinors $s\tilde{\psi}\varphi$ is neutral in the cases $(p, q) \neq (0, n)$. There is an exception $(p, q) = (1, 0)$, where the spinor space is 1-dimensional with an identically vanishing scalar product and invariance group $GL(1, \mathbb{R})$, see Lounesto 1981, p. 733, the Table VB, line $p + q = 1$. On p. 178, l. -1 the author claims that the scalar product of spinors $s\tilde{\psi}\varphi$ is neutral when $(p, q) \neq (n, 0)$. Again, there are exceptions $(p, q) = (0, 1), (0, 2), (0, 3)$, where the scalar product cannot be neutral, while it is 1-dimensional over $\mathbb{C}, \mathbb{H}, {}^2\mathbb{H}$ (or $\mathbb{H}$), with invariance groups $O(1, \mathbb{C}), SO^*(2), GL(1, \mathbb{H})$, see Lounesto 1981, p. 733, the Table VA, lines $p + q = 1, 2, 3$. ∎

Baum 1981. In Sektion 1.5: *Bilinearformen auf dem Spinormodul* on page 68 Satz 1.12 is incorrect. Here the author discusses the complexified Clifford algebra $\mathbb{C} \otimes C\ell_{k,n-k}$ [author's notation $C_{n,k}^c$] and a spinor space, which could equally well be represented, without loss of generality, within the Clifford algebra $\mathbb{C} \otimes C\ell_{k,n-k}$ as a minimal left ideal

$$S = (\mathbb{C} \otimes C\ell_{k,n-k})f$$

where f is a primitive idempotent of $\mathbb{C} \otimes C\ell_{k,n-k}$. Baum introduces a complex valued positive definite scalar product (ψ, φ) on S, distinguishes a special element

$$\mathbf{b} = \begin{cases} \mathbf{e}_1\mathbf{e}_2 \ldots \mathbf{e}_k & \text{if } k = 0, 1 \bmod 4 \\ i\mathbf{e}_1\mathbf{e}_2 \ldots \mathbf{e}_k & \text{if } k = 2, 3 \bmod 4 \end{cases}$$

and defines a new scalar product $\langle \psi, \varphi \rangle = (\psi, \mathbf{b}\varphi)$ on S. Then Baum proves in Satz 1.12 that the scalar product $\langle \psi, \varphi \rangle$ is indefinite, but this is false, because the case $k = 0$ is not excluded. A counter-example satisfying all the assumptions comes in $\mathbb{C} \otimes C\ell_{0,3}$ where for any non-zero $\psi \in S = (\mathbb{C} \otimes C\ell_{0,3})f$ we have $\langle \psi, \psi \rangle > 0$ because $\mathbf{b} = 1$. Another series of counter-examples could be constructed in the cases $k = n$, with n odd, because also then $\langle \psi, \varphi \rangle$ is positive-definite [but it should be noted that Baum has actually excluded the cases $k > [n/2]$ on p. 51, l. 5].

Baum also proves in Satz 1.12 that the scalar product $\langle \psi, \varphi \rangle$ is invariant under the group $\mathbf{Spin}_+(k, n - k)$. This statement is true, but it is not the best result in the sense that $\mathbf{Spin}_+(k, n - k)$ is not the largest group preserving $\langle \psi, \varphi \rangle$. The largest groups are much bigger, and they are seen to be the following, in a few lower-dimensional cases,

n	$k = 0$	$k > 0$
3	$U(2) \times U(2)$	$U(1,1) \times U(1,1)$
4	$U(4)$	$U(2,2)$
5	$U(4) \times U(4)$	$U(2,2) \times U(2,2)$
6	$U(8)$	$U(4,4)$

A complete list of these maximal groups can be found, together with other groups not reachable by Baum's approach, in the Table VII on page 735 of Lounesto 1981.

The above observations point out genuine errors, or at least inaccuracies. The following observation is more important, although no error will be pointed out; only inadequate treatment. The approach presented in Sektion 1.5: *Bilinearformen auf dem Spinormodul* gives access only to half of the groups naturally invoked in conjunction with scalar products of spinors [induced by anti-involutions of Clifford algebras]. First, one can divide the set of good scalar products or correlations into two equivalence classes, see Porteous 1969, 1981, pp. 203 and 207. Secondly, all good scalar products of spinors have been determined, and there are 4 types of them on the complexified spinor spaces $S = (\mathbb{C} \otimes C\ell_{p,q})f$, whereas Baum's method reaches only 2 classes, see Porteous p. 272. The situation with real spinors is more involved: there are 32 classes of scalar products on the real spinor spaces $C\ell_{p,q}f$, see Porteous p. 271 [Porteous' tables may seem cryptic because they condense also other information, but a more accessible source to the scalar products on real spinor spaces might be the Table V on page 733 of Lounesto 1981.

I will make explicit what scalar products on complexified spinor spaces cannot be reached by Baum's approach. First, recall that the two anti-involutions, reversion and Clifford-conjugation, of $u \in C\ell_{p,q}$ are denoted in this paper by $\tilde{u}$ and $\bar{u}$; and that $\tilde{\mathbf{x}} = \mathbf{x}$ and $\bar{\mathbf{x}} = -\mathbf{x}$ for $\mathbf{x} \in \mathbb{R}^{p,q}$. Extend these anti-automorphisms as complex linear to $\mathbb{C} \otimes C\ell_{p,q}$. Introduce in $\mathbb{C} \otimes C\ell_{p,q}$ a complex conjugation so that $u^* = a - ib$ for $u = a + ib$ with $a, b \in C\ell_{p,q}$. Then define in the spinor space $S = (\mathbb{C} \otimes C\ell_{p,q})f$ two scalar products

$$s\tilde{\psi}^*\varphi \quad \text{and} \quad s\bar{\psi}^*\varphi,$$

where $s \in \mathbb{C} \otimes C\ell_{p,q}$ is an invertible element chosen so that the scalar product has values in $f(\mathbb{C} \otimes C\ell_{p,q})f \simeq \mathbb{C}$. In a few lower dimensional cases we find then for a non-zero $\psi \in S$ and arbitrary $\varphi \in S$

$$
\begin{array}{l|lll}
\mathbb{C} \otimes C\ell_{0,2} & f = \frac{1}{2}(1 + ie_1) & \underline{\bar{\psi}^*\psi > 0} & e_2\tilde{\psi}^*\psi \gtreqless 0 \\
\mathbb{C} \otimes C\ell_{1,1} & f = \frac{1}{2}(1 + e_1) & ie_2\bar{\psi}^*\psi \gtreqless 0 & \tilde{\psi}^*\psi \gtreqless 0 \\
\mathbb{C} \otimes C\ell_2 & f = \frac{1}{2}(1 + e_1) & ie_2\bar{\psi}^*\psi \gtreqless 0 & \tilde{\psi}^*\psi > 0 \\
\end{array}
$$

$$
\begin{array}{l|lll}
\mathbb{C} \otimes C\ell_{0,3} & f = \frac{1}{2}(1 + ie_1)\frac{1}{2}(1 + ie_2e_3) & \underline{\bar{\psi}^*\psi > 0} & \tilde{\psi}^*\varphi = 0 \\
\mathbb{C} \otimes C\ell_{1,2} & f = \frac{1}{2}(1 + e_1)\frac{1}{2}(1 + ie_2e_3) & \bar{\psi}^*\varphi = 0 & \tilde{\psi}^*\psi \gtreqless 0 \\
\mathbb{C} \otimes C\ell_{2,1} & f = \frac{1}{2}(1 + e_1)\frac{1}{2}(1 + e_2e_3) & ie_2\bar{\psi}^*\psi \gtreqless 0 & \tilde{\psi}^*\varphi = 0 \\
\mathbb{C} \otimes C\ell_3 & f = \frac{1}{2}(1 + e_1)\frac{1}{2}(1 + ie_2e_3) & \bar{\psi}^*\varphi = 0 & \tilde{\psi}^*\psi > 0 \\
\end{array}
$$

$$
\begin{array}{l|lll}
\mathbb{C} \otimes C\ell_{0,4} & f = \frac{1}{2}(1 + ie_1)\frac{1}{2}(1 + ie_2e_3) & \bar{\psi}^*\psi > 0 & e_4\tilde{\psi}^*\psi \gtreqless 0 \\
\mathbb{C} \otimes C\ell_{1,3} & f = \frac{1}{2}(1 + e_1)\frac{1}{2}(1 + ie_2e_3) & ie_4\bar{\psi}^*\psi \gtreqless 0 & \tilde{\psi}^*\psi \gtreqless 0 \\
\mathbb{C} \otimes C\ell_{2,2} & f = \frac{1}{2}(1 + e_1)\frac{1}{2}(1 + e_2e_3) & ie_2\bar{\psi}^*\psi \gtreqless 0 & ie_2e_4\bar{\psi}^*\psi \gtreqless 0 \\
\mathbb{C} \otimes C\ell_{3,1} & f = \frac{1}{2}(1 + e_1)\frac{1}{2}(1 + ie_2e_3) & ie_4\bar{\psi}^*\psi \gtreqless 0 & \tilde{\psi}^*\psi \gtreqless 0 \\
\mathbb{C} \otimes C\ell_4 & f = \frac{1}{2}(1 + e_1)\frac{1}{2}(1 + ie_2e_3) & \underline{e_2e_3e_4\bar{\psi}^*\psi \gtreqless 0} & \tilde{\psi}^*\psi > 0 \\
\end{array}
$$

The automorphism groups of these scalar products can be found in the lines 2,3,4 of the Table VII on page 735 of Lounesto 1981. To relate these scalar products to Baum's products I will list a similar table with the element

$$
\mathbf{b} = \begin{cases} e_1e_2 \ldots e_p & \text{if} \quad p = 0, 1 \bmod 4 \\ ie_1e_2 \ldots e_p & \text{if} \quad p = 2, 3 \bmod 4 \end{cases}
$$

inserted between the factors of the scalar product, and look for the cases when the product so obtained will be positive definite [and coincides with Baum's (ψ, φ)]. Then we find that

$$
\begin{array}{l|lll}
\mathbb{C} \otimes C\ell_{0,2} & \bar{\psi}^*\mathbf{b}\psi > 0 & e_2\tilde{\psi}^*\mathbf{b}\psi \gtreqless 0 & \overline{\mathbf{x}\tilde{\psi}}^*\varphi + \bar{\psi}^*\mathbf{x}\varphi = 0 \\
\mathbb{C} \otimes C\ell_{1,1} & e_2\bar{\psi}^*\mathbf{b}\psi \gtreqless 0 & \tilde{\psi}^*\mathbf{b}\psi \gtreqless 0 & \overline{\mathbf{x}\tilde{\psi}}^*\varphi - \bar{\psi}^*\mathbf{x}\varphi = 0 \\
\mathbb{C} \otimes C\ell_2 & ie_2\bar{\psi}^*\mathbf{b}\psi \gtreqless 0 & \tilde{\psi}^*\mathbf{b}\psi > 0 & \overline{\mathbf{x}\tilde{\psi}}^*\varphi + \bar{\psi}^*\mathbf{x}\varphi = 0 \\
\end{array}
$$

$$
\begin{array}{l|lll}
\mathbb{C} \otimes C\ell_{0,3} & \bar{\psi}^*\mathbf{b}\psi > 0 & \tilde{\psi}^*\mathbf{b}\varphi = 0 & \overline{\mathbf{x}\tilde{\psi}}^*\varphi + \bar{\psi}^*\mathbf{x}\varphi = 0 \\
\mathbb{C} \otimes C\ell_{1,2} & \bar{\psi}^*\mathbf{b}\varphi = 0 & \tilde{\psi}^*\mathbf{b}\psi \gtreqless 0 & \overline{\mathbf{x}\tilde{\psi}}^*\varphi - \tilde{\psi}^*\mathbf{x}\varphi = 0 \\
\mathbb{C} \otimes C\ell_{2,1} & ie_2\bar{\psi}^*\mathbf{b}\psi \gtreqless 0 & \tilde{\psi}^*\mathbf{b}\varphi = 0 & \overline{\mathbf{x}\tilde{\psi}}^*\varphi + \bar{\psi}^*\mathbf{x}\varphi = 0 \\
\mathbb{C} \otimes C\ell_3 & \bar{\psi}^*\mathbf{b}\varphi = 0 & \tilde{\psi}^*\mathbf{b}\psi > 0 & \overline{\mathbf{x}\tilde{\psi}}^*\varphi - \bar{\psi}^*\mathbf{x}\varphi = 0 \\
\end{array}
$$

$$
\begin{array}{l|lll}
\mathbb{C} \otimes C\ell_{0,4} & \bar{\psi}^*\mathbf{b}\psi > 0 & e_4\tilde{\psi}^*\mathbf{b}\psi \gtreqless 0 & \overline{\mathbf{x}\tilde{\psi}}^*\varphi + \bar{\psi}^*\mathbf{x}\varphi = 0 \\
\mathbb{C} \otimes C\ell_{1,3} & e_4\bar{\psi}^*\mathbf{b}\psi \gtreqless 0 & \tilde{\psi}^*\mathbf{b}\psi \gtreqless 0 & \overline{\mathbf{x}\tilde{\psi}}^*\varphi - \tilde{\psi}^*\mathbf{x}\varphi = 0 \\
\mathbb{C} \otimes C\ell_{2,2} & ie_2\bar{\psi}^*\mathbf{b}\psi \gtreqless 0 & ie_2e_4\tilde{\psi}^*\mathbf{b}\psi \gtreqless 0 & \overline{\mathbf{x}\tilde{\psi}}^*\varphi + \bar{\psi}^*\mathbf{x}\varphi = 0 \\
\mathbb{C} \otimes C\ell_{3,1} & e_4\bar{\psi}^*\mathbf{b}\psi \gtreqless 0 & \tilde{\psi}^*\mathbf{b}\psi \gtreqless 0 & \overline{\mathbf{x}\tilde{\psi}}^*\varphi - \bar{\psi}^*\mathbf{x}\varphi = 0 \\
\mathbb{C} \otimes C\ell_4 & e_2e_3e_4\bar{\psi}^*\mathbf{b}\psi \gtreqless 0 & \tilde{\psi}^*\mathbf{b}\psi > 0 & \overline{\mathbf{x}\tilde{\psi}}^*\varphi + \bar{\psi}^*\mathbf{x}\varphi = 0 \\
\end{array}
$$

The last column is for checking that my interpretation of Baum's results coincides with the case (3) on page 69 [since Baum restricts herself to the indefinite metrics at the top of the page, her results are this time correct]. With this backward tracking we can find that the underlined cases in the first table above correspond to Baum's positive definite scalar product (ψ, φ), and that half of my scalar products were not reached by her approach. Baum delimited the topic to include only certain scalar products, but as mentioned earlier, the scalar products can be collected into equivalence classes, and she missed in each signature one of the two classes.

When I informed Baum about the above errors, she told me that also Satz 2.2, p. 71, Lemma 2.7, p. 77, and Folgerung 2.1, p. 78, are wrong, and that correct information can be found in Karoubi 1968, pp. 174-175, Proposition 1.1.26. I hope that her frankness and sincere attitude would be shared by a larger portion of scientists. ∎

Published articles on scalar products of spinors are invariably incomplete. To the best of my knowledge, there is once again an incomplete article in press. Complete information about scalar products of spinors can be found in Porteous 1969, 1981 pp. 271-272, Lounesto 1981, and Porteous 1995.

Discussion of Budinich & Trautman 1988. First, I will give a counter-example to the Theorem 7.3 on p. 104 [a mistake about scalar products of spinors]. The restriction of the reversion to $C\ell_{0,3}^+ \simeq \mathbb{H}$ [the authors denote the reversion anti-automorphism by β and the even Clifford algebra by $C\ell_0(0,3)$] is the quaternion conjugation, which makes h of the formula (7.35) positive definite and not neutral.

Secondly, I will straighten out a confusion on pp. 90–95 [concerning the charge conjugation of the Dirac equation and a possible existence of eigenspinors of the charge conjugation, the Majorana spinors, in dimension 8].

Minkowski space $\mathbb{R}^{1,3}$ with $C\ell_{1,3} \simeq \text{Mat}(2, \mathbb{H})$. In the signature $\mathbb{R}^{1,3}$ we have the Dirac equation $i\partial \psi - e\mathbf{A}\psi = m\psi$ or $\gamma^\mu(i\partial_\mu - eA_\mu)\psi = m\psi$ where

$$\gamma_0 = \gamma^0 = \begin{pmatrix} I & 0 \\ 0 & -I \end{pmatrix} \quad \text{and} \quad \gamma_k = -\gamma^k = \begin{pmatrix} 0 & -\sigma_k \\ \sigma_k & 0 \end{pmatrix} \quad \text{for} \quad k = 1, 2, 3$$

and where σ_1, σ_2, σ_3 are the Pauli spin matrices (page 3 line 4 formula 1.5). Usually $\psi \in \mathbb{C}^4$ is a column spinor but I shall regard the spinor ψ as a 4×4-matrix with only the first column being non-zero, that is, $\psi \in \text{Mat}(4, \mathbb{C})f$ where $f = \frac{1}{2}(1 + \gamma_0)\frac{1}{2}(1 + i\gamma_1\gamma_2)$. Majorana spinors ψ are eigenspinors $\mathcal{C}(\psi) = \pm\psi$ of the charge conjugation operator $\mathcal{C}$ which is defined by

$$\mathcal{C}(\psi) = -i\gamma_2\psi^* \qquad \text{complex conjugate } \psi^* \text{ taken in } \text{Mat}(4, \mathbb{C}) \text{ or}$$
$$\mathcal{C}(\psi) = \hat{\psi}^*\gamma_1 \qquad \text{complex conjugate } \psi^* \text{ taken in } \mathbb{C} \otimes C\ell_{1,3}$$

[as before $u \to \hat{u}$ means the grade involution]. The charge conjugated spinor $\psi_\mathcal{C} = \mathcal{C}(\psi)$ satisfies the Dirac equation $i\partial\psi_\mathcal{C} + e\mathbf{A}\psi_\mathcal{C} = m\psi_\mathcal{C}$. The charge conjugation is anti-linear $\mathcal{C}i = -i\mathcal{C}$, involutory $\mathcal{C}^2 = I$ and satisfies $\mathcal{C}(\mathbf{A}\psi) = -\mathbf{A}\mathcal{C}(\psi)$. In the sequel I will *abbreviate* the last equation into the form $\mathcal{C}\mathbf{A} = -\mathbf{A}\mathcal{C}$.

The bilinear covariants associate to the spinor ψ a vector $u = u_0\gamma^0 + u_1\gamma^1 + u_2\gamma^2 + u_3\gamma^3$, $u^2 = u_0^2 - u_1^2 - u_2^2 - u_3^2$, which is time-like, that is $u^2 > 0$ or $u_0^2 > u_1^2 + u_2^2 + u_3^2$. When normalized $u^2 = 1$ this vector is related to the velocity $v < c$ of a real particle by

$$u_0 = \frac{1}{\sqrt{1 - \dfrac{v^2}{c^2}}}.$$

For an imaginary particle, or superluminary tachyon, $v > c$ and u is purely imaginary so that iu is space-like $(iu)^2 = -1 < 0$.

Opposite metric $\mathbb{R}^{3,1}$ with $C\ell_{3,1} \simeq \mathrm{Mat}(4,\mathbb{R})$. In the opposite metric $\mathbb{R}^{3,1}$ the Dirac equation is $\partial\psi + ie\mathbf{A}\psi = m\psi$. This form of the Dirac equation guarantees that for a real particle the unit velocity $u = u_0\mathbf{e}^0 + u_1\mathbf{e}^1 + u_2\mathbf{e}^2 + u_3\mathbf{e}^3$, $u^2 = -u_0^2 + u_1^2 + u_2^2 + u_3^2 = -1$, is real and time-like, that is $u^2 < 0$ or $u_0^2 > u_1^2 + u_2^2 + u_3^2$. The matrix representation

$$\mathbf{e}_0 = -\mathbf{e}^0 = \begin{pmatrix} i & 0 \\ 0 & -i \end{pmatrix} \quad \text{and} \quad \mathbf{e}_k = \mathbf{e}^k = \begin{pmatrix} 0 & \sigma_k \\ \sigma_k & 0 \end{pmatrix} \quad \text{for } k = 1,2,3$$

is obtained by the choice $f = \frac{1}{2}(1 - i\mathbf{e}_0)\frac{1}{2}(1 - i\mathbf{e}_1\mathbf{e}_2)$ and results in the charge conjugation

$$\begin{aligned} \mathcal{C}(\psi) &= -i\mathbf{e}_0\mathbf{e}_2\psi^* & \text{complex conjugate } \psi^* \text{ taken in } \mathrm{Mat}(4,\mathbb{C}) \quad \text{or} \\ \mathcal{C}(\psi) &= \psi^*\mathbf{e}_0\mathbf{e}_1 & \text{complex conjugate } \psi^* \text{ taken in } \mathbb{C} \otimes C\ell_{3,1}. \end{aligned}$$

It is possible to represent $C\ell_{3,1}$ by real matrices as follows

$$\mathbf{e}_1 = \begin{pmatrix} \sigma_3 & 0 \\ 0 & -\sigma_3 \end{pmatrix}, \quad \mathbf{e}_2 = \begin{pmatrix} \sigma_1 & 0 \\ 0 & \sigma_1 \end{pmatrix}, \quad \mathbf{e}_3 = \begin{pmatrix} 0 & \sigma_3 \\ \sigma_3 & 0 \end{pmatrix},$$

$$\mathbf{e}_4 = \begin{pmatrix} -i\sigma_2 & 0 \\ 0 & -i\sigma_2 \end{pmatrix}$$

obtained by the choice $f = \frac{1}{2}(1 + \mathbf{e}_1)\frac{1}{2}(1 + \mathbf{e}_2\mathbf{e}_4)$. This results in $\mathcal{C}(\psi) = \psi^*$ for both the real structures $\mathbb{C} \otimes \mathrm{Mat}(4,\mathbb{R}) \simeq \mathrm{Mat}(4,\mathbb{C})$ and $\mathbb{C} \otimes C\ell_{3,1}$.

For either matrix representation the charge conjugation is anti-linear $\mathcal{C}i = -i\mathcal{C}$, involutory $\mathcal{C}^2 = I$ and satisfies $\mathcal{C}\mathbf{A} = \mathbf{A}\mathcal{C}$.

Conjugations in general. Regard the spinor space S as a minimal left ideal $S = (\mathbb{C} \otimes C\ell_{p,q})f$ where the idempotent f is primitive in the complexified Clifford algebra $\mathbb{C} \otimes C\ell_{p,q}$. A real linear transformation $\mathcal{C} : S \to S$ is called a *conjugation* if it is anti-linear $\mathcal{C}i = -i\mathcal{C}$ and satisfies $\mathcal{C}^2 = \pm I$. In principle, a charge conjugation could be either involutory $\mathcal{C}^2 = I$ or satisfy $\mathcal{C}^2 = -I$. However, in the latter case anti-linearity implies $\frac{1}{2}(1 \pm i\mathcal{C})^2 = \pm i\mathcal{C}$, and so the operations $\frac{1}{2}(1 \pm i\mathcal{C})$ do not project complex subspaces of the spinor space S.

For this reason it is appropriate to omit the 'quaternionic' charge conjugations $\mathcal{C}^2 = -I$; I will exclude them, since Budinich & Trautman attach Majorana spinors only to the involutory charge conjugations $\mathcal{C}^2 = I$ (this is implied by page 95 lines 5, 8 and 22).

On pages 90-97 the authors have adopted two different approaches to charge conjugation which are not compatible with each other. In order to make a distinction between the two concepts of charge conjugation of Budinich & Trautman, I will call one of them from now on the *electric conjugation* $\mathcal{E}$ and the other the *Majorana conjugation* $\mathcal{M}$ [$\mathcal{E}$ swaps the sign of the charge of the Dirac equation and for $\mathcal{M}$ the eigenspinors are Majorana spinors].

Electric conjugation in higher dimensions. On page 94 the authors consider the *Dirac equation* not only in the Minkowski space $\mathbb{R}^{3,1}$ or $\mathbb{R}^{1,3}$ but also in higher dimensions. They refer to 'real particles' of 'momentum p' whereby they focus onto the Lorentz signatures $\mathbb{R}^{n-1,1}$ and $\mathbb{R}^{1,n-1}$. [The momentum p of a particle is apparently a time-like vector in a space-time with Lorentz signature (one-dimensional time).]

The electric conjugation $\mathcal{E}$ should transform the Dirac equation (page 94 formulas 7.25 and 7.25i)

$$\mathbb{R}^{n-1,1} \qquad\qquad \mathbb{R}^{1,n-1}$$
$$\partial\psi + ie\mathbf{A}\psi = m\psi \qquad\qquad i\partial\psi - e\mathbf{A}\psi = m\psi$$

to the following form (page 95 lines 1-3)

$$\mathbb{R}^{n-1,1} \qquad\qquad \mathbb{R}^{1,n-1} \qquad .$$
$$\partial\psi_C - ie\mathbf{A}\psi_C = m\psi_C \qquad\qquad i\partial\psi_C + e\mathbf{A}\psi_C = m\psi_C .$$

This change of sign is assumed to be accomplished by a real linear transformation of spinors $\mathcal{E}: S \to S$, $\psi \to \psi_C$ which is anti-linear $\mathcal{E}(i\psi) = -i\mathcal{E}(\psi)$ and involutory $\mathcal{E}(\mathcal{E}(\psi)) = \psi$. [7] An anti-linear transformation swaps the signs as designated above if and only if the following is satisfied (consequences of page 95 lines 1-3)

$$\mathbb{R}^{n-1,1} \qquad\qquad \mathbb{R}^{1,n-1}$$
$$\mathcal{E}(\mathbf{A}\psi) = \mathbf{A}\mathcal{E}(\psi) \qquad\qquad \mathcal{E}(\mathbf{A}\psi) = -\mathbf{A}\mathcal{E}(\psi).$$

To summarize, Budinich & Trautman are looking for higher-dimensional analogs of the Dirac particle – by implication in the Lorentz signatures with a time-like momentum $p = i\partial$ – and refer to an electric conjugation which is anti-linear $\mathcal{E}i = -i\mathcal{E}$, involutory $\mathcal{E}^2 = I$ and such that (gauge $\partial \to \partial + ie\mathbf{A}$)

$$
\begin{array}{cc}
C\ell_{n-1,1} & C\ell_{1,n-1} \\
p^2 + m^2 = 0 & p^2 - m^2 = 0 \\
(\partial^2 - m^2)\psi = 0 & (\partial^2 + m^2)\psi = 0 \\
\partial\psi + ie\mathbf{A}\psi = m\psi & i\partial\psi - e\mathbf{A}\psi = m\psi \\
\boxed{\mathcal{E}\mathbf{A} = \mathbf{A}\mathcal{E}} & \boxed{\mathcal{E}\mathbf{A} = -\mathbf{A}\mathcal{E}}
\end{array}
$$

The Dirac equation could be generalized from the Lorentz spaces to arbitrary signatures. It is not essential whether or not the authors confine themselves to the Lorentz signatures. The essential point is that they *characterize* the electric conjugation by the above *properties* in the case of the Lorentz signatures.

---

[7] Note that e is a scalar for which $\mathcal{E}(e\psi) = e\mathcal{E}(\psi)$, that is, $\mathcal{E}e = e\mathcal{E}$.

<u>Majorana conjugation and its Majorana spinors</u>. The authors define the Majorana conjugation on page 91 line 1 formula 7.19, page 91 lines 6-7, page 95 lines 5/8 formulas 7.26/7.26i and page 95 line -5. Budinich & Trautman's definition results in a real linear transformation $\mathcal{M}$ which is anti-linear $\mathcal{M}i = -i\mathcal{M}$, involutory $\mathcal{M}^2 = I$ and such that [for short denote $\mathbb{D}(d) = \mathrm{Mat}(d, \mathbb{D})$]

			$Cl_{n-1,1}$		$Cl_{1,n-1}$			
		$p-q$		n		$p-q$		
$\mathcal{M}\mathbf{A} = \pm\mathbf{A}\mathcal{M}$	$\mathcal{M} \simeq C, C\Gamma$	0	$\mathbb{R}(2)$	2	$\mathbb{R}(2)$	0	$\mathcal{M} \simeq C, C\Gamma$	$\mathcal{M}\mathbf{A} = \pm\mathbf{A}\mathcal{M}$
$\mathcal{M}\mathbf{A} = \mathbf{A}\mathcal{M}$	$\mathcal{M} \simeq C$	2	$\mathbb{R}(4)$	4	$\mathbb{H}(2)$	6	$\mathcal{M} \simeq C\Gamma$	$\mathcal{M}\mathbf{A} = -\mathbf{A}\mathcal{M}$
		4	$\mathbb{H}(4)$	6	$\mathbb{H}(4)$	4		
$\boxed{\mathcal{M}\mathbf{A} = -\mathbf{A}\mathcal{M}}$	$\mathcal{M} \simeq C\Gamma$	6	$\mathbb{H}(8)$	8	$\mathbb{R}(16)$	2	$\mathcal{M} \simeq C$	$\boxed{\mathcal{M}\mathbf{A} = \mathbf{A}\mathcal{M}}$

The notation $\mathcal{M} \simeq C$ and $\mathcal{M} \simeq C\Gamma$ above means $\mathcal{M}(\psi) = [C(\psi)]^*$ (page 95 line 5 formula 7.26) and $\mathcal{M}(\psi) = [C\Gamma(\psi)]^*$ (page 95 line 8 formula 7.26i). The commutation relations on page 91 line 1 formula 7.19 together with page 95 line 5 (and 8) mean that

$$[C(\mathbf{A}\psi)]^* = C^*\mathbf{A}^*\psi^* = \mathbf{A}C^*\psi^* = \mathbf{A}[C(\psi)]^* \quad \text{in} \quad \mathrm{Mat}(d, \mathbb{C})$$

which together with $\mathbf{A}\Gamma = -\Gamma\mathbf{A}$ imply (the case $p - q = 0$ involves both signs)

$$\mathcal{M}\mathbf{A} = \mathbf{A}\mathcal{M} \quad \text{for} \quad \mathbf{A} \in \mathbb{R}^{p,q}, \quad p - q = 0, 2 \bmod 8$$
$$\mathcal{M}\mathbf{A} = -\mathbf{A}\mathcal{M} \quad \text{for} \quad \mathbf{A} \in \mathbb{R}^{p,q}, \quad p - q = 0, 6 \bmod 8$$

after recalling that the authors define on page 95 lines 8,-5,-1 the Majorana conjugation to be $\mathcal{M} \simeq C\Gamma$ in the signature $p - q = 6 \bmod 8$.

Majorana spinors are eigenspinors of the Majorana conjugation $\mathcal{M}$ with eigenvalues ± 1, $\mathcal{M}(\psi) = \pm\psi$; as a consequence, for Majorana spinors to exist the Majorana conjugation must be involutive $\mathcal{M}^2 = I$. In the signatures $p - q = 4 \bmod 8$ there is a Majorana conjugation which is quaternionic $\mathcal{M}^2 = -I$, and/but there are no Majorana spinors (page 97 line 12), so exclude $p - q = 4 \bmod 8$.

<u>Electric conjugation $\neq$ Majorana conjugation</u>. The above formulas in the boxes ($=$ consequences of page 91 line 1 formula 7.19 and page 95 line -5) show that the Majorana conjugation $\mathcal{M}$ of the authors is not in accordance with their electric conjugation $\mathcal{E}$ in dimension 8. The Majorana conjugation transforms the Dirac equation as follows [the operator ∂ commutes like the vector $\mathbf{A}$; denote $\psi_\mathcal{M} = \mathcal{M}(\psi)$]

$\mathbb{R}^{7,1}$	$\mathbb{R}^{1,7}$
$\partial\psi + ie\mathbf{A}\psi = m\psi$	$i\partial\psi - e\mathbf{A}\psi = m\psi$
$-\partial\psi_\mathcal{M} + ie\mathbf{A}\psi_\mathcal{M} = m\psi_\mathcal{M}$	$-i\partial\psi_\mathcal{M} - e\mathbf{A}\psi_\mathcal{M} = m\psi_\mathcal{M}$
$[\partial\psi_\mathcal{M} - ie\mathbf{A}\psi_\mathcal{M} = -m\psi_\mathcal{M}]$	$[i\partial\psi_\mathcal{M} + e\mathbf{A}\psi_\mathcal{M} = -m\psi_\mathcal{M}]$.

The electric charge is not conjugated according to the characterizations on page 94 and top three lines on page 95, which contradicts Budinich & Trautman's theory of (electric) charge conjugation [of the Dirac equation] having (supposedly) Majorana spinors as its eigenspinors.

I have asked the authors to provide me with an explicit conjugation swapping the sign of the electric charge of the Dirac equation in dimension 8. While the authors have not presented such an involutive antilinear conjugation, I remain confident that no electric (charge) conjugation exists in dimension 8. [If assumptions/definitions are altered, it is possible to have a charge conjugation, which swaps the signs of the electric charge and the *mass*, like the Majorana conjugation $\mathcal{M}$, but discussion about changes possibly rescuing a theory is beyond the scope of this paper: locating inconsistencies in scientific publications – thus offering the authors a chance to enhance their theories.]   ∎

Rodriguez-Romo & Viniegra & Keller 1992. The authors discuss the bilinear covariants of Dirac spinors and give quadratic relations between the bilinear covariants (also other than the Fierz identities). The computations are carried out in matrix formalism, without taking advantage of the multivector structure of Clifford algebras. In a matrix approach some details are blurred.

For multivector approach to bilinear covariants, see Lounesto 1993.

Recall that a column spinor $\psi \in \mathbb{C}^4$ can be replaced by a square-matrix spinor $\psi \in \mathrm{Mat}(4, \mathbb{C})$, more explicitly,

$$\psi = \begin{pmatrix} \psi_1 \\ \psi_2 \\ \psi_3 \\ \psi_4 \end{pmatrix} \in \mathbb{C}^4 \quad \text{by} \quad \psi = \begin{pmatrix} \psi_1 & 0 & 0 & 0 \\ \psi_2 & 0 & 0 & 0 \\ \psi_3 & 0 & 0 & 0 \\ \psi_4 & 0 & 0 & 0 \end{pmatrix} \in \mathrm{Mat}(4, \mathbb{C})f$$

where f is a primitive idempotent

$$f = \begin{pmatrix} 1 & 0 & 0 & 0 \\ 0 & 0 & 0 & 0 \\ 0 & 0 & 0 & 0 \\ 0 & 0 & 0 & 0 \end{pmatrix}$$

of $\mathrm{Mat}(4, \mathbb{C})$. Taking account of the isomorphism $\mathrm{Mat}(4, \mathbb{C}) \simeq \mathbb{C} \otimes C\ell_{1,3}$ we can regard the square-matrix spinor $\psi \in \mathrm{Mat}(4, \mathbb{C})f$ as a Clifford-algebraic spinor

$$\psi = \psi_1 f_1 + \psi_2 f_2 + \psi_3 f_3 + \psi_4 f_4 \in (\mathbb{C} \otimes C\ell_{1,3})f$$

expressed in the basis

$$\begin{aligned}
f_1 &= \tfrac{1}{4}(1 + \gamma_0 + i\gamma_{12} + i\gamma_{012}) &&= f \\
f_2 &= \tfrac{1}{4}(-\gamma_{13} + i\gamma_{23} - \gamma_{013} + i\gamma_{023}) &&= -\gamma_{13}f \\
f_3 &= \tfrac{1}{4}(\gamma_3 - \gamma_{03} + i\gamma_{123} - i\gamma_{0123}) &&= -\gamma_{03}f \\
f_4 &= \tfrac{1}{4}(\gamma_1 - i\gamma_2 - \gamma_{01} + i\gamma_{02}) &&= -\gamma_{01}f
\end{aligned}$$

of the minimal left ideal $S = (\mathbb{C} \otimes C\ell_{1,3})f$,

$$f = \frac{1}{2}(1 + \gamma_0)\frac{1}{2}(1 + i\gamma_1\gamma_2).$$

The bilinear covariants are now, in terms of both the column spinor $\psi \in \mathbb{C}^4$ and the Clifford-algebraic spinor $\psi \in (\mathbb{C} \otimes C\ell_{1,3})f$,

$$\sigma = \psi^\dagger \gamma_0 \psi = 4\langle \tilde{\psi}^* \psi \rangle_0$$
$$J_\mu = \psi^\dagger \gamma_0 \gamma_\mu \psi = 4\langle \tilde{\psi}^* \gamma_\mu \psi \rangle_0$$
$$S_{\mu\nu} = \psi^\dagger \gamma_0 i\gamma_{\mu\nu} \psi = 4\langle \tilde{\psi}^* i\gamma_{\mu\nu} \psi \rangle_0$$
$$K_\mu = \psi^\dagger \gamma_0 i\gamma_{0123}\gamma_\mu \psi = 4\langle \tilde{\psi}^* i\gamma_{0123}\gamma_\mu \psi \rangle_0$$
$$\omega = -\psi^\dagger \gamma_0 \gamma_{0123} \psi = -4\langle \tilde{\psi}^* \gamma_{0123} \psi \rangle_0 \qquad\qquad [\gamma_{0123} = \gamma_0\gamma_1\gamma_2\gamma_3].$$

The Dirac current density is the vector

$$\mathbf{J} = \gamma^\mu J_\mu = \gamma^\mu 4\langle \gamma_\mu \psi \psi^\dagger \gamma_0 \rangle_0 \qquad\qquad \psi \in \mathrm{Mat}(4,\mathbb{C})f$$
$$= \langle 4\psi\tilde{\psi}^* \rangle_1 \qquad\qquad \psi \in (\mathbb{C} \otimes C\ell_{1,3})f$$

and the electromagnetic moment density is the bivector

$$\mathbf{S} = \tfrac{1}{2}\gamma^{\mu\nu} S_{\mu\nu} = \tfrac{1}{2}\gamma^{\mu\nu} 4\langle i\gamma_{\mu\nu} \psi \psi^\dagger \gamma_0 \rangle_0 \qquad\qquad \psi \in \mathrm{Mat}(4,\mathbb{C})f$$
$$= -i\langle 4\psi\tilde{\psi}^* \rangle_2 \qquad\qquad \psi \in (\mathbb{C} \otimes C\ell_{1,3})f.$$

Thus, we have bilinear covariants

$$\sigma \in \mathbb{R} \qquad\qquad \text{a scalar}$$
$$\mathbf{J} \in \mathbb{R}^{1,3} \qquad\qquad \text{a vector}$$
$$\mathbf{S} \in {\textstyle\bigwedge}^2 \mathbb{R}^{1,3} \qquad\qquad \text{a bivector}$$
$$\mathbf{K}\gamma_{0123} \in {\textstyle\bigwedge}^3 \mathbb{R}^{1,3} \qquad\qquad \text{a 3-vector}$$
$$\omega\gamma_{0123} \in {\textstyle\bigwedge}^4 \mathbb{R}^{1,3} \qquad\qquad \text{a 4-volume element}$$

all of which are real [contrary to the statement of Grandy 1991, p. 96, l. -1, his $\psi^\dagger \gamma_0(-i\gamma_{0123})\psi$ is not real].

It should be noted that the square-matrix spinor and the Clifford-algebraic spinor differ by their real structures/parts; for the Clifford-algebraic spinor the real part is

$$\mathrm{Re}(\psi) = \frac{1}{2} \begin{pmatrix} \psi_1 & -\psi_2^* & 0 & 0 \\ \psi_2 & \psi_1^* & 0 & 0 \\ \psi_3 & \psi_4^* & 0 & 0 \\ \psi_4 & -\psi_3^* & 0 & 0 \end{pmatrix} \quad \text{in} \quad \mathbb{C} \otimes C\ell_{1,3}.$$

Denote $\Phi = 4\,\mathrm{Re}(\psi)$, and define the *spinor operator*

$$\Psi = \mathrm{even}(\Phi) = \begin{pmatrix} \psi_1 & -\psi_2^* & \psi_3 & \psi_4^* \\ \psi_2 & \psi_1^* & \psi_4 & -\psi_3^* \\ \psi_3 & \psi_4^* & \psi_1 & -\psi_2^* \\ \psi_4 & -\psi_3^* & \psi_2 & \psi_1^* \end{pmatrix}$$

from which we may re-obtain the original Dirac spinor as

$$\psi = \Psi \frac{1}{2}(1 + \gamma_0)\frac{1}{2}(1 + i\gamma_{12}).$$

In terms of the spinor operator the bilinear covariants are

$$\boldsymbol{\Psi}\tilde{\boldsymbol{\Psi}} = \sigma + \omega\gamma_{0123}$$
$$\boldsymbol{\Psi}\gamma_0\tilde{\boldsymbol{\Psi}} = \mathbf{J}$$
$$\boldsymbol{\Psi}\gamma_{12}\tilde{\boldsymbol{\Psi}} = \mathbf{S}, \qquad \boldsymbol{\Psi}\gamma_{03}\tilde{\boldsymbol{\Psi}} = -\mathbf{S}\gamma_{0123}$$
$$\boldsymbol{\Psi}\gamma_3\tilde{\boldsymbol{\Psi}} = \mathbf{K}, \qquad \boldsymbol{\Psi}\gamma_{012}\tilde{\boldsymbol{\Psi}} = \mathbf{K}\gamma_{0123}$$

that is, $\boldsymbol{\Psi}$ operates like a Lorentz transformation composed with a dilation (and possibly a duality transformation), in the case $\boldsymbol{\Psi}\tilde{\boldsymbol{\Psi}} \neq 0$.

Define $Z = 4\psi\psi^\dagger\gamma_0 = 4\psi\tilde{\psi}^*$ to find that

$$Z = \sigma + \mathbf{J} + i\mathbf{S} + i\mathbf{K}\gamma_{0123} + \omega\gamma_{0123}.$$

If $\sigma+\omega\gamma_{0123} \neq 0$, the Fierz identities imply $Z^2 = 4\sigma Z$. Define $W = 4\psi\psi_C^\dagger\gamma_0 = 4\psi\tilde{\psi}_C^*$ (the last factor is the Dirac adjoint of the charge conjugate of ψ). Then

$$W = \mathcal{K} - \mathcal{S}\gamma_{0123} \quad \text{where} \quad \mathcal{K} = \mathbf{K}_1 + i\mathbf{K}_2, \ \mathcal{S} = \mathbf{S}_1 + i\mathbf{S}_2,$$

when $\mathbf{K}_k = \boldsymbol{\Psi}\gamma_k\tilde{\boldsymbol{\Psi}}$, $\mathbf{S}_k = \boldsymbol{\Psi}\gamma_{ij}\tilde{\boldsymbol{\Psi}}$ (ijk cycl.). We may find that $W^2 = 0$, in particular

$$\langle W^2 \rangle_0 = \mathcal{K}^2 - \mathcal{S} \lrcorner \mathcal{S} = 0,$$

which is the same as Rodriguez-Romo & Viniegra & Keller 1992, p. 492, (3.12), although they did not observe that separately $\mathcal{K}^2 = 0$, $\mathcal{S}^2 = 0$. Furthermore, we find $WZ = 0$, in particular

$$\langle WZ \rangle_0 = \mathcal{K} \cdot \mathbf{J} - i(\gamma_{0123}\mathcal{S}) \lrcorner \mathbf{S} = 0,$$

which corresponds to Rodriguez-Romo & Viniegra & Keller 1992, p. 492, (3.13), although they did not notice that separately $\mathcal{K} \cdot \mathbf{J} = 0$; and $ZW = 4\sigma W$, in particular

$$\langle ZW \rangle_3 = -\mathbf{J} \wedge (\gamma_{0123}\mathcal{S}) + i\mathbf{S} \wedge \mathcal{K} + \gamma_{0123}\omega\mathcal{K} + i\mathbf{K} \wedge \mathcal{S} = 0,$$

which rectifies an error in Rodriguez-Romo & Viniegra & Keller 1992, p. 496, (3.22), where the last term is missing [observe that the real and imaginary parts both vanish separately]. ∎

CLICAL was effectively used to untangle Budinich & Trautman and Rodriguez-Romo & Viniegra & Keller. In both cases, some details did not fit together, and computations seemed too tedious for hand calculations. The picture was gradually clarified with CLICAL.

6. Decomposing conformal transformations

Recall that, in dimensions $n \geq 3$, sense preserving conformal mappings are just restrictions of the Möbius transformations, that is, compositions of rotations, translations, dilations and transversions (called special conformal transformations, in physics). A Möbius transformation of $\mathbb{R}^{p,q}$ can be written in the form

$$\mathbf{x} \to (a\mathbf{x} + b)(c\mathbf{x} + d)^{-1},$$

where $a, b, c, d \in C\ell_{p,q}$, and represented by a Vahlen matrix $\begin{pmatrix} a & b \\ c & d \end{pmatrix} \in \mathrm{Mat}(2, C\ell_{p,q})$.
The elements of a Vahlen matrix must satisfy the following conditions

     1. a, b, c, d are products of vectors,
     2. $a\tilde{d} - b\tilde{c} = \pm 1$,
     3. $a\tilde{b}, \tilde{b}d, d\tilde{c}, \tilde{c}a$ are vectors,

see Lounesto & Springer 1989. Rotations, translations, dilations and transversions
will be represented as follows

$$ax a^{-1} \qquad a \in \mathbf{Spin}_+(p, q) \qquad \begin{pmatrix} a & 0 \\ 0 & a \end{pmatrix}$$

$$\mathbf{x} + \mathbf{b} \qquad \mathbf{b} \in \mathbb{R}^{p,q} \qquad \begin{pmatrix} 1 & \mathbf{b} \\ 0 & 1 \end{pmatrix}$$

$$\mathbf{x}\delta \qquad \delta > 0 \qquad \begin{pmatrix} \sqrt{\delta} & 0 \\ 0 & 1/\sqrt{\delta} \end{pmatrix}$$

$$\frac{\mathbf{x} + \mathbf{x}^2 \mathbf{c}}{1 + 2\mathbf{x} \cdot \mathbf{c} + \mathbf{x}^2 \mathbf{c}^2} \qquad \mathbf{c} \in \mathbb{R}^{p,q} \qquad \begin{pmatrix} 1 & 0 \\ \mathbf{c} & 1 \end{pmatrix}.$$

Note that a product, where each of these four matrices appears just once, always has
an invertible entry in its diagonal (there are $4! = 24$ such products). For instance,
in the product

$$\begin{pmatrix} a & 0 \\ 0 & a \end{pmatrix} \begin{pmatrix} 1 & \mathbf{b} \\ 0 & 1 \end{pmatrix} \begin{pmatrix} \sqrt{\delta} & 0 \\ 0 & 1/\sqrt{\delta} \end{pmatrix} \begin{pmatrix} 1 & 0 \\ \mathbf{c} & 1 \end{pmatrix}$$
$$= \begin{pmatrix} a\sqrt{\delta} + a\mathbf{b}\mathbf{c}/\sqrt{\delta} & a\mathbf{b}/\sqrt{\delta} \\ a\mathbf{c}/\sqrt{\delta} & a/\sqrt{\delta} \end{pmatrix}$$

the lower right-hand diagonal element $a/\sqrt{\delta}$ is invertible.

Now, consider the Minkowski space $\mathbb{R}^{3,1}$ and its Clifford algebra $C\ell_{3,1}$ generated
by $\mathbf{e}_1, \mathbf{e}_2, \mathbf{e}_3, \mathbf{e}_4$ satisfying $\mathbf{e}_1^2 = \mathbf{e}_2^2 = \mathbf{e}_3^2 = 1$, $\mathbf{e}_4^2 = -1$. The Vahlen matrix

$$W = \frac{1}{2} \begin{pmatrix} 1 - \mathbf{e}_{14} & -\mathbf{e}_1 + \mathbf{e}_4 \\ \mathbf{e}_1 + \mathbf{e}_4 & 1 + \mathbf{e}_{14} \end{pmatrix}$$

with entries in the Clifford algebra $C\ell_{3,1} \simeq \mathrm{Mat}(4, \mathbb{R})$ is such that all its entries
are non-invertible. [8] [This example is due to J. Maks 1989, p. 41.] Thereby,
the Vahlen matrix W is not a product of just one rotation, one translation, one
dilation and one transversion (in any order). However, this Vahlen matrix W is
in the identity component of the conformal group of the Minkowski space, since its
diagonal elements a, d are even and its pseudo-determinant $a\tilde{d} - b\tilde{c}$ equals 1, or
path-wise, since it can be written as a product of a transversion, a translation and
a transversion

$$W = \begin{pmatrix} 1 & 0 \\ \frac{1}{2}(\mathbf{e}_1 + \mathbf{e}_4) & 1 \end{pmatrix} \begin{pmatrix} 1 & \frac{1}{2}(-\mathbf{e}_1 + \mathbf{e}_4) \\ 0 & 1 \end{pmatrix} \begin{pmatrix} 1 & 0 \\ \frac{1}{2}(\mathbf{e}_1 + \mathbf{e}_4) & 1 \end{pmatrix}.$$

---

[8] In the Euclidean spaces $\mathbb{R}^n$ we could replace the first condition for the Vahlen matrices by
1. $a, b, c, d \in \Gamma_n \cup \{0\}$. This cannot be done in the indefinite metrics, as shown by the matrix W.
Exercise: locate this error in the literature.

Here we have a counter-example to Hestenes 1991, p. 91, ll. 10-12, in particular formula (5.89). [This mistake is a small variation of an earlier mistake in Hestenes & Sobczyk 1984/1987, p. 218, ll. 11-13, formula (5.50).]

In 1991, Hestenes was aware of Maks' counter-example, but misinterpreted it, as can be seen from the last two lines on page 90 and the first line on page 91. Hestenes writes in 1991, p. 91, ll. 1-2, that his *diversion is a more fundamental counter-example* than Maks' counter-example. It is true that W factors into a product of two 'diversions' as follows

$$W = \frac{1}{\sqrt{2}} \begin{pmatrix} 1 & -e_1 \\ e_1 & 1 \end{pmatrix} \frac{1}{\sqrt{2}} \begin{pmatrix} 1 & e_4 \\ e_4 & 1 \end{pmatrix}$$

[see p. 90, l. -4, formula (5.88), where the last term has an obvious misprint in it, and should be $(1 - \mathbf{ab})e_- e_1$]. However, it is not true that a 'diversion' cannot be expressed as a product of just one transversion, one dilation, one translation and one rotation. This can be seen by expressing the first factor above as a product of just one transversion, one dilation and one translation as follows

$$\frac{1}{\sqrt{2}} \begin{pmatrix} 1 & -e_1 \\ e_1 & 1 \end{pmatrix} = \begin{pmatrix} 1 & 0 \\ e_1 & 1 \end{pmatrix} \begin{pmatrix} 1/\sqrt{2} & 0 \\ 0 & \sqrt{2} \end{pmatrix} \begin{pmatrix} 1 & -e_1 \\ 0 & 1 \end{pmatrix}.$$

Inserting the identity rotation as the last factor we see that a 'diversion' is a product of just one transversion, one dilation, one translation and one rotation, and as such cannot serve as a *more fundamental counter-example*, contrary to the claim of Hestenes 1991, p. 91, ll. 1-2.

7. Exterior algebra and characteristic 2

Witt 1937 noticed the connection between quadratic forms and Clifford algebras. At that time mathematicians tried to develop theories over arbitrary fields. Witt's work had left open the case of characteristic 2. Chevalley 1954 gave two definitions for the Clifford algebra,

$$C\ell(Q) \subset \mathrm{End}(\textstyle\bigwedge V) \quad \text{and} \quad C\ell(Q) = \otimes V/\mathcal{I}_Q.$$

The first definition works for all fields, even in characteristic 2, and the second definition is the most general, being valid for commutative rings.

Chevalley's theory incorporating characteristic 2 differed essentially from Clifford's original approach: it used the endomorphism algebra $\mathrm{End}(\bigwedge V)$ of the exterior algebra $\bigwedge V$ and was based on non-symmetric bilinear forms, the symmetric part being the symmetric scalar product (in characteristic $\neq 2$). Chevalley's approach has caused confusion among those who have forgotten his motives: the inclusion of characteristic 2.

Lawson & Michelsohn 1989, p. 10, claim that there is a canonical isomorphism of linear spaces $\bigwedge V \to C\ell(Q)$, and Crumeyrolle 1990, p. 292, claims that there is no such canonical isomorphism. In other words, Lawson & Michelsohn and Crumeyrolle have opposite opinions about a possible existence of a multivector structure in Clifford algebras. However, both are wrong: Lawson & Michelsohn because they allow characteristic 2, and Crumeyrolle because he excludes characteristic 2. [Roughly speaking, Lawson & Michelsohn claim that two numbers x, y in a field $\mathbb{F}$ always have a mean $\frac{1}{2}(x + y)$ within that field $\mathbb{F}$, but this is false in the field of two elements $\mathbb{F} = \{0, 1\}$, whereas Crumeyrolle claims that no such mean exists, but he is wrong since he excludes characteristic 2.]

Before discussing the confusion in detail, I will point out a related mistake of Lawson & Michelsohn 1989. On page 10 lines 15-16 the authors claim that the defining relations $\mathbf{xy} + \mathbf{yx} = 2g(\mathbf{x}, \mathbf{y})$ have the form $\mathbf{xy} + \mathbf{yx} = 0$ in characteristic 2 [because $2 = 0$]. This shows that the authors were unaware of the interesting, non-linear quadratic forms, induced by non-symmetric non-diagonalizable bilinear forms: Take a linear space V of dimension 2 over the field $\mathbb{F} = \{0, 1\}$, with quadratic form $Q(x_1 e_1 + x_2 e_2) = x_1 x_2$ on V. There are only three non-zero vectors in V, the basis vectors e_1, e_2 and their sum $e_1 + e_2$, for which $Q(e_1) = 0$, $Q(e_2) = 0$ and $Q(e_1 + e_2) = 1$. In the Clifford algebra $C\ell(Q)$ one gets $e_1^2 = 0$, $e_2^2 = 0$ and $(e_1 + e_2)^2 = 1$, which implies $e_1 e_2 + e_2 e_1 = 1 \neq 0$.

In fact, the above Clifford algebra $C\ell(Q)$ serves as maximal counter-example in the sense that for all non-zero vectors $\mathbf{v}, \mathbf{w} \in V$ such that $\mathbf{v} \neq \mathbf{w}$ we have $\mathbf{vw} + \mathbf{wv} = 1$. To see this, work out the multiplication table of $C\ell(Q)$. First, there are only two linear isomorphisms $C\ell(Q) \to \bigwedge V$ which preserve the even-odd parity and which are identity mappings when restricted to $\mathbb{F} \oplus V$ [the image of $e_2 e_1$ being either $e_1 \wedge e_2$ or $1 + e_1 \wedge e_2$]. Respectively, there are two different representations of the Clifford algebra $C\ell(Q)$ in the exterior algebra $\bigwedge V$ with the following multiplication tables

	e_1	e_2	$e_1 \wedge e_2$
e_1	0	$1 + e_1 \wedge e_2$	$-e_1$
e_2	$-e_1 \wedge e_2$	0	0
$e_1 \wedge e_2$	0	$-e_2$	$-e_1 \wedge e_2$

	e_1	e_2	$e_1 \wedge e_2$
e_1	0	$e_1 \wedge e_2$	0
e_2	$1 - e_1 \wedge e_2$	0	e_2
$e_1 \wedge e_2$	e_1	0	$e_1 \wedge e_2$

expressed in terms of the basis $1, e_1, e_2, e_1 \wedge e_2$ for $\bigwedge V$.

We may conclude that *there are no canonical linear isomorphisms* $\bigwedge V \to C\ell(Q)$, that is, neither of the above multiplication tables can be preferred over the other. In other words, $\bigwedge^2 V$ cannot be canonically embedded in $C\ell(Q)$, that is, Clifford algebras do not have a natural multivector structure in characteristic 2.

The above discussion shows that Lawson & Michelsohn pp. 10-11 are wrong, since they allow characteristic 2, and claim that there is a canonical linear isomorphism $\bigwedge V \to C\ell(Q)$. To show that the opposite statement of Crumeyrolle is also wrong, it is sufficient to present in characteristic $\neq 2$ a construction of a linear isomorphism $C\ell(Q) \to \bigwedge V$, which is canonical/natural (= independent of a particular choice). Such a linear isomorphism was presented by M. Riesz 1958/1993, pp. 61-67. Riesz reobtained the exterior product of $\mathbf{x} \in V$ and $u \in C\ell(Q)$ by the formula [9]

$$\mathbf{x} \wedge u = \frac{1}{2}(\mathbf{x}u + \hat{u}\mathbf{x});$$

thus the subspace of k-vectors is constructed recursively via [10]

$$\mathbf{x} \wedge \mathbf{a} = \frac{1}{2}(\mathbf{x}\mathbf{a} + (-1)^k \mathbf{a}\mathbf{x}) \in \overset{k}{\bigwedge} V \quad \text{for} \quad \mathbf{a} \in \overset{k-1}{\bigwedge} V.$$

Salamon 1989. The author claims on p. 170, l. 1, formula (12.2), that

$$\mathbf{a}\mathbf{b} = \mathbf{a} \wedge \mathbf{b} + \mathbf{a} \lrcorner \mathbf{b} \quad \text{for} \quad \mathbf{a} \in \overset{k}{\bigwedge} \mathbb{R}^{0,n}, \ \mathbf{b} \in \overset{l}{\bigwedge} \mathbb{R}^{0,n}, \ k < l,$$

where $\mathbf{a} \lrcorner \mathbf{b} \in \bigwedge^{l-k} \mathbb{R}^{0,n}$. A counter-example, satisfying all the assumptions, is given by $\mathbf{a} = \mathbf{e}_{12} \in \bigwedge^2 \mathbb{R}^{0,5}$ and $\mathbf{b} = \mathbf{e}_{134} \in \bigwedge^3 \mathbb{R}^{0,5}$ [as assumed $k < l$ while $k = 2$, $l = 3$], for which

$$\mathbf{a}\mathbf{b} = \mathbf{e}_{12}\mathbf{e}_{134} = -\mathbf{e}_1^2 \mathbf{e}_{234} = \mathbf{e}_{234} \notin \overset{5}{\bigwedge} \mathbb{R}^{0,5} \oplus \mathbb{R}^{0,5}$$

while $\mathbf{e}_{234} \in \bigwedge^3 \mathbb{R}^{0,5}$. ∎

Oziewicz 1986. Oziewicz p. 252, l. -1, formula (26), tried to generalize the Clifford algebra of a quadratic form Q on V. He proposed to replace the symmetric bilinear form g, $Q(\mathbf{x}) = g(\mathbf{x}, \mathbf{x})$, by a *not necessarily symmetric* bilinear form B, $B(\mathbf{x}, \mathbf{x}) = Q(\mathbf{x})$. However, this 'generalization' does not result in a new algebra, in the sense that the algebra so constructed is isomorphic, as an associative algebra, to the Clifford algebra of the symmetric part g of B [in characteristic $\neq 2$]. Only the multivector structure varies along with the antisymmetric part A of B, so that the new exterior product is given by

$$\mathbf{x} \overset{.}{\wedge} \mathbf{y} = \mathbf{x} \wedge \mathbf{y} + A(\mathbf{x}, \mathbf{y}) \quad \text{for} \quad \mathbf{x}, \mathbf{y} \in V.$$

On the same page 252, lines 2-4, formula (23), Oziewicz presented a definition of his Clifford algebra, but it is not necessary to assume his condition 1, which is

---

[9] Chevalley had already earlier in 1946 found that completely antisymmetric Clifford-products of vectors correspond to simple elements in the exterior algebra

$$\mathbf{x}_1 \wedge \mathbf{x}_2 \wedge \ldots \wedge \mathbf{x}_k = \frac{1}{k!} \sum_{\sigma \in S_k} \text{sign}(\sigma) \mathbf{x}_{\sigma(1)} \mathbf{x}_{\sigma(2)} \cdots \mathbf{x}_{\sigma(k)},$$

but this correspondence required the ground field to be of characteristic 0.

[10] Riesz also found that the derivation, or contraction by a vector $\mathbf{x} \in V$, can be reobtained from the Clifford product by

$$\mathbf{x} \lrcorner \mathbf{a} = \frac{1}{2}(\mathbf{x}\mathbf{a} - (-1)^k \mathbf{a}\mathbf{x}) \in \overset{k-1}{\bigwedge} V \quad \text{for} \quad \mathbf{a} \in \overset{k}{\bigwedge} V.$$

a consequence of his conditions 2 and 3 (in the case $Q \neq 0$). This can be seen as follows. Take $\mathbf{x} \in V$ and $u \in \bigwedge V$. Then [11]

$$
\begin{aligned}
(\mathbf{xx})u &= \mathbf{x}(\mathbf{x}u) && \text{by 3.} \\
&= \mathbf{x} \wedge (\mathbf{x} \wedge u + \mathbf{x} \lrcorner\, u) + \mathbf{x} \lrcorner\, (\mathbf{x} \wedge u + \mathbf{x} \lrcorner\, u) && \text{by 2.} \\
&= \mathbf{x} \wedge \mathbf{x} \wedge u + \mathbf{x} \wedge (\mathbf{x} \lrcorner\, u) + \mathbf{x} \lrcorner\, (\mathbf{x} \wedge u) + \mathbf{x} \lrcorner\, (\mathbf{x} \lrcorner\, u) \\
&= (\mathbf{x} \lrcorner\, \mathbf{x})u,
\end{aligned}
$$

since $\mathbf{x} \wedge \mathbf{x} \wedge u = 0$ by p. 247, l. -9, $\mathbf{x} \lrcorner\, (\mathbf{x} \lrcorner\, u) = 0$ by p. 248, l. -11, and $\mathbf{x} \lrcorner\, (\mathbf{x} \wedge u) = (\mathbf{x} \lrcorner\, \mathbf{x}) \wedge u + \hat{\mathbf{x}} \wedge (\mathbf{x} \lrcorner\, u)$ by p. 249, l. 1. ∎

Lounesto 1995. The following related mistake of mine was brought to my attention by Rafał Abłamowicz, who pointed out that from

$$
xu = \mathbf{x} \wedge u + \mathbf{x} \lrcorner\, u \tag{1}
$$

on page 138 one cannot conclude that

$$
\hat{u}\mathbf{x} = \mathbf{x} \wedge u - \mathbf{x} \lrcorner\, u \tag{2}
$$

in the case of a non-symmetric bilinear form B.

The error can be corrected [12] either by restricting to a symmetric B or by making explicit that the contraction depends on B. In the latter case, denote by $B^{\mathrm{op}} = -B^{\mathsf{T}}$ or

$$
B^{\mathrm{op}}(\mathbf{x}, \mathbf{y}) = -B(\mathbf{y}, \mathbf{x})
$$

the bilinear form inducing the opposite metric, that is, $B^{\mathrm{op}}(\mathbf{x}, \mathbf{x}) = Q^{\mathrm{op}}(\mathbf{x}) = -Q(\mathbf{x})$, $Q(\mathbf{x}) = B(\mathbf{x}, \mathbf{x})$. Then, we have

$$
xu = \mathbf{x} \wedge u + \mathbf{x} \lrcorner_B\, u \tag{1$'$}
$$

and, since $\mathrm{op}(\mathbf{x}u) = \hat{u}\mathbf{x}$, it follows that

$$
\hat{u}\mathbf{x} = \mathbf{x} \wedge u + \mathbf{x} \lrcorner_{B^{\mathrm{op}}}\, u \tag{2$'$}
$$

which, in the case of a symmetric B, can be denoted as (1) and (2), which in turn imply (the formulas on page 139)

$$
\mathbf{x} \wedge u = \frac{1}{2}(xu + \hat{u}\mathbf{x}) \tag{3}
$$

---

[11] We will replace the 'interior' product of Oziewicz $\bigwedge V^* \times \bigwedge V \to \bigwedge V$, $(u, v) \to i_u(v)$ by a product computed just within one algebra $\bigwedge V$, namely the left contraction $\bigwedge V \times \bigwedge V \to \bigwedge V$, $(u, v) \to u \lrcorner\, v$ determined uniquely also in the degenerate case by

  1. $\mathbf{x} \lrcorner\, \mathbf{y} = B(\mathbf{x}, \mathbf{y})$
  2. $\mathbf{x} \lrcorner\, (u \wedge v) = (\mathbf{x} \lrcorner\, u) \wedge v + \hat{u} \wedge (\mathbf{x} \lrcorner\, v)$
  3. $(u \wedge v) \lrcorner\, w = u \lrcorner\, (v \lrcorner\, w)$

for $\mathbf{x}, \mathbf{y} \in V$, $u, v, w \in \bigwedge V$.

[12] As the author who made the mistake, I am entitled to propose a correction. In general, I try to avoid presenting corrections, because the authors might have (had) different goals in mind. I merely point out interior inconsistencies and justify my claims by counter-examples.

$$\mathbf{x} \lrcorner u = \frac{1}{2}(\mathbf{x}u - \hat{u}\mathbf{x}) \tag{4}$$

where it is not necessary to make explicit the symmetric B.

I would like to use this occasion to express my gratitude to Rafał Abłamowicz for giving me an opportunity to learn more about Clifford algebras.   ∎

8. Cayley-Dickson process

For the Cayley-Dickson algebras, see Hagmark & Lounesto 1986 and Schafer 1954. Micali 1990 introduces in $C\ell(Q)$ a deformed product $x \circ y$ given by

$$(x_0 + x_1) \circ (y_0 + y_1) = (x_0 y_0 + \alpha \tilde{y}_1 x_1) + (y_1 x_0 + x_1 \tilde{y}_0)$$

where $x_0, y_0 \in C\ell^+(Q)$, $x_1, y_1 \in C\ell^-(Q)$, $\alpha \in \mathbb{F}$; as before the reversion is denoted $u \to \tilde{u}$. On pp. 237-238, the author says that each Cayley-Dickson algebra $\mathbb{F}(\alpha_1, \alpha_2, ..., \alpha_n)$ is a deformation of some Clifford algebra $C\ell(Q, V)$ of a non-degenerate Q on V, such that $\dim V = n$, over the field $\mathbb{F}$. This does not hold for the following reasons: First, the order of the scalars $\alpha_1, \alpha_2, ..., \alpha_n$ in $\mathbb{F}(\alpha_1, \alpha_2, ..., \alpha_n)$ matters, whereas for the Clifford algebra of $Q(\mathbf{x}) = \alpha_1 x_1^2 + \alpha_2 x_2^2 + ... + \alpha_n x_n^2$ (in characteristic $\neq 2$) the order of the terms is irrelevant (that is, there are more Cayley-Dickson algebras than Clifford algebras). Secondly, the deformations of Clifford algebras are not flexible after $n \geq 5$. This can be seen in the Clifford algebra $C\ell_5 \simeq \mathrm{Mat}(2, \mathbb{H}) \oplus \mathrm{Mat}(2, \mathbb{H})$ by considering the elements $a = \mathbf{e}_1 + \mathbf{e}_{12}$ and $b = \mathbf{e}_{345}$ for which Micali's deformed product with $\alpha = -1$ satisfies $(a \circ b) \circ a = -2\mathbf{e}_{2345} \neq a \circ (b \circ a) = 2\mathbf{e}_{2345}$.

9. Not the best result

In the Clifford algebras of definite spaces one can introduce a norm in a natural way

$$|u| = \sqrt{\langle u\tilde{u} \rangle_0} \quad \text{for} \quad u \in C\ell_n = C\ell_{n,0},$$
$$|u| = \sqrt{\langle u\bar{u} \rangle_0} \quad \text{for} \quad u \in C\ell_{0,n}.$$

Constales 1989, p. 6, gives an estimate for the norm

$$|uv| \leq 2^{n/2}|u||v| \quad \text{in the case} \quad u, v \in C\ell_{0,n}$$

and admits that the constant $2^{n/2}$ is not optimal, and refers to a better value [Constales' misprint corrected: missing last parenthesis]

$$H_{0,n} = \sqrt{2^{n-1}(1 + 2^{(1-n)/2} \cos((n+1)\pi/4))}$$

given by Habetha 1985. However, even this is not optimal. Compute Habetha's estimates in lower dimensions

n	1	2	3	4	5	6	7	8
$H_{0,n}$	1	1	$\sqrt{2}$	$\sqrt{6}$	4	6	$6\sqrt{2}$	$2\sqrt{34}$

and compare them to the best estimates

n	1	2	3	4	5	6	7	8
$K_{0,n}$	1	1	$\sqrt{2}$	$\sqrt{2}$	2	$2\sqrt{2}$	4	4

given by Hile & Lounesto 1988. This table continues with recursion $K_{0,n+8} = 4K_{0,n}$. Hile & Lounesto 1988 also give the best estimate for the norm of the positive definite Clifford algebras $C\ell_n$

n	1	2	3	4	5	6	7	8
K_n	$\sqrt{2}$	$\sqrt{2}$	$\sqrt{2}$	$\sqrt{2}$	2	2	$2\sqrt{2}$	4

with recursion $K_{n+8} = 4K_n$.

10. Misinterpretation of a reader

The following is not a mistake in the sense that from the text it cannot be concluded that the author has/had a misconception. However, some readers have misinterpreted the text.

Lam 1973, p. 126, determines all the Clifford algebras of the type

$$C\ell_{p,q}(\mathbb{F}) = C\ell(p\langle 1\rangle \perp q\langle -1\rangle, \mathbb{F})$$

over arbitrary fields $\mathbb{F}$ of characteristic $\neq 2$. [Lam does not claim to classify all Clifford algebras (in any characteristic $\neq 2$).] Here

$$p\langle 1\rangle \perp q\langle -1\rangle = \langle \underbrace{1,1,\ldots,1}_{p}, \underbrace{-1,\ldots,-1}_{q}\rangle$$

is a diagonal quadratic form

$$x_1^2 + x_2^2 + \ldots + x_p^2 - x_{p+1}^2 - \ldots - x_{p+q}^2, \quad p + q = n.$$

This indeed classifies all the Clifford algebras in the case when $\mathbb{F} = \mathbb{R}$, since the Witt ring of $\mathbb{R}$ is $W(\mathbb{R}) = \mathbb{Z}$, and so the Clifford algebras $C\ell_{p,q} = C\ell_{p,q}(\mathbb{R})$ can be identified by the dimension $n = p + q$ and the signature $p - q \in \mathbb{Z}$. However, all the Clifford algebras, over arbitrary fields $\mathbb{F}$, cannot be reached by the notation $C\ell_{p,q}(\mathbb{F})$. For instance, the Clifford algebras $C\ell_{1,0}(\mathbb{F}_5)$ and $C\ell_{0,1}(\mathbb{F}_5)$ both split and are isomorphic to the double-ring ${}^2\mathbb{F}_5 = \mathbb{F}_5 \times \mathbb{F}_5$, which is not a field, whereas the Clifford algebra $C\ell(\langle 2\rangle, \mathbb{F}_5)$, of the quadratic form $x \to 2x^2$ on $\mathbb{F}_5$, is isomorphic to the field extension $\mathbb{F}_5(\sqrt{2}) \simeq \mathbb{F}_{25}$. In other words, in the case of the 1-dimensional vector space over $\mathbb{F}_5$ neither of the Clifford algebras of $x \to \pm x^2$ is isomorphic to the Clifford algebra of $x \to 2x^2$ [because 2 is not a square in $\mathbb{F}_5 = \{0,1,2,3,4\}$].

11. Discussion

In trying to get a picture of what is new in a published work, I have checked formulas with CLICAL. Evaluating the left hand side with arbitrary arguments satisfying all the assumptions, and comparing the result to the right hand side, reveals sometimes a discrepancy. The next step is to find the simplest non-trivial counter-example, in the lowest dimension and degree and with the smallest number of components. In discussions with authors about the fine points of their works, CLICAL has helped me to follow, verify or disqualify, the arguments presented, and to penetrate into the topic during the conversation.

The role of counter-examples in mathematics has been discussed by Lakatos 1976, Dubnov 1963 and Hauchecorne 1988. Lakatos focuses on the historical development of mathematics and Dubnov on various levels of abstraction. Both restrict themselves to a specific topic within mathematics (like this paper). Hauchecorne gives counter-examples in almost all branches of mathematics. He also elaborates on virtues of counter-examples in teaching and in research: A theorem often necessitates several hypotheses – to chart out its domain of applications it is important to become convinced about the relevancy of each hypothesis. This can be done by dropping one assumption at a time, and giving a counter-example to each new 'theorem'. Counter-examples cannot be ignored on the basis that 'they do not treat the general case'. Counter-examples are not 'exceptions that confirm the rule'. In mathematical research, the negation of a theorem, affirmation that it is false, is demonstrated by existence of a case, where all the hypotheses are verified without the conclusion being valid. The mathematical justification for the falsity of a theorem is completed by presenting a counter-example; after verification of such a presentation further study in the same line at whatever generality is useless and erroneous activity.

There are several books listing counter-examples in various branches of mathematics: Capobianco & Molluzo 1978 (graph theory), Gelbaum & Olstead 1964 (analysis), Khaleelulla 1982 (topological vector spaces), Romano & Siegel 1986 (probability and statistics), Steen & Seebach 1970 (topology) and Stoyanov 1987 (probability). Similarly as Lakatos, Dubnov and Hauchecorne these authors do not point out errors of contemporary mathematicians. The present work differs from those studies in that respect: counter-examples are given to the works of living mathematicians, who can participate in a public debate about possible correctness of counter-examples presented in this paper.

Some scientists refrain from participating in discussions about errors in published works, presumably because they anticipate a misinterpretation on the part of the author. Some scientists refrain from a public debate on errors, because of their mistaken belief that the peer refereeing system guarantees correctness of published works. Often scientists cannot come up with a suggestion on how to evaluate details of their works [other than the peer review system in abstract journals and academic appointments] – in this paper such a method has been suggested/revived: public scrutiny focusing on interior consistency of details [of publications available in scientific libraries – thus guaranteeing everybody's access to a public debate].

References

H. Baum: *Spin-Strukturen und Dirac-Operatoren über pseudoriemannschen Mannigfaltigkeiten.* Teubner, Leipzig, 1981.

I. M. Benn, R. W. Tucker: *An Introduction to Spinors and Geometry with Applications in Physics.* Adam Hilger, Bristol, 1987.

N. Bourbaki: *Algèbre, Chapitre 9, Formes sesquilinéaires et formes quadratiques*. Hermann, Paris, 1959.

P. Budinich, A. Trautman: *The Spinorial Chessboard*. Springer, Berlin, 1988.

M. Capobianco, J. Molluzo: *Examples and Counterexamples in Graph Theory*. North Holland, Amsterdam, 1978.

C. Chevalley: *Theory of Lie Groups*. Princeton University Press, Princeton, 1946.

C. Chevalley: *The Algebraic Theory of Spinors*. Columbia University Press, New York, 1954.

Y. Choquet-Bruhat, C. DeWitt-Morette: *Analysis, Manifolds and Physics, Part II*. North Holland, Amsterdam, 1989.

D. Constales: *The relative position of L^2 domains in complex and Clifford analysis*. Thesis, Univ. Gent, 1989.

R. Coquereaux: Spinors, reflections and Clifford algebras: a review, pp. 135-190 in A. Trautman, G. Furlan (eds.): *Spinors in Physics and Geometry (Trieste, 1986)*. World Scientific, Singapore, 1988.

A. Crumeyrolle: *Orthogonal and Symplectic Clifford Algebras, Spinor Structures*. Kluwer, Dordrecht, 1990.

L. Dabrowski: *Group Actions on Spinors*. Bibliopolis, Napoli, 1988.

R. Deheuvels: *Formes quadratiques et groupes classiques*. Presses Universitaires de France, Paris, 1981.

R. Deheuvels: *Tenseurs et spineurs*. Presses Universitaires de France, Paris, 1993.

R. Delanghe, F. Sommen and V. Souček: *Clifford Algebra and Spinor-Valued Functions, A Function Theory for the Dirac Operator*. Kluwer, Dordrecht, 1992.

G.M. Dixon: *Division Algebras: Octonions, Quaternions, Complex Numbers and the Algebraic Design of Physics*. Kluwer, Dordrecht, 1994.

C. Doran: *Geometric Algebra and its Applications to Mathematical Physics*. Thesis, Univ. Cambridge, 1994.

Ya.S. Dubnov: *Mistakes in Geometric Proofs*. Heath, Boston, 1963.

V. Figueiredo: Clifford algebra approach to Cayley-Klein matrices, pp. 230-236 in P.S. Letelier, W.A. Rodrigues (eds.): *Gravitation: The Space-Time Structure, SILARG VIII. Proc. 8th Latin American Symposium on Relativity and Gravitation (Brazil, 1993)*. World Scientific, Singapore, 1994.

B.R. Gelbaum, J.H.M. Olstead: *Counterexamples in Analysis*. Holden-Day, San Francisco, 1964.

J. Gilbert, M. Murray: *Clifford Algebras and Dirac Operators in Harmonic Analysis*. Cambridge Studies in Advanced Mathematics, Cambridge University Press, Cambridge, 1991.

H.P. Ginsburg, S. Opper: *Piaget's Theory of Intellectual Development*. Prentice Hall, Englewood Cliffs, NJ, 1988.

M. Göckeler, Th. Schücker: *Differential Geometry, Gauge Theories, and Gravity*. Cambridge University Press, Cambridge, 1987.

W.T. Grandy: *Relativistic Quantum Mechanics of Leptons and Fields*. Kluwer, Dordrecht, 1991.

K. Habetha: Einige geometrischen Aussagen in der Cliffordalgebra und die Subharmonizität des Betrages einer Regulären Funktion, pp. 79-83 in *Mathematical Papers given on occasion of Ernst Mohr's 75th birthday (TU Berlin, 1985)*.

P.-E. Hagmark, P. Lounesto: Walsh functions, Clifford algebras and Cayley-Dickson process, pp. 531-540 in J.S.R. Chisholm, A.K. Common (eds.): *Clifford Algebras and their Applications in Mathematical Physics (Canterbury, 1985)*. Reidel, Dordrecht, 1986,

F.R. Harvey: *Spinors and Calibrations*. Academic Press, San Diego, 1990.

B. Hauchecorne: *Les contre-exemples en mathématiques*. Ellipses, Paris, 1988.

D. Hestenes, G. Sobczyk: *Clifford Algebra to Geometric Calculus*. Reidel, Dordrecht, 1984, 1987.

D. Hestenes: *New Foundations for Classical Mechanics*. Reidel, Dordrecht, 1986, 1987.

D. Hestenes: The design of linear algebra and geometry. *Acta Applic. Math.* **23** (1991), 65-93.

G.N. Hile, P. Lounesto: Inequalities for spinor norms in Clifford algebras, pp. 285-297 in A. Trautman, G. Furlan (eds.): *Spinors in Physics and Geometry (Trieste, 1986)*. World Scientific, Singapore, 1988.

M. Karoubi: Algèbres de Clifford et K-théorie. *Ann. scient. Éc. Norm. Sup.* 4e série, t. 1 (1968), 161-270.

S.M. Khaleelulla: *Counterexamples in Topological Vector Spaces*. Springer, Berlin, 1982.

M.-A. Knus: *Quadratic and Hermitian Forms over Rings*. Springer, Berlin, 1991.

I. Lakatos: *Proofs and Refutations*. Cambridge UP, Cambridge, 1976.

T.Y. Lam: *The Algebraic Theory of Quadratic Forms*. Benjamin, Reading, MA, 1973, 1980.

H.B. Lawson, M.-L. Michelsohn: *Spin Geometry*. Universidade Federal do Ceará, Brazil, 1983. Princeton University Press, Princeton, NJ, 1989.

R. Lipschitz: Principes d'un calcul algébrique qui contient comme espèces particulières le calcul des quantités imaginaires et des quaternions. *C. R. Acad. Sci. Paris* **91** (1880), 619-621, 660-664. Reprinted in *Bull. Soc. Math.* (2) **11** (1887), 115-120.

R. Lipschitz: *Untersuchungen über die Summen von Quadraten*. Max Cohen und Sohn, Bonn, 1886, pp. 1-147. The first chapter of pp. 5-57 translated into French by J. Molk: Recherches sur la transformation, par des substitutions réelles, d'une somme de deux ou troix carrés en elle-même. *J. Math. Pures Appl.* (4) **2** (1886), 373-439. French résumé of all three chapters in *Bull. Sci. Math.* (2) **10** (1886), 163-183.

R. Lipschitz (signed): Correspondence. *Ann. of Math.* **69** (1959), 247-251.

P. Lounesto: Scalar products of spinors and an extension of Brauer-Wall groups. *Found. Phys.* **11** (1981), 721-740.

P. Lounesto, A. Springer: Möbius transformations and Clifford algebras of Euclidean and anti-Euclidean spaces, pp. 79-90 in J. Ławrynowicz (ed.): *Deformations of Mathematical Structures*. Kluwer, Dordrecht, 1989.

P. Lounesto: Clifford algebras and Hestenes spinors. *Found. Phys.* **23** (1993), 1203-1237.

P. Lounesto: Crumeyrolle's bivectors and spinors, pp. 137-166 in R. Abłamowicz, P. Lounesto (eds.): *Clifford Algebras and Spinor Structures, A Special Volume Dedicated to the Memory of Albert Crumeyrolle (1919-1992)*. Kluwer, Dordrecht, 1995.

J. Maks: *Modulo (1,1) periodicity of Clifford algebras and the generalized (anti-)Möbius transformations*. Thesis, Technische Universiteit Delft, 1989.

A. Micali: Progrés récents dans la théorie des algèbres de Clifford, pp. 232-248 in *Proc. Int. Symp. Math. Theor. Phys. (Guarajá, 1989)*. Symp. Gaussiana Ser. A Math. Theor. Phys. 1. Inst. Gaussianium, Toronto, Ont., 1990.

R. Moresi: A remark on the Clifford group of a quadratic form, pp. 621-626 in *Stochastic Processes, Physics and Geometry*. Ascona/Locarno, 1988.

Z. Oziewicz: From Grassmann to Clifford, pp. 245-255 in J.S.R. Chisholm, A.K. Common (eds.): *Clifford Algebras and their Applications in Mathematical Physics (Canterbury, 1985)*. Reidel, Dordrecht, 1986.

K. Popper: *Objective Knowledge*. Oxford UP, Oxford, 1972.

I.R. Porteous: *Topological Geometry*. Van Nostrand Reinhold, London, 1969. Cambridge University Press, Cambridge, 1981.

I.R. Porteous: *Clifford Algebras and the Classical Groups*. Cambridge University Press, Cambridge, 1995.

M. Riesz: *Clifford Numbers and Spinors*. The Institute for Fluid Dynamics and Applied Mathematics, Lecture Series No. **38**, University of Maryland, 1958. Reprinted as facsimile (eds.: E.F. Bolinder, P. Lounesto) by Kluwer, 1993.

S. Rodrigues-Romo, F. Viniegra, J. Keller: Geometrical contents of the Fierz identities, pp. 479-497 in A. Micali et al. (eds.): *Clifford Algebras and their Applications in Mathematical Physics (Montpellier, 1989)*. Kluwer, Dordrecht, 1992.

J.P. Romano, A.F. Siegel: *Counterexamples in Probability and Statistics*. Wadsworth & Brooks, Monterey, CA, 1986.

S. Salamon: *Riemannian Geometry and Holonomy Groups*. Longman Scientific, Essex, 1989.

R.D. Schafer: On algebras formed by the Cayley-Dickson process. *Amer. J. Math.* **76** (1954), 435-446.

L.A. Steen, J.A. Seebach: *Counterexamples in Topology*. Holt, Rinehart and Winston, New York, 1970. Springer, New York, 1978.

J. Stoyanov: *Counterexamples in Probability*. John Wiley, Chichester, 1987.

E. Witt: Theorie der quadratischen Formen in beliebigen Körpern. *J. Reine Angew. Math.* **176** (1937), 31-44.

2

Differential Geometry, Quantum Mechanics, Spinors and Conformal Group

THE USE OF COMPUTER ALGEBRA AND CLIFFORD ALGEBRA IN TEACHING MATHEMATICAL PHYSICS

JAYME VAZ, JR.
Department of Applied Mathematics - IMECC
State University at Campinas (UNICAMP)
CP 6065, 13081-970 Campinas, S.P., Brazil
e-mail: vaz@ime.unicamp.br

Abstract. In this paper we give a collection of examples of how computer algebra can be used within Clifford algebras in teaching of Mathematical Physics. These examples cover elementary and advanced topics, and are performed using REDUCE.

Key words: Computer algebra, Clifford algebra.

1. Introduction

If carefully and wisely used, a computer may be an important ally in teaching. In fact, looking to computers nowadays we see that they are an essential element of modern society, sometimes replacing humans in some activities, sometimes being a very helpful tool for the developing human activities. In teaching, however, one cannot accept that a computer could replace the teacher. Notwithstanding they can help him. It is true that in some subjects one can freely use computers, but this is not the case in Physics and Mathematics, and in particular in Mathematical Physics, for in these cases much of the apprenticeship comes from a training which models the way of thinking about these subjects.

The advent of algebraic computation has introduced a new element, and a new perspective, in the teaching of Mathematical Physics. It is true that one cannot think of using an already made package to teach something, but one can use a computer algebra system in a constructive way in order to help with teaching. Once we accept this, we arrive at a fundamental question: the choice of using a specific computer algebra system for doing so.

We have several different computer algebra systems like MAPLE, MATHEMATICA, REDUCE, etc., and each one has some advantages over another in some specific cases. It is our interest that must define the system to be used. For example: if one wants to solve a system of differential equations MAPLE is a good choice; if one wants to draw graphics, MATHEMATICA and MAPLE are good choices; if one wants to work with differential forms REDUCE is better, and so on.

Our objective in this paper is to describe use of computer algebra and Clifford algebras in the teaching of Mathematical Physics. We want therefore a system which has some algebraic facilities. Moreover, according to our ideas about the use of a computer in teaching, this system is expected to have some programming facilities. The programming facilities must be such that a student who has not worked with it before still can quickly learn basics of programming. In fact, the student is expected to learn Clifford algebras and its applications, and not computer algebra system programming. Due to these reasons, we believe that the best choice is REDUCE.

This paper has been conceived to be a collection of examples. We do not intend (and do not have the pretension) to say how to teach Clifford algebras and its applications. We prefer rather to give some examples of applications of computer algebra in teaching with hope that these examples will motivate the professor/lecturer not used to work with computers to start doing so, and also the student for using it in connection with the lectures. In this way, our examples are not expected to be very elaborate ones since our idea is to show a wide range of applications. The examples are also supposed to be directed to a beginner user of computer algebra. We suppose therefore that the reader has only a very basic knowledge of computer algebra systems, in our case REDUCE, and we try to avoid elaborate algorithms and some tricks of experienced users.

We have divided our examples into three levels. At the first level (Section 2), we discuss examples directed to a Clifford algebra beginner whom we expect to be a beginning undergraduate student. In this section we first introduce the Clifford algebra of the Euclidean plane and try to extract from this very simple example some very important concepts. The idea of extracting important concepts like spinors, spin group, etc., from a trivial example is that it would be understandable for, say, a first year undergraduate student who could then become aware of certain concepts which will be introduced to him only after some years in an usually abstract way. We also discuss in Section 2 applications to classical differential geometry of curves and to Lorentz transformations. The latter is of particular interest since we obtain the magnetic moment of the electron with correct gyromagnetic factor.

At the second level (Section 3), our examples are a little more advanced and are directed to an advanced undergraduate student. In this section we introduce Dirac operator and discuss its applications to calculations in differential geometry and electromagnetism. We also give an example of the use of spinors in electromagnetism. Finally, the third level (Section 4) is directed to a graduate student. At this level, the student is expected to be familiar with a system like REDUCE and with the calculus of differential forms, so that the powerful package **EXCALC** can be used. In this section we show the use of Clifford algebra valued differential forms in differential geometry and gauge fields.

Some parts of our approach in this paper to Clifford algebras and its applications are based on lectures given by us at the University of Brasília (UnB) in the first part of 1994 academic year, and some of those lectures have been inspired by Professor Lounesto lectures at UNICAMP in 1993. We follow in this paper the notations suggested by (Ablamowicz *et al.*, 1991).

2. Clifford algebras within REDUCE

Defining Clifford product in REDUCE is a trivial thing. First we take the vectors as operators and then tell REDUCE that it is a non-commutative operator:

```
operator e;
noncom e;
```

Then we have $e(1)$ as the vector e_1, $e(2)$ as the vector e_2, and so on. The command

```
factor e;
```

will cause REDUCE to factor the vectors in a given multivectorial expression. The Clifford product is defined by a list of assignment statements; for example:

```
e(1)*e(1):=1;
e(2)*e(2):=1;
e(2)*e(1):=-e(1)*e(2);
...
```

People who work with Clifford algebras will agree that it is more efficient tool work in terms of an orthonormal basis. This is in fact what we are doing here. If we have not used an orthonormal basis, the Clifford product could not have been defined the way it has been since then there is no trivial way of doing so. The package **EXCALC** can be used in this case since it has the contraction ⌐ and the exterior product ∧, but now we have to define a correlation between a vector space V and its dual $V^\star$, and this involves unnecessary complications which are outside the scope of our objectives.

We also observe that within our orthonormal basis (or in a Witt basis) the deformations of Clifford algebras (Lawrynowicz *et al.*, 1995) can be easily handled:

```
e(2)*e(1):=-q*e(1)*e(2);
```

where q is (in general) a complex parameter.

Since REDUCE recognizes sum of operators, it is trivial to work out a multivector. The multivector $a_0 + a_1 e_1 + a_{13} e_{13}$ is

```
ex := a0 + a1*e(1) + a13*e(1)*e(3);
```

Let us make a pause for a brief comment. Typical undergraduate students are not familiar with non-commutative algebras, so it is important before going on into Clifford algebras and its applications working out some "experiments" with the Clifford product. Moreover, the student is familiar with the concept of vectors, but not of multivectors. The fact that

```
(e(1)*e(2))*(e(1)*e(2));
```

gives -1 is the opportunity of introducing the concept of a 2-vector. By means of other examples we introduce the concept of a k-vector, and then the need for the concept of multivectors will become evident after performing products of k-vectors of different grades k. After being familiar with the concept of multivector, the student is then motivated to introduce some operations, namely the involutions.

The morphisms of reversion ~, main involution ^, and conjugation ¯ can be introduced by procedures. Such procedures have been written by Brackx and Constales, and are found in a floppy disk which accompanies the book of Delanghe *et al.* (1992). It is not difficult to write them for the cases discussed here. They are explained in Brackx *et al.* (1987). These procedures are **rev**, **mainvol** and **bar**.

The k-vector part of a multivector is given by the procedure **grade**. For example, a_0 is the scalar part of **ex** above, given by

```
grade(ex,0);
```

We can now introduce to a Clifford algebra beginner some examples of their applications in Mathematical Physics.

2.1. ROTATIONS

Rotations in the Euclidean plane are a trivial example, but it is an example which can be used to introduce several important concepts. These four examples are illustrative:[1]

```
trigrules:={sin(~x)*cos(~x) => (1/2)*sin(2*x),
sin(~x)^2 => (1/2)*(1-cos(2*x)),
cos(~x)^2 => (1/2)*(1+cos(2*x))};
let trigrules;
ex1 := (cos(th)+sin(th)*e(1)*e(2))*(v1*e(1)+v2*e(2))*
(cos(th)-sin(th)*e(1)*e(2));
ex2 := (cos(th)+sin(th)*e(1)*e(2))*(v1*e(1)+v2*e(2))*
(cos(th)+sin(th)*e(1)*e(2));
ex3 := (1+a*e(1)+b*e(1)*e(2))*(v1*e(1)+v2*e(2))*
(1+a*e(1)-b*e(1)*e(2));
ex4 := (a+b*e(1)*e(2))*(v1*e(1)+v2*e(2))*(a-b*e(1)*e(2));
```

The first example shows that **ex1** is just the rotation of the vector $v = v_1 e_1 + v_2 e_2$ by an angle 2θ. The second example shows that rotations are given by the operation $v \mapsto Rv\tilde{R}$ and not by $v \mapsto RvR$. The third example shows that the form of R is not arbitrary, but that it must be an even multivector. The result of **ex4** will not be clear to the student at a first sight; however, after doing the substitutions

```
sub(a=rho*cos(th),b=rho*sin(th),ex4);
```

the student will see that it is just the rotation of v, that is **ex1**, plus its dilation by a factor ρ^2. It will be clear that not every operation $v \mapsto Rv\tilde{R}$ consists of only a rotation when R is an even multivector. The examples

```
ex5 := (cos(th)+sin(th)*e(1)*e(2))*(cos(th)-sin(th)*e(1)*e(2));
ex6 := (a+b*e(1)*e(2))*(a-b*e(1)*e(2));
```

show that $R_1 = \cos\theta + \sin\theta e_{12}$ is a unitary multivector, and that $\tilde{R}_1 = R_1^{-1}$, while R_2 is not a unitary multivector since it has the norm $\rho^2 = a^2 + b^2$. Two important conclusions appear:
(i)   multivectors may have an inverse under the Clifford product;
(ii) rotations are given by $v \mapsto Rv\tilde{R}$ where R is a unitary even multivector.
The next example:[2]

```
ex7 := (cos(th1)+sin(th1)*e(1)*e(2))*(cos(th2)+sin(th2)*e(1)*e(2));
```

---

[1] Computer algebra systems like REDUCE do not "know" trigonometric rules; since they are essential in some Clifford algebra calculations, it is better to define then explicitly.

[2] Here it is convenient to add to our trigonometric rules formulas like $\cos x \cos y = (1/2)(\cos(x+y) + \cos(x-y))$, or after the calculations, introduce them by assignments.

shows that unitary even multivectors form a group. Since both $(\cos\theta + \sin\theta\mathbf{e}_{12})$ and $-(\cos\theta + \sin\theta\mathbf{e}_{12})$ give the same rotation by an angle 2θ, the student is directed to conclude that this group is not SO(2), and we can now introduce the concept of double covering of groups, and define $\mathbf{Spin}(2) = \{R \in C\ell_{2,0}^+ \mid R\tilde{R} = \tilde{R}R = 1\}$ such that $\mathbf{Spin}(2)/\{\pm 1\} = \mathrm{SO}(2)$.

These examples can be further exploited in order to introduce the important concept of spinors. At this stage it should be clear to the student that $C\ell_{2,0}^+ \simeq \mathbb{C}$. We define a spinor operator as an even multivector[3]. From our examples we have that $v \mapsto \psi v\tilde{\psi}$ is a rotation and a dilation of v if ψ is a spinor operator. Complex numbers are therefore the first example of spinor operators we can give to the student who will not be surprised when we introduce the Weyl spinor in the Clifford algebra of the three dimensional Euclidean space and the Dirac spinor in the Clifford algebra of Minkowski spacetime.

Before we continue it is interesting to make some comments about two other facts. The first is that a rotation in a different sense, obtained by replacing $\theta \mapsto -\theta$ in the above examples, equals the rotation by θ if we replace $\mathbf{e}_{12} \mapsto -\mathbf{e}_{12}$. The student sees therefore the question of orientation of space and the definition of its volume element τ, in this $\tau = \mathbf{e}_{12}$ or $\tau = -\mathbf{e}_{12}$. When the three dimensional case is to be considered, it will be clear to the student that a right-handed orientation of the space is defined by the volume element $\tau = \mathbf{e}_{123}$, while a left-handed orientation is given by the volume element $\tau = -\mathbf{e}_{123}$.

The second fact is that Clifford algebras for vector spaces with the same dimension but different signatures are not isomorphic. First of all, our starting point $\mathbf{e}_1^2 = 1$ and $\mathbf{e}_2^2 = 1$ was in no way arbitrary. It would sound strange to most beginning undergraduate students that a vector may have a negative square. However, after manipulating examples with multivectors in $C\ell_{2,0}$ the student will see that multivectors, and then vectors, may have a negative square. It would be appropriate therefore to compare $C\ell_{2,0}$ and $C\ell_{0,2}$, and conclude that they are not isomorphic. The student must be motivated to work with $C\ell_{0,2}$ since it is isomorphic to the quaternions. It would be interesting to consider in more details the algebra $C\ell_{1,1}$ since it is related to conformal transformations (Lounesto and Latvamaa 1980, Hestenes 1991), but it is better to discuss it later since the structure of $C\ell_{1,1}$ is better appreciated when we work with a null (Witt) basis.

It is natural now to go to the three dimensional Euclidean space. We shall not discuss examples of the generalization to that space of what we discussed here since it is too trivial, and our objective was to show how an interactive computer algebra session can be used in order to introduce the student to Clifford algebras. Concepts like Weyl spinors, Cayley-Klein parameters, etc., will appear in $C\ell_{3,0}$ as natural generalizations of what we have discussed.

2.2. THEORY OF SPACE CURVES

The applications of Clifford algebras to the theory of curves is based on the method of moving frames. A moving frame $\{\boldsymbol{f}_i = \boldsymbol{f}_i(t)\}$ is defined in REDUCE as:

---

[3] This definition is only partially correct, but it suffices for our purposes. For more details and a correct definition see Rodrigues *et al.* (1995). We also observe that in higher dimensions $v \mapsto \psi v\tilde{\psi}$, where ψ is an arbitrary even multivector, does not necessarily give another vector.

```
operator f;
noncom f;
for j:=1:3 do f(j)*f(j) := 1;
for all j,k such that j>k let f(j)*f(k) = -f(k)*f(j);
for j:=1:3 do depend f(j),t;
factor f;
```

Note that we have identified t with the arc length since $\boldsymbol{f}_1^2 = 1$. The Frenet equations are then given in terms of the Darboux 2-vector Ω:[4]

```
om := k1*f(2)*f(1) + k2*f(3)*f(2);
depend k1,t;
depend k2,t;
```

The Frenet equations can be obtained as follows:

```
operator aux;
for j:=1:3 do aux(j) := df(f(j),t)-(1/2)*(om*f(j)-f(j)*om);
```

that is: $\text{aux}(1) = \dot{\boldsymbol{f}}_1 - k_1\boldsymbol{f}_2 = 0$, $\text{aux}(2) = \dot{\boldsymbol{f}}_2 + k_1\boldsymbol{f}_1 - k_2\boldsymbol{f}_3 = 0$ and $\text{aux}(3) = \dot{\boldsymbol{f}}_3 + k_2\boldsymbol{f}_2 = 0$, where k_1 and k_2 are, respectively, the curvature and the torsion of the space curve.

Now, questions like to find a curve whose curvature and torsion satisfies certain equations are replaced by questions like to find a curve whose Darboux 2-vector satisfies a certain equation. As an example, let this equation be $\dot{\Omega} = k\Omega$, where k may depend on t. In order to write it we must first express $\dot{\boldsymbol{f}}_i$ using Frenet equations, that is:

```
for j:=1:3 do df(f(j),t) := rhs first solve(aux(j),df(f(j),t));
aux(4) := df(om,t) - k*om;
```

Thus, we find $\text{aux}(4) = 0 = \boldsymbol{f}_{23}(-\dot{k}_2 + kk_2) + \boldsymbol{f}_{12}(-\dot{k}_1 + kk_1)$, which we can then separate into two equations:[5]

```
aux(5) := grade(f(3)*f(2)*aux(4),0);
aux(6) := grade(f(2)*f(1)*aux(4),0);
```

Now we may use package **ODESOLVE** to solve these differential equations. Note that regardless of the form of $k = k(t)$, their solutions will be helixes since $k_1/k_2 = \text{const.}$, which characterizes a helix. The case $k = 0$ defines a circular helix. It is easy to verify by generalizing our approach to an arbitrary dimension that $\dot{\Omega} = k\Omega$ is the equation of a helix in any dimension since one obtains $k_i/k_j = \text{const.}$ $(i \neq j)$.

Although the package **ODESOLVE** can be used to solve aux(5) and aux(6), it will not help us to integrate Frenet equations since it deals (at the present time) only with a single scalar first-order differential equation (and of a simple type – like aux(5)

---

[4] In fact, the Frenet frame is defined as a moving frame in which the Darboux 2-vector assumes that form.

[5] Note that the procedure **grade** has to be defined now in terms of the operator **f** and not in terms of the operator **e** as before.

and aux(6)). In order to integrate Frenet equations we need in this case to solve a system of nine ordinary differential equations. REDUCE can help us to find this system by using operators e(i) for a fixed frame e_i, and we can save this system and later solve it, for example, by using "dsolve" in MAPLE.

The example we gave in this subsection can be extended to spaces of arbitrary dimension and signature. We only need to take care in the case of pseudo-Euclidean spaces of some signs since the Darboux 2-vector must be expressed in terms of the reciprocal frame $\{\boldsymbol{f}^i\}$. An example of application to the study of the spinning particle model of Barut and Zhangi (1984) can be found in Vaz (1995).

2.3. LORENTZ TRANSFORMATIONS

Let us see some examples of the use of Clifford algebras within REDUCE to Lorentz transformations and its applications. Let $C\ell_{1,3}$ be the spacetime algebra, that is, the Clifford algebra of Minkowski vector space $\mathbb{R}^{1,3}$.

```
operator e;
noncom e;
e(0)*e(0) := 1;
for j:=1:3 do e(j)*e(j) := -1;
for all j,k such that j>k let e(j)*e(k) = -e(k)*e(j);
factor e;
```

We have seen that a rotation is given in Clifford algebras by $v \mapsto Rv\tilde{R}$ with $R\tilde{R} = 1$. In the case of a Lorentz transformation this rotation is a hyperbolic one, that is, R is of the form $R = \cosh\alpha + \mathbf{B}\sinh\alpha$ with $\mathbf{B}$ being a 2-vector such that $\mathbf{B}^2 = 1$. It can be shown (Hestenes, 1966) that $R = (1 + \gamma + \gamma v e_{01})/\sqrt{2(1+\gamma)}$ for the Lorentz transformation along x_1 where $\gamma = 1/\sqrt{1-v^2}$. In fact:

```
x   := x0*e(0) + x1*e(1) + x2*e(2) + x3*e(3);
lor := (1 + ga + ga*v*e(0)*e(1))/sqrt(2*(1+ga));
let 1 - v^2 = 1/(ga^2);
y   := lor*x*rev(lor);
```

The result is $y_0 = \gamma(x_0 - vx_1)$, $y_1 = \gamma(x_1 - vx_0)$, $y_2 = x_2$ and $y_3 = x_3$, which are the usual expressions for the Lorentz transformation when $c = 1$. In a similar way we obtain expressions for transformations of electric and magnetic fields. The electromagnetic field is described by a 2-vector F (the Faraday 2-vector) with $F_{0i} = e_i$ and $F_{ij} = \epsilon_{ijk}h^k$. We have

```
far := f01*e(0)*e(1) + f02*e(0)*e(2) + f03*e(0)*e(3)
+ f12*e(1)*e(2) + f13*e(1)*e(3) + f23*e(2)*e(3);
farpr := lor*f*rev(lor);
```

which gives $F'_{01} = F_{01}$, $F'_{02} = \gamma(F_{02} - vF_{12})$, $\cdots$, as expected.

As a non-trivial example, let us calculate the magnetic moment of a moving point particle. The current density 1-vector J in the rest frame is $J = \rho e_0$, where ρ is the charge density $\rho = q\delta(\vec{x} - \vec{x}_0)$. Under a Lorentz transformation we have

```
curr := rho*e(0);
currpr := lor*curr*rev(lor);
```

that is: $J' = \gamma\rho e_0 - \gamma\rho v e_1$, in such a way that $q = \int \rho d^3 x = \int (\gamma\rho) d^3 y$ is an invariant. The electromagnetic moment density 2-vector M is $M = x \wedge J$, i.e.:

```
emm  := grade(x*curr,2);
mel  := (1/2)*(m - e(0)*m*e(0));
mmag := (1/2)*(m + e(0)*m*e(0));
```

We obtain $M_{el} = \rho(x_1 e_{10} + x_2 e_{20} + x_3 e_{30}) = \rho\vec{x}$ and $M_{mag} = 0$, as expected. If the particle is moving with velocity $\vec{v} = v\vec{\sigma}_1 = v e_{10}$ we have:

```
lorentzrules:={x0 = ga*(y0 + v*y1),
x1 = ga*(y1 + v*y0),
x2 = y2,
x3 = y3};
let lorentzrules;
emmpr  := lor*emm*rev(lor);
mprel  := (1/2)*(emmpr - e(0)*emmpr*e(0));
mprmag := (1/2)*(emmpr + e(0)*emmpr*e(0));
```

We obtain $M'_{el} = \gamma\rho((y_1 + v y_0)e_{10} + y_2 e_{20} + y_3 e_{30})$ and $M'_{mag} = \gamma\rho(v y_3 e_{13} + v y_2 e_{12})$. Note that $(c = 1)$

$$M'_{mag} = \gamma\rho v e_{10}(y_3 e_{03} + y_2 e_{02}) = \gamma\rho\vec{y} \wedge \vec{v},$$

and since ρ is a Dirac delta distribution:

$$m' = \int d^3 y(\vec{y} \wedge \vec{v})\gamma\rho = e\vec{y} \wedge \vec{v} = \frac{e}{m_0}\vec{y} \wedge \vec{p},$$

where $\vec{p} = m_0\vec{v}$. This is the expression for the magnetic dipole moment of a point particle with the correct gyromagnetic factor $g = 2$. This definition of M as $M = x \wedge J$ and the fact that it gives $g = 2$ as shown above was noticed by Rego (1963).

3. The Dirac operator

The fundamental differential operator acting on sections of the Clifford algebra bundle is the so-called Dirac operator. In this section we shall introduce a procedure to calculate the action of the Dirac operator and show some applications. We shall be restricted to Lorentzian spacetimes.

Let $\{g_\mu = \frac{\partial}{\partial x^\mu}\}$ and $\{e_\mu\}$ be, respectively, a coordinate basis and an orthonormal basis for the tangent space $T_x M \simeq C\ell_{1,3}$. We have $e_\mu = h_\mu{}^\nu g_\nu$ and $g_\nu = h^\mu{}_\nu e_\mu$ with $h^\mu{}_\sigma h_\nu{}^\sigma = \delta_\nu^\mu$, $e_\mu \cdot e_\nu = \eta_{\mu\nu} = \mathrm{diag}(1,-1,-1,-1)$, $g_\mu \cdot g_\nu = g_{\mu\nu}$. For the reciprocal basis $e^\mu = h^\mu{}_\nu g^\nu$, $g^\nu = h_\mu{}^\nu e^\mu$, $e^\mu \cdot e_\nu = \delta_\nu^\mu$, $g^\mu \cdot g_\nu = \delta_\nu^\mu$. The Dirac operator is

$$\partial = g^\mu \nabla_{g_\mu} = e^\mu \nabla_{e_\mu},$$

where ∇_{g_μ} [resp. ∇_{e_μ}] is the covariant derivative in the direction of g_μ [resp. e_μ].

Now, we have to work in terms of an orthonormal basis. The expression for ∇_{e_μ} is

$$\nabla_{e_\mu} A = \nabla_{e_\mu}(A^\nu e_\nu) = d_\mu(A^\nu)e_\nu + A^\nu \Gamma^\sigma_{\mu\nu} e_\sigma$$

where $d_\mu(A^\nu) = h_\mu{}^\sigma \partial_\sigma(A^\nu)$ is the Pfaff derivative (Choquet-Bruhat *et al.*, 1982) and $\Gamma^\sigma_{\mu\nu}$ are the connexion coefficients defined in terms of $\{e_\mu\}$: $\nabla_{e_\mu} e_\nu = \Gamma^\sigma_{\mu\nu} e_\sigma$. For a connexion which is metric compatible we have that $\gamma_{\mu\sigma\nu} = -\Gamma_{\nu\sigma\mu}$, where $\Gamma_{\mu\sigma\nu} = \eta_{\mu\rho} \Gamma^\rho_{\sigma\nu}$. Then we can define a 2-vector $\gamma_\sigma = \frac{1}{2} \Gamma_{\mu\sigma\nu} e^{\mu\nu}$ in such a way that

$$\nabla_{e_\mu} e_\nu = \frac{1}{2}[\Gamma_\mu, e_\nu] = \frac{1}{2}(\Gamma_\mu e_\nu - e_\nu \Gamma_\mu).$$

We have therefore

$$\nabla_{e_\mu} A = d_\mu A + \frac{1}{2}[\Gamma_\mu, A]$$

where d_μ is called *fiducial derivative* (along e_μ), with $d_\mu(A^\nu e_\nu) = (d_\mu A^\nu)e_\nu = h_\mu{}^\sigma(\partial_\sigma A^\nu)e_\nu$.

Let us write a procedure to calculate the Dirac operator when $\{g_\mu = \frac{\partial}{\partial x^\mu}\}$ is an orthogonal basis and the connexion is Riemannian. In this case $h_\mu{}^\nu = h_{(\mu)} \delta_\mu^\nu$ (the index between parenthesis means that it is not to be summed – note that $h_{(\alpha)} \delta_\mu^\alpha \delta_\alpha^\nu = h_{(\mu)} \delta_\mu^\nu$), and $\Gamma_{\mu\nu\sigma} = (1/2)(C_{\mu\nu\sigma} + C_{\nu\mu\sigma} + C_{\sigma\mu\nu})$, where $C_{\mu\nu\sigma} = \eta_{\mu\rho} C^\rho_{\nu\sigma}$ with $\{C^\rho_{\nu\sigma}\}$ being the structure coefficients of the basis $\{e_\mu\}$ – i.e.: $[e_\mu, e_\nu] = C^\sigma_{\mu\nu} e_\sigma$, where $[e_\mu, e_\nu] = d_\mu e_\nu - d_\nu e_\mu$ is the Lie bracket. It is easy to see that

$$C_{\sigma\mu\nu} = \eta_{\sigma\nu} h_{(\nu)}^{-1}(\partial_\mu h_{(\nu)})h_{(\mu)} - \eta_{\sigma\mu} h_{(\mu)}^{-1}(\partial_\nu h_{(\mu)})h_{(\nu)},$$

and then

$$\Gamma_\mu = h_{(\mu)} h_{(\nu)} (\partial_\nu h_{(\mu)}^{-1}) e_\mu{}^\nu.$$

The Dirac operator ∂ becomes

$$\partial\phi = h_{(\mu)} e^\mu \left[\partial_\mu \phi + \frac{1}{2}[e_\mu{}^\nu h_{(\nu)} \partial_\nu h_{(\mu)}^{-1}, \phi] \right].$$

Let us introduce a simple procedure to calculate the Dirac operator. First, define $C\ell_{1,3}$ as in Section 2.3. Since we need the reciprocal basis, we need to use $\eta_{(\mu)} = \eta_{\mu\mu} = \eta^{\mu\mu}$ in $e^\mu = \eta_{(\mu)} e_\mu$.

```
operator eta;
eta(0) := 1;
for j:=1:3 do eta(j) := -1;
operator h;
operator co;
co(0) := x0;
co(1) := x1;
co(2) := x2;
co(3) := x3;
procedure auxdirac(var1,var2);
begin;
z := 0;
for k:=0:3 do z := z + eta(k)*h(k)*df(1/h(var1),co(k))*
(1/2)*(e(var1)*e(k)*var2 - var2*e(var1)*e(k));
return z end;
```

```
procedure dirac(ex);
begin;
y := 0;
for j:=0:3 do y := y+eta(j)*h(j)*e(j)*(df(ex,co(j))+auxdirac(j,ex));
return y end;
```

Note that it remains to specify h_0, h_1, h_2 and h_3, that is, to specify the coordinates (x_0, x_1, x_2, x_3). As an example, consider the coordinates (t, r, θ, ϕ). Then:

```
h(0) := 1;
h(1) := 1;
h(2) := 1/x1;
h(3) := 1/(x1*sin(x2));
depend psi,x0,x1,x2,x3;
dirac(psi);
```

For the scalar field ψ we obtain

$$\partial \psi = \operatorname{grad} \psi = \mathbf{e}_0 \frac{\partial \psi}{\partial t} - \mathbf{e}_1 \frac{\partial \psi}{\partial r} - \mathbf{e}_2 \frac{1}{r} \frac{\partial \psi}{\partial \theta} - \mathbf{e}_3 \frac{1}{r \sin \theta} \frac{\partial \psi}{\partial \phi}.$$

The Laplace-Beltrami operator $\Box$ is $\Box = \partial^2$.

```
dirac(dirac(psi));
```

It gives

$$\Box \psi = \frac{\partial^2 \psi}{\partial t^2} - \left(\frac{\partial^2 \psi}{\partial r^2} + \frac{2}{r} \frac{\partial \psi}{\partial r} + \frac{1}{r^2} \frac{\partial^2 \psi}{\partial \theta^2} + \frac{\cot \theta}{r^2} \frac{\partial \psi}{\partial \theta} + \frac{1}{r^2 \sin^2 \theta} \frac{\partial^2 \psi}{\partial \phi^2} \right),$$

and the terms between parentheses give the usual Laplacian in spherical coordinates.

Expressions for the divergent ($\partial \cdot$) and for the gradient ($\partial \wedge$) of arbitrary multivector fields can be trivially obtained in a similar way.

3.1. CALCULATIONS IN DIFFERENTIAL GEOMETRY

It was shown by Hestenes (1986) that the use of Clifford algebras simplifies calculations with differential forms. We shall in this section illustrate with our computational machinery the example given by Hestenes (1986).

For a metric-compatible Riemannian connexion, the connexion 2-vectors Γ_μ are given by (Hestenes, 1986)

$$\Gamma_\mu = \frac{1}{2} (\mathbf{e}^\nu \wedge \partial \wedge \mathbf{e}_\nu) \cdot \mathbf{e}_\mu - \partial \wedge \mathbf{e}_\mu.$$

If our coordinate basis is orthogonal the above reduces to

$$\Gamma_\mu = -\partial \wedge \mathbf{e}_\mu,$$

which can be calculated by

```
procedure omega(ex);
begin;
ome := - grade(dirac(e(ex)),2);
return ome end;
```

For the Schwarzschild metric we have[6]

```
h(0) := exp(-psi1);
h(1) := exp(-psi2);
h(2) := 1/x1;
h(3) := 1/(x1*sin(x2));
depend psi1,x0,x1;
depend psi2,x0,x1;
```

Now we can calculate the connexion 2-vectors. For example:

```
omega(0);
```

gives $\Gamma_0 = -\partial_r \psi_1 e^{-\psi_2} e_{01}$ for the connexion coefficient along $e_0 = e^{-\psi_1} \partial_t$. For Γ_t we have $\Gamma_t = -\partial_r \psi_1 e^{\psi_2 - \psi_1} e_{01}$. However, calculating the connexion 2-vectors Γ_μ using the **dirac** procedure and the **grade** procedure is not too efficient in this case. We can start using the fact that $\Gamma_\mu = h_{(\mu)} h_{(\nu)} (\partial_\nu h_{(\mu)}^{-1}) e_\mu{}^\nu$, and define the following procedure:

```
procedure connexion(ex);
begin;
conn := 0;
for k:=0:3 do conn := conn + h(ex)*h(k)*df(1/h(ex),co(k))*
eta(k)*(1/2)*(e(ex)*e(k)-e(k)*e(ex));
return conn end;
```

Calculations of Γ_0 using REDUCE 3.5 on a 433DX/D PS/ValuePoint IBM microcomputer took 380 ms using the procedure **omega** and 50 ms using the procedure **connexion**, and the calculations of all 2-vectors Γ_μ took 1540 ms using **omega** and 160 ms using **connexion**. Anyway, it is not our objective here to look for the most efficient procedure, and in this way **omega** does its pedagogical job.

The Riemannian curvature 2-vector $\Omega_{\mu\nu}$ is defined by

$$[\nabla_{e_\mu}, \nabla_{e_\nu}]A = \frac{1}{2}[\Omega_{\mu\nu}, A].$$

We obtain that

$$\Omega_{\mu\nu} = d_\mu \Gamma_\nu - d_\nu \Gamma_\mu + \frac{1}{2}[\Gamma_\mu, \Gamma_\nu].$$

We remark that here d is the fiducial derivative, and we have to express those quantities in terms of a coordinate basis in order to calculate $\Omega_{\mu\nu}$ using the command **df**. One possible procedure is:

---

[6] Our definition of $h_{(\mu)}$ is just the inverse of the one of Hestenes (1986).

```
procedure riemann(var1,var2);
begin;
riem := h(var1)*h(var2)*(df(connexion(var2)/h(var2),co(var1)) -
df(connexion(var1)/h(var1),co(var2))) +
(1/2)*(connexion(var1)*connexion(var2) -
connexion(var2)*connexion(var1));
return riem end;
```

We obtain, for example,

$$\Omega_{01} = - \left[\left(\partial_{tt}\psi_2 + (\partial_t\psi_2)^2 - (\partial_t\psi_2)(\partial_t\psi_1) \right) e^{-2\psi_1} \right.$$
$$\left. - \left(\partial_{rr}\psi_1 + (\partial_r\psi_1)^2 - (\partial_r\psi_1)(\partial_r\psi_2) \right) e^{-2\psi_2} \right] e_{01}.$$

The Ricci 1-vectors R_μ are $R_\mu = e^\nu \cdot \Omega_{\nu\mu}$, which can be easily calculated using **grade**. The empty space condition $R_\mu = 0$ gives the equations $\partial_t\psi_1 = \partial_t\psi_2 = 0$, $\psi_2 = -\psi_1$ and $r\partial_r\psi_2 + e^{-\psi_2} - 1 = 0$, which can be integrated giving $e^{-\psi_2} = e^{\psi_1} = 1 - 2m/r$.

The Dirac operator acting on a scalar field ψ, for example, is:

```
depend psi,x0,x1,x2,x3;
dirac(psi);
```

which gives

$$\partial\psi = e_0 \frac{1}{1-2m/r} \frac{\partial\psi}{\partial t} - e_1(1-2m/r)\frac{\partial\psi}{\partial r} - e_2\frac{1}{r}\frac{\partial\psi}{\partial\theta} - e_3\frac{1}{r\sin\theta}\frac{\partial\psi}{\partial\phi}.$$

The Laplace-Beltrami operator $\Box$ is given by

```
dirac(dirac(psi));
```

which gives, after some rearrangements:

$$\Box\psi = \frac{1}{(1-2m/r)^2}\frac{\partial^2\psi}{\partial t^2} - \left[(1-2m/r)^2\frac{\partial^2\psi}{\partial r^2} + (1-2m/r)\frac{2}{r}\frac{\partial\psi}{\partial r} \right.$$
$$\left. + \frac{1}{r^2}\frac{\partial^2\psi}{\partial\theta^2} + \frac{\cot\theta}{r^2}\frac{\partial\psi}{\partial\theta} + \frac{1}{r^2\sin^2\theta}\frac{\partial^2\psi}{\partial\psi^2} \right].$$

Similar expressions can be easily obtained for any multivector field.

3.2. APPLICATIONS IN ELECTROMAGNETISM

The Clifford algebra of spacetime is a powerful tool for studying electromagnetism (Hestenes 1966, Jancewicz 1988). In terms of the spacetime algebra Maxwell equations assume a very elegant and concise expression. Let us work, for simplicity, with rectangular coordinates, i.e., we take $h_{(\mu)} = 1$ ($\mu = 0, 1, 2, 3$). Let us define the Faraday 2-vector F and the current density 1-vector J:

```
far := e1*e(0)*e(1) + e2*e(0)*e(2) + e3*e(0)*e(3)
+ b3*e(1)*e(2) + b2*e(3)*e(1) + b1*e(2)*e(3);
curr := -(rho*e(0) + j1*e(1) + j2*e(2) + j3*e(3));
depend e1,x0,x1,x2,x3;
```

```
depend e2,x0,x1,x2,x3;
...
depend j3,x0,x1,x2,x3;
max := dirac(far) - curr;
maxin := grade(max,1);
maxhom := grade(max,3);
```

We have that `maxin` are the inhomogeneous Maxwell equations:

$$\mathbf{e}_3 \left(\frac{\partial b_1}{\partial x_2} - \frac{\partial b_2}{\partial x_1} + \frac{\partial e_3}{\partial t} + j_3 \right) + \mathbf{e}_2 \left(\frac{\partial b_3}{\partial x_1} - \frac{\partial b_1}{\partial x_3} + \frac{\partial e_2}{\partial t} + j_2 \right) +$$

$$\mathbf{e}_1 \left(\frac{\partial b_2}{\partial x_3} - \frac{\partial b_3}{\partial x_2} + \frac{\partial e_1}{\partial t} + j_1 \right) - \mathbf{e}_0 \left(\frac{\partial e_1}{\partial x_1} + \frac{\partial e_2}{\partial x_2} + \frac{\partial e_3}{\partial x_3} - \rho \right) = 0,$$

and `maxhom` are the homogeneous Maxwell equations:

$$\mathbf{e}_{012} \left(\frac{\partial b_3}{\partial t} + \frac{\partial e_2}{\partial x_1} - \frac{\partial e_1}{\partial x_2} \right) + \mathbf{e}_{013} \left(-\frac{\partial b_2}{\partial t} - \frac{\partial e_1}{\partial x_3} + \frac{\partial e_3}{\partial x_1} \right) +$$

$$\mathbf{e}_{023} \left(\frac{\partial b_1}{\partial t} - \frac{\partial e_2}{\partial x_3} + \frac{\partial e_3}{\partial x_2} \right) - \mathbf{e}_{123} \left(\frac{\partial b_1}{\partial x_1} + \frac{\partial b_2}{\partial x_2} + \frac{\partial b_3}{\partial x_3} \right) = 0$$

This computation shows that in terms of the spacetime algebra, Maxwell equations assume the form $\partial F = J$. We can also demonstrate explicitly that $F = \partial \wedge A$, or $F = \partial A$ in Lorenz gauge $\partial \cdot A = 0$.

By direct computation we show that

```
grade(far^2,0);
grade(far^2,4);
```

are the invariants of the electromagnetic field $\vec{e}^2 - \vec{b}^2$ and $2\vec{e} \cdot \vec{b}$, respectively. Again, by direct computation we show that

```
(1/2)*far*e(0)*rev(far);
```

is the energy-momentum 1-vector $S_0 = \mathbf{e}_0(\vec{e}^2 + \vec{b}^2)/2 + \mathbf{e}_i(\vec{e} \times \vec{b})^i$. It is a natural guess that $T_{ij} = \mathbf{e}_i \cdot S_j$, with $S_j = (1/2)F\mathbf{e}_j\tilde{F}$ $(i,j = 1,2,3)$ are the components of Maxwell strain tensor, as it can be easily verified.

Several explicit examples can now be constructed. We shall not do this here since one can find interesting examples in the book of Jancewicz (1988). Here we prefer rather to exploit a very interesting relationship between spinors and electromagnetism.

We have proved elsewhere (Vaz and Rodrigues, 1993) that any electromagnetic field described by the Faraday 2-vector F can be written in the form $F = \psi \mathbf{e}_{21} \tilde{\psi}$ (in appropriate units). Let us illustrate this result for the case of the Coulomb field.

A Dirac-Hestenes spinor is an element of $Cl_{1,3}^{+}$[7]. Consider a Dirac-Hestenes spinor field ψ of the form

$$\psi = \frac{1 + \mathbf{e}_{0123}}{\sqrt{2(\rho^2 + z^2)}} \left(\cos \phi/2 - \mathbf{e}_{13} \sin \phi/2 \right)$$

---

[7] This definition is only partly correct. See (Rodrigues *et al.*, 1995).

where we are using cylindrical spatial coordinates ($e_1 = \partial_\rho$, $e_2 = (1/\rho)\partial_\theta$, $e_3 = \partial_z$), and ϕ, the azimuthal angle of spherical coordinates, is given by

$$\tan \phi = \frac{\rho}{z}.$$

Let us show that $F = \psi e_{21} \tilde{\psi}$ is indeed the Coulomb field.

```
cos(phi) := x3/sqrt(x1^2 + x3^2);
sin(phi) := x1/sqrt(x1^2 + x3^2);
psi := (1/sqrt(2*(x1^2 + x3^2)))*(1 + e(0)*e(1)*e(2)*e(3))*
(cos(phi/2) - e(1)*e(3)*sin(phi/2));
let trigrules;
far := psi*e(2)*e(1)*rev(psi);
```

The result is

$$F = \frac{\rho e_{01} + z e_{03}}{(\rho^2 + z^2)^{3/2}}$$

which is just the Coulomb field $-\vec{r}/r^3$ in terms of cylindrical coordinates. Also

```
h(0) := 1;
h(1) := 1;
h(2) := 1/x1;
h(3) := 1;
dirac(far);
```

gives $\partial F = 0$ (of course computer algebra systems still don't know topology!). Note that the electromagnetic potential A can be written in this example as

```
pot := -psi*e(0)*rev(psi)/sqrt(-psi*rev(psi)*e(0)*e(1)*e(2)*e(3));
```

which is easy to verify since

```
far - dirac(pot);
```

gives zero.

The electromagnetic field of a moving charged particle can be obtained in the same way as $\psi' e_{21} \tilde{\psi}'$, where now ψ' is

```
psipr := lor*psi;
```

where `lor` describes a Lorentz transformation as in Section 2.3. This illustrates the fact that although $\psi \in C\ell_{1,3}^+$, it does not transforms under the action of $\mathbf{Spin}_+(1,3)$ as other elements of $C\ell_{1,3}^+$. This fact has lead us to define a Dirac-Hestenes spinor field as a certain equivalence class of even multivectors (Rodrigues et $al.$, 1995).

4. Clifford algebra valued differential forms

The main disadvantage of working the way we have with Clifford algebras is the dependence on the choice of a basis. It would be better if we could define a new non-commutative product and work in terms of it, but this is not a trivial thing to do. On the other hand, this apparent disadvantage can be exploited into a powerful mathematical tool.

Given a set of p-forms A^μ we can define a Clifford algebra valued differential form (clifform, for short) $A = A^\mu \otimes e_\mu$. For example: $\theta = dx^\mu \otimes e_\mu$, where $\{e_\mu\}$ need not to be identified with $\{\partial_\mu\}$, that is, $\{e_\mu\}$ may be a basis of an abstract vector space. Similarly, for a set of p-forms $A^{\mu\nu}$ such that $A^{\mu\nu} = -A^{\nu\mu}$ we define a clifform $A = (1/2)A^{\mu\nu} \otimes e_{\mu\nu}$. For example: the connexion 1-form is given by $\Gamma_{\mu\nu} = \Gamma_{\mu\sigma\nu}dx^\sigma$ such that in terms of an orthonormal coframe $\{\theta^\mu\}$ a metric compatible connexion satisfies $\Gamma_{\mu\nu} = -\Gamma_{\nu\mu}$; we can define therefore a 2-vector valued 1-form $\Gamma = (1/2)\Gamma^{\mu\nu} \otimes e_{\mu\nu}$. One can find in (Dimakis and Müller-Hoissen 1991) several interesting applications of clifforms to Mathematical Physics.

Since a clifform is defined in terms of the generators of the Clifford algebra, our approach does not show any deficiency in this case. On the other hand, REDUCE has a very powerful package for leading with exterior forms called **EXCALC** (Schrüfer 1994). In this section we shall illustrate some examples of applications of clifforms using **EXCALC**[8].

4.1. STRUCTURAL EQUATIONS

The Cartan structural equations

$$\Theta^\mu = d\theta^\mu + \Gamma^\mu{}_\nu \wedge \theta^\nu, \quad \Omega^\mu{}_\nu = d\Gamma^\mu{}_\nu + \Gamma^\mu{}_\sigma \wedge \Gamma^\sigma{}_\nu$$

define the torsion 2-form Θ^μ and the curvature 2-form $\Omega^\mu{}_\nu$. Here $\{\theta^\mu\}$ is the coframe field and d is the exterior derivative. For a metric compatible connexion we have $\Gamma^{\mu\nu} = -\Gamma^{\nu\mu}$. If we define in this case

$$\theta = \theta^\mu \otimes e_\mu, \quad \Gamma = \frac{1}{2}\Gamma^{\mu\nu} \otimes e_{\mu\nu},$$

$$\Theta = \Theta^\mu \otimes e_\mu, \quad \Omega = \frac{1}{2}\Omega^{\mu\nu} \otimes e_{\mu\nu},$$

then Cartan structural equations read

$$\Theta = d\theta + \Gamma \wedge \theta + \theta \wedge \Gamma,$$

$$\Omega = d\Gamma + \Gamma \wedge \Gamma.$$

The Bianchi identities are

$$d\Theta + \Gamma \wedge \Theta - \Theta \wedge \Gamma = \Omega \wedge \theta - \theta \wedge \Omega,$$

$$d\Omega + \Gamma \wedge \Omega - \Omega \wedge \Gamma = 0.$$

As in the example, let us work out again the Schwarzschild solution of the gravitational field equations, and see how easier it is to work with clifforms.

---

[8] For REDUCE 3.5 it is necessary to define some patches in order to **EXCALC** work with clifforms. These patches are reproduced in the appendix below so that the reader could reproduce our results.

```
load_package excalc;
operator e;
noncom e;
e(0)*e(0) := 1;
...
pform {psi1,psi2}=0, {theta,conn(k,l),gamma}=1;
fdomain psi1=psi1(t,r), psi2=psi2(t,r);
coframe o(0) = exp(psi1)*d t, o(1) = exp(psi2)*d r, o(2) = r*d th,
o(3) = r*sin(th)*d phi with signature (1,-1,-1,-1);
theta := o(0)*e(0) + o(1)*e(1) + o(2)*e(2) + o(3)*e(3);
gamma := (1/2)*(conn(0,1)*e(0)*e(1) + conn(0,2)*e(0)*e(2)
+ conn(0,3)*e(0)*e(3) + conn(1,2)*e(1)*e(2)
+ conn(1,3)*e(1)*e(3) + conn(2,3)*e(2)*e(3));
```

Now we have to impose the condition of Riemannian connexion, that is, of vanishing torsion: $d\theta + \Gamma \wedge \theta + \theta \wedge \Gamma = 0$. However, **EXCALC** has a command **riemannconx** which gives us the Riemannian connexion.

```
riemannconx conn;
d theta + gamma^theta + theta^gamma;
```

We see that $\Theta = 0$ indeed. For the curvature we have

```
pform curv=2;
curv := d gamma + gamma^gamma;
```

The Ricci 1-vector 1-form is given by $R = w \lrcorner \Omega$, where $w = \mathbf{e}^\mu \otimes v_\mu$, with $\{v_\mu\}$ being the frame dual to $\{\theta^\mu\}$.

```
frame v;
w := e(0)*v(-0) - e(1)*v(-1) - e(2)*v(-2) - e(3)*v(-3);
pform ricci=1;
ricci := w _| curv;
```

The result is

$$
\begin{aligned}
R = R^\mu{}_\nu \mathbf{e}_\mu \otimes \theta^\nu = {}& \mathbf{e}_0 \otimes \theta^0 \left[\tfrac{1}{2} e^{-2\psi_2} \left(\partial_{rr}\psi_1 + (\partial_r\psi_1)^2 - \partial_r\psi_1 \partial_r\psi_2 \right) \right. \\
& \left. - \tfrac{1}{2} e^{-2\psi_1} \left(\partial_{tt}\psi_2 + (\partial_t\psi_2)^2 - \partial_t\psi_1\partial_t\psi_2 \right) + \tfrac{1}{r} e^{-2\psi_2} \partial_r\psi_1 \right] \\
& + \mathbf{e}_0 \otimes \theta^1 \left(\tfrac{1}{r} e^{-(\psi_1+\psi_2)} \partial_t\psi_2 \right) + \mathbf{e}_1 \otimes \theta^1 \left(-\tfrac{1}{r} e^{-(\psi_1+\psi_2)} \partial_t\psi_2 \right) \\
& + \mathbf{e}_1 \otimes \theta^1 \left[\tfrac{1}{2} e^{-2\psi_2} \left(\partial_{rr}\psi_1 + (\partial_r\psi_1)^2 - \partial_r\psi_1\partial_r\psi_2 \right) \right. \\
& \left. - \tfrac{1}{2} e^{-2\psi_1} \left(\partial_{tt}\psi_2 + (\partial_t\psi_2)^2 - \partial_t\psi_1\partial_t\psi_2 \right) - \tfrac{1}{r} e^{-2\psi_2} \partial_r\psi_2 \right] \\
& + \mathbf{e}_2 \otimes \theta^2 \left[\tfrac{1}{2r^2} e^{-2\psi_2} \left(1 - e^{2\psi_2} + r(\partial_r\psi_1 - \partial_r\psi_2) \right) \right] \\
& + \mathbf{e}_3 \otimes \theta^3 \left[\tfrac{1}{2r^2} e^{-2\psi_2} \left(1 - e^{2\psi_2} + r(\partial_r\psi_1 - \partial_r\psi_2) \right) \right],
\end{aligned}
$$

and from $R^\mu{}_\nu = 0$ we obtain the Schwarzschild solution

$$
e^{\psi_1} = \sqrt{1 - 2m/r}, \quad e^{\psi_2} = 1/\sqrt{1 - 2m/r}
$$

which gives

$$\Omega = \frac{1}{2} \left[\left(\frac{-2m}{r^3} \right) e_{01} \otimes (\theta^0 \wedge \theta^1) + \left(\frac{m}{r^3} \right) e_{02} \otimes (\theta^0 \wedge \theta^2) + \left(\frac{m}{r^3} \right) e_{03} \otimes (\theta^0 \wedge \theta^3) \right.$$
$$\left. + \left(\frac{m}{r^3} \right) e_{12} \otimes (\theta^1 \wedge \theta^2) + \left(\frac{m}{r^3} \right) e_{13} \otimes (\theta^1 \wedge \theta^3) + \left(\frac{-2m}{r^3} \right) e_{23} \otimes (\theta^2 \wedge \theta^3) \right].$$

One can construct several other interesting examples. For example, one particularly interesting is the derivation of the Reissner-Nordström solution from the Schwarzschild one given by (Puntigam *et al.*, 1995). In the same reference we can also find the derivation of the Kerr-Newman solution from Kerr one.

4.2. GAUGE FIELDS

Clifforms can also be used in the description of gauge fields. The gauge (or Yang-Mills) potential is a Lie algebra valued 1-form. But a Lie algebra is isomorphic (Doran *et al.*, 1993) to the sub-algebra of 2-vectors with the commutator product of a Clifford algebra. The gauge potential could therefore be taken as a 2-vector valued 1-form.

Let us consider, as a concrete example, the SU(2) gauge theory over $\mathbb{R}^4$. The Clifford algebra of $\mathbb{R}^4$ is $C\ell_{4,0} \simeq \mathbb{H}(2)$. The generators of $su(2)$ can be identified with $\{e_{12}, e_{23}, e_{13}\}$ with the product $A \times B = (1/2)[A, B]$, as we shall see. Note that in order to describe $su(2)$ we don't need to use $C\ell_{4,0}$; we can also work, for example, with $C\ell_{3,0} \in C\ell_{4,0}$ or $C\ell_{1,3} \simeq C\ell_{4,0}$. The choice of $C\ell_{3,0}$ is the simplest one, so we shall work with it.

```
operator e;
noncom e;
for k:=1:3 do e(k)*e(k) := 1;
for all j,k such that j>k let e(j)*e(k) = -e(k)*e(j);
factor e;
(1/2)*((e(1)*e(2))*(e(2)*e(3))-(e(2)*e(3))*(e(1)*e(2)));
(1/2)*((e(2)*e(3))*(e(1)*e(3))-(e(1)*e(3))*(e(2)*e(3)));
(1/2)*((e(1)*e(3))*(e(1)*e(2))-(e(1)*e(2))*(e(1)*e(3)));
```

We see that $(1/2)[e_{12}, e_{23}] = e_{13}$, $(1/2)[e_{23}, e_{13}] = e_{12}$ and $(1/2)[e_{13}, e_{12}] = e_{23}$, that is, $\{e_{12}, e_{23}, e_{13}\}$ are the generators of $su(2)$. Let us define the SU(2) gauge potential.

```
load_package excalc;
pform a(1) = 1;
indexrange 1 = {1,2,3};
a := a(1)*e(1)*e(2) + a(2)*e(2)*e(3) + a(3)*e(1)*e(3);
```

The field strength F is given by the covariant derivative of A.

```
pform f = 2;
f := d a + a^a;
```

This gives

$$F = (dA^1 + 2A^2 \wedge A^3)e_{12} + (dA^2 + 2A^3 \wedge A^1)e_{23} + (dA^3 + 2A^1 \wedge A^2)e_{13}.$$

Note that the field strength F satisfies the Bianchi identity $DF = dF + A \wedge F - F \wedge A = 0$.

```
d f + a^f - f^a;
```

The field strength F has to satisfy, besides $DF = 0$, also the field equation $D \star F = 0$, where $\star$ denotes the Hodge duality operator. However, for those fields which satisfy $F = \star F$ or $F = - \star F$ (resp. called self dual and anti-self dual fields) the equation $D \star F = 0$ is automatically satisfied. In order to find these solutions (called *instantons*) we write F in terms of A and require $F = \star F$ (or $F = - \star F$), which will give an equation to be solved for A. Consider $V \in C\ell_{3,0}^{+}$:

```
v:= v0 + v1*e(1)*e(2) + v2*e(2)*e(3) + v3*e(1)*e(3);
```

In order for $V \in \mathbf{Spin}(3)$, we must have $V\tilde{V} = 1$.

```
v*rev(v);
```

We see that $V\tilde{V} = v_0^2 + v_1^2 + v_2^2 + v_3^2$. So, $V/\sqrt{V\tilde{V}} \in \mathbf{Spin}(3)$. We know that if $V \in \mathbf{Spin}(3)$ then $\tilde{V}\partial_\mu V = -(\partial_\mu \tilde{V})V \in spin(3) \simeq su(2)$. Let us take $U \in \mathbf{Spin}(3)$ given by

```
u := (x0 + x1*e(1)*e(2) + x2*e(2)*e(3)
+ x3*e(1)*e(3))/sqrt(x0*x0 + x1*x1 + x2*x2 + x3*x3);
```

Since $\tilde{U}\partial_\mu U \in spin(3)$ we shall suppose that A is of the form $A = g\tilde{U}\partial_\mu U$, where g is a function to be defined.

```
coframe o(0)=d x0, o(1)=d x1, o(2)=d x2, o(3)=d x3
with signature (1,1,1,1);
pform g=0;
a:= g*rev(u)*d u;
```

This gives

$$
\begin{aligned}
A = \ &\frac{g}{(x_0^2 + x_1^2 + x_2^2 + x_3^2)} \left[(-\theta^0 x_1 + \theta^1 x_0 + \theta^2 x_3 - \theta^3 x_2)e_{12} \right. \\
&+ \left. (-\theta^0 x_2 - \theta^1 x_3 + \theta^2 x_0 + \theta^3 x_1)e_{23} + (-\theta^0 x_3 + \theta^1 x_2 - \theta^2 x_1 + \theta^3 x_0)e_{13} \right].
\end{aligned}
$$

Now $F = dA + A \wedge A$, and we shall impose $F = \star F$.

```
f := d a + a^a;
auxf := f - #f;
```

Looking to the expression for $F - \star F = 0$ it is suggested that a possible solution could comes from the ansatz $g = g(|\, x \,|)$, where $|\, x \,| = \sqrt{x_0^2 + x_1^2 + x_2^2 + x_3^2}$. It is convenient, however, to work with $\rho = x_0^2 + x_1^2 + x_2^2 + x_3^2$. We have therefore:

```
pform rho=0;
fdomain g=g(rho);
fdomain rho=rho(x0,x1,x2,x3);
rho:= x0*x0 + x1*x1 + x2*x2 + x3*x3;
auxf;
```

The result is:

$$F - \star F = \frac{1}{\rho} \left\{ 2\mathbf{e}_{12} \left[(\theta^0 \wedge \theta^1)(\frac{dg}{d\rho}\rho - g^2 + g) - (\theta^2 \wedge \theta^3)(\frac{dg}{d\rho}\rho - g^2 + g) \right] \right.$$

$$+ \ 2\mathbf{e}_{23} \left[(\theta^0 \wedge \theta^2)(\frac{dg}{d\rho}\rho - g^2 + g) + (\theta^1 \wedge \theta^3)(\frac{dg}{d\rho}\rho - g^2 + g) \right]$$

$$\left. + \ 2\mathbf{e}_{13} \left[(\theta^0 \wedge \theta^3)(\frac{dg}{d\rho}\rho - g^2 + g) - (\theta^1 \wedge \theta^2)(\frac{dg}{d\rho}\rho - g^2 + g) \right] \right\}.$$

The equation to be solved is

$$\frac{dg}{d\rho}\rho - g^2 + g = 0,$$

and this can be done with ODESOLVE.

```
load_package odesolve;
depend(g,rho);
odesolve(df(g,rho)*rho - g*g + g,g,rho);
```

which gives

```
arbconst(1) + log(g-1) - log(g) - log(rho)=0
```

or

$$g = \frac{1}{1 + \kappa^2 \rho},$$

where κ^2 is an arbitrary constant.

Finally, we observe that characteristic classes can also be written using clifforms, and therefore calculated with EXCALC. For example, the second Pontrijagin character, which is used to define the instanton number, is related to the scalar part (calculated using **grade**) of $F\tilde{F}$.

5. Conclusions

In this paper we tried to show how computer algebra can be a powerful tool in the teaching of Clifford algebras and its applications. In order to show this we took well known examples in Mathematical Physics, expressed these examples in terms of Clifford algebras and used REDUCE in order to manipulate them. We tried to focus attention on Clifford algebras and computer algebra together, and not on each one separately, so that only a basic knowledge of each was required. By showing examples from different areas of Mathematical Physics we expected to motivate the professor/lecturer and the student to work out more elaborate examples in areas of their interest. By showing examples from elementary to advanced levels we also expected to suggest that computer algebra and Clifford algebras can work together also in the research activities.

Acknowledgements

We are grateful to E. Schrüfer for discussions about **EXCALC** and for the patches needed to work with clifforms, to W. A. Rodrigues, W. Seixas and Q. A. G. Souza for years of discussions on computers, computer algebra systems and Clifford algebras, and to R. Abłamowicz, P. Lounesto and J. M. Parra for their kind invitation to collaborate in this volume.

Appendix

This appendix consists of two parts: in Part A one can find rules for manipulating with the Clifford algebra of spacetime and the Dirac operator (see text). Part B contains patches (written by E. Schrüfer to whom we are grateful for consenting to their reproduction here) that are needed to work with Clifford algebra valued differential forms package **EXCALC** from REDUCE 3.5 (see Section 4). These patches are in no way in final form and some problems still can be found (for example, if one wants to use the command **vardf** of **EXCALC** in order to calculate variational derivatives, some other patches are needed). The patches must be read after loading **EXCALC** in REDUCE 3.5. to reproduce results of Section 4.

A. Clifford algebra of Spacetime

```
operator e;
noncom e;

e(0)*e(0):=1;
for j:=1:3 do e(j)*e(j):=-1;
for all j,k such that j>k let e(j)*e(k)=-e(k)*e(j);

factor e;

operator eta;      % choice of signature
eta(0):=1;
for j:=1:3 do eta(j):=-1;

operator h;        % specification of coordinate system
% h(0), h(1), h(2) and h(3) must be
% specified according to the definition
% on the text

operator co;
co(0):=x0;
co(1):=x1;
co(2):=x2;
co(3):=x3;
```

```
% procedure to calculate the Dirac operator

procedure auxdirac(ex1,ex2);
begin;
z:=0;
for k:=0:3 do z := z + eta(k)*h(k)*df(1/h(ex1),co(k))*(1/2)*
(e(ex1)*e(k)*ex2 - ex2*e(ex1)*e(k));
return z end;

procedure dirac(ex);
begin;
y:=0;
for j:=0:3 do y := y + eta(j)*h(j)*e(j)*(df(ex,co(j)) +
auxdirac(j,ex));
return y end;

;end;
```

B. Patches for EXCALC version 2

The patches shown below are necessary for **EXCALC** version 2 only, and must be read after loading **EXCALC** in REDUCE 3.5. Please contact Eberhard Schrüefer (schruefer@gmd.de) for the status of **EXCALC** version 3.

```
%--- patch

symbolic procedure multpfs(u,v);
if null u or null v then nil
else if ldpf u = 1 then multsqpf(lc u,v)
else if ldpf v = 1 then multpfsq(u,lc v)
else addpf(addpf(multttpf(lt u,lt v),multpfs(red u,v)),
multpfs(lt u .+ nil,red v));

symbolic procedure multpfsq(u,v);
if null u or null numr v then nil
else ldpf u .* multsq(lc u,v) .+ multpfsq(red u,v);

symbolic procedure multsqpf(u,v);
if null v or null numr u then nil
else ldpf v .* multsq(u,lc v) .+ multsqpf(u,red v);

symbolic procedure partitsq!* u;
%U is a standardquotient. Partitfunction for *sq's.
%Leaves unexpanded structure if possible;
(if null x then nil
else if domainp x then 1 .* u .+ nil
else addpf(if sfp mvar x and sfexform1p lt mvar x
```

```
then multpfsq(exptpf(partitsq!*(mvar x ./ 1),
ldeg x),
cancel(lc x ./ y))
else if null sfp mvar x and deg!*form mvar x
then mvar x .* cancel(lc x ./ y) .+ nil
else multsqpf(!*p2q lpow x,partitsq!*(lc x ./ y)),
partitsq!*(red x ./ y)))
where x = numr u, y = denr u;

symbolic procedure exdff0 u;
if domainp u then nil
else addpf(addpf(multsqpf(!*p2q lpow u,exdff0 lc u),
multpfsq(exdfp0 lpow u,lc u ./ 1)),
exdff0 red u);

symbolic procedure partitop u;
begin scalar x,alglist!*;
return
if atom u then if x := get(u,'avalue)
then partitsq!* simp!* cadr x
else if get!*fdeg u then mkupf u
else if numr(x := simp!* u)
then 1 .* x .+ nil
else nil
else if x := get(car u,'partitfn)
then if flagp(car u,'full) then apply1(x,u)
else apply1(x,cdr u)
else if car u eq '!*sq then partitsq!* simp!* u
else if car u eq 'plus then
<<for each j in cdr u do
x := addpf(partitop j,x); x>>
else if car u eq 'minus then negpf partitop cadr u
else if car u eq 'difference then
addpf(partitop cadr u,
negpf partitop caddr u)
else if car u eq 'times then
<<x := partitop cadr u;
for each j in cddr u do
x := multpfs(x,partitop j);
x>>
else if car u eq 'quotient then
multpfsq(partitop cadr u,simprecip cddr u)
else if car u eq 'recip then
1 .* simprecip cdr u .+ nil
else if numr(x := simp!* u)
then 1 .* x .+ nil
```

```
else nil
end;
```

`%--- end patch ---`

`; end;`

References

Ablamowicz, R., Lounesto, P. and Maks, J.: 1991, "Conference Report: Second Workshop on "Clifford Algebras and their applications in mathematical physics"", *Foundations of Physics* **21**, pp. 735–748.

Barut, A. O. and Zanghi, N.: 1984, "Classical theory of Dirac electron", *Physical Review Letters* **52**, pp. 2009–2012.

Brackx, F., Constales, D., Delanghe, R. and Serras, H.: 1987, "Clifford algebra with Reduce", *Supplemento ai Rendiconti del Circolo Matemático di Palermo*, **16**, serie II, pp. 11–19.

Choquet-Bruhat, Y., DeWitt-Morette, C. and Dillard-Bleick, M.: 1982, *Analysis, Manifolds and Physics*, revised edition, North-Holland Publishing Company, Amsterdam, The Netherlands.

Delanghe, R., Sommen, F. and Souček, V.: 1992, *Clifford Algebra and Spinor-Valued Functions*, Kluwer Academic Publishers, Dordrecht, The Netherlands.

Dimakis, A. and Müller-Hoissen, F.: 1991, "Clifform calculus with applications to classical field theory", *Classical and Quantum Gravity*, **8**, pp. 2093–2132.

Doran, C., Hestenes, D., Sommen, F. and Van Acker, N.: 1993, "Lie groups as spin groups", *Journal of Mathematical Physics*, **34**, pp. 3642–3669.

Hestenes, D.: 1966, *Spacetime Algebra*, Gordon & Breach, New York.

Hestenes, D.: 1986, "Curvature calculations with spacetime algebra", *International Journal of Theoretical Physics*, **25**, pp. 581–588.

Hestenes, D.: 1991, "The design of linear algebra and geometry", *Acta Applicandae Mathematicae*, **23**, pp. 65–93.

Jancewicz, B.: 1988, *Multivectors and Clifford Algebra in Electrodynamics*, World Scientific, Singapore.

Ławrynowicz, J., Papaloucas, L. C. and Rembieliński, J.: 1995, "Quantum Braided Clifford Algebras", in *Clifford Algebras and Spinor Structures*, Abłamowicz, R. and Lounesto, P. (eds.), Kluwer Academic Publishers, Dordrecht, pp. 387–395.

Lounesto, P. and Latvamaa, E.: 1980, "Conformal Transformations and Clifford algebras", *Proceedings of the American Mathematical Society*, **79**, p. 533.

Puntigam, R. A., Schrüfer, E. and Hehl, F. W.: 1995, "The use of computer algebra in Maxwell's Theory", to appear in *Computer Algebra in Science and Engineering – Proc. of the ZiF-Workshop*, Fleischer, J. et al. (eds.), World Scientific, Singapore.

Rego, G. B.: 1963, "On the definition of some electromagnetic quantities", in *Proceedings of the Symposium on Electromagnetic Theory and Antennas*, Copenhagen 1962, Pergamon Press, pp. 997–1007.

Rodrigues, Jr., W. A., Souza, Q. A. G. and Vaz, Jr., J.: 1995, "Spinor fields and superfields as equivalence classes of exterior algebra fields", in *Clifford Algebras and Spinor Structures*, Abłamowicz, R. and Lounesto, P. (eds.), Kluwer Academic Publishers, Dordrecht, pp. 177–198.

Schrüfer, E.: 1994, **EXCALC**: *A System for Doing Calculations in the Calculus of Modern Differential Geometry*, GMD, Institut I1, D-53757 St. Augustin, Germany.

Vaz, Jr., J. and Rodrigues, Jr., W. A.: 1993, "On the equivalence of Maxwell and Dirac equations, and quantum mechanics" *International Journal of Theoretical Physics*, **32**, pp. 945–958.

Vaz, Jr., J.: 1995, "The Barut and Zanghi Model, and some generalizations", *Physics Letters* **B 344**, pp. 149–157.

GENERAL CLIFFORD ALGEBRA AND RELATED DIFFERENTIAL GEOMETRY CALCULATIONS WITH *MATHEMATICA*

JOSEP M. PARRA * and LLORENÇ ROSELLÓ †
Laboratori de Física Matemàtica,I.E.C.
Dept. de Física Fonamental, Universitat de Barcelona
Diagonal 647, E-08028, Barcelona, Spain
e-mail: jmparra@hermes.ffn.ub.es

Abstract. In this paper the syntax of the functions of the packages Clifford.m and Nabla.m is explained by means of simple examples. These files together with some examples mostly taken from physical applications are meant to be an integral part of the software developed by the authors. Also included is a *MatLab* appendix referring to another specific set of files.

Key words: Clifford algebra – Mathematica – Nabla operator – Differential forms

1. Introduction

A specific Clifford Algebra package has been developed to perform in *Mathematica* some calculations related to this algebra. The greatest technical difficulty encountered while programming is the scarce facilities that *Mathematica* offers to define a *"non commutative product"* using its function *,Times or its function **, NonCommutativeMultiply. For instance, if we want $e_1 e_2 = e_{12}$ one could think that it can be modified this way:

```
Unprotect[Times]
Times[e[i_], e[j_]] := e[i, j]
```

But with this definition the outcome of the product:

```
e[2] * e[1]
```

turns out to be:

```
e[1,2]
```

The problem is that it is very difficult (and we don't know now if it is possible) to force *Mathematica* not to put the factors in order before operating with them. Other kind of difficulties arise when trying to use **. They have led us to write a complete new function product CP specific to the Clifford algebra needs. This procedure, in fact a direct implementation of the product definition given in (Artin, 1957, p.186), causes some loss of "naturality" that can be remedied with the definition of an

---

* The author acknowledges financial support from the Spanish Ministry of Education under contract No. PB93-1050
† Dept. L.S.I., U. Politècnica de Catalunya, e-mail: lm38124917@quartz.upc.es

optional `PreReadOn[]` procedure that allows a substitution of the prefix operator symbol `CP` by the infix operator symbol `**`.

The other most important function inside the program is `Nabla`. It is the Clifford algebra straightforward generalization of Hamilton's ∇ original operator. This is the true square root of the Laplacian over arbitrary exterior algebra fields or differential forms. Dirac found it in a disguised complex matrix form applied upon complex entities of dubious interpretation. Naming it after Dirac makes no justice to Maxwell whose set of equations in empty space constitute a paradigmatic example of its importance in physics. Indeed, Maxwell's vacuum equations are in $Cl_{1,3}$ or $Cl_{3,1}$: `Nabla[BivectorField]=0`. It is our opinion that its use in arbitrary orthogonal coordinates, in any number of dimensions, and with any signature of the metric, may foster a more extensive use of Clifford analysis both in teaching and research, enhancing and enlarging the revival of the Grassmann-Cartan differential form calculus advocated some years ago in (Regge, 1984)

> "The first step, or rather advice, in the group manifold approach is to take the calculus of forms seriously and use it instead of the conventional tensor calculus. The chief advantage is that computations are much simpler and their geometrical and physical interpretation more transparent. ... the calculus of forms is basically so much simpler than the standard tensor calculus that it should win in the long run. Also the relationship between geometry, group theory and also physics is much more obvious."

As the works of Marcel Riesz, David Hestenes and many others have shown, all these connections and even the power of this differential calculus are greatly increased when: a) one allows linear combinations of forms of different degree, especially those that can be called even or odd; b) the exterior or Grassmann product is combined with the inner product of the metric structure to produce the single associative **geometric** product, as Clifford named it. A *Mathematica* tool designed to facilitate the use of Clifford algebra and analysis while keeping all the time the characteristic geometric nature of the quantities at sight –an outstanding feature of `CLICAL` (Lounesto, 1987)–, can be extensively useful to overcome the natural reservations and actively promote the geometric algebra usage among those used to the traditional vectorial and matrix methods.

2. Description of `Clifford.m` functions

- `e`

  `e[`$i_1, i_2, \ldots, i_n$`]` is a basis multivector, obtained by the exterior product or the Clifford product of orthogonal basis vectors. The correspondence between the standard and *Mathematica* representations is as follows:

$$1 \longmapsto \texttt{e[]}$$
$$\mathrm{e}_{i_1,\ldots,i_n} \longmapsto \texttt{e[}i_1,\ldots,i_n\texttt{]}$$

  Although the order of the indices doesn't matter in the input and is fixed in the output, it is very important to take care **not to repeat any index inside**

the brackets, e.g.:

$$\cos x + \sin x\, e_{12} \longmapsto \text{Cos[x]} \ \text{e[]} \ + \ \text{Sin[x]} \ \text{e[1,2]}$$

- **CliffordAlgebra, CA**
  CliffordAlgebra$[e_0^2, e_1^2, \ldots, e_n^2]$ is used to introduce the specific Clifford algebra with which we want to work. Contrary to other programs we are allowed to change it at any time keeping the value of the variables, thus making full use of the universality character of the Clifford algebra. As the package has been designed to suit the needs of physicists for whom the index 0 is used to denote the time coordinate, it turns out to be **always** necessary to include it and specify e_0^2, even when no use of this dimension is intended. **CA[1,1,1,1]**, **CA[1,-1-1-1]**, **CA[-1,1,1,1]** define respectively the Clifford algebras $C\ell_{4,0}, C\ell_{1,3}, C\ell_{3,1}$. The first and the last can be indistinctly chosen to work out $C\ell_{3,0}$ with basis vectors e_1, e_2, e_3.

- **form**
  form is the list that contains the list of the values e_i^2. It may be useful to inform us about what Clifford algebra are we using.

- **CP, ****
  CP[*cliffor*$_1, \ldots,$ *cliffor*$_n$] Gives the product of *cliffor*$_1 \ldots$ *cliffor*$_n$, where *cliffor* will stand from now on as a generic name for an arbitrary element of the Clifford algebra, i. e., a general linear combination of the basis multivectors $e_{i_1,\ldots,i_r}$. This name has been repeatedly proposed by B. Jancewicz (Jancewicz, 1989) as belonging to the same 'family names' of some of the most used quantities in physics: vector, tensor, spinor, affinor. It has been objected that being Clifford a person name it does not seem a 'correct choice'. But due to the reluctance to call *form* and *differential form* the linear combinations of antisymmetric tensors of different degrees, and also to admit that in metric spaces and manifolds they are "naturally subjected" to the Clifford product in no lesser degree than they are subjected to the exterior or Grassmann product, Jancewicz's proposal has the merit of avoiding any ambiguity. It makes clear that all objects considered here are in the domain of Clifford calculus, something that challenges the vector, tensor or spinor calculus.
  For instance, a *cliffor* could be:

$$\text{Log[z] e[1,2,3] + a e[1] - c e[3,0]}$$

but we cannot write in *Mathematica*

$$\text{Log[z] e[1,2] e[3] + a e[1] (e[] - c e[1,3,0])}$$

Instead, we have to make explicit use of the Clifford product. For instance we will translate $(\cos x + \sin x\, e_{12})(ae_1 + be_2)(\cos x - \sin x\, e_{12})$ in *Mathematica* as:
```
CP[Cos[x] e[] + Sin[x] e[1,2], a e[1] + b e[2],
Cos[x] e[] - Sin[x] e[1,2]k]
```

or, alternatively, if we have previously established
`PreReadOn[]`
`(Cos[x] e[] + Sin[x] e[1,2], a e[1] + b e[2]) **`
`(Cos[x] e[] - Sin[x] e[1,2])`
Full associativity and distributivity with respect the sum is implemented in
either of the two ways of expressing the Clifford product of *cliffors*, without any
limitation in the number of terms. This means that we can use ** for *cliffors* in
the same way as we use the ordinary product. Notice that because the names
for a *cliffor* are not distinguished from the names of numeric variables we have:
`3 * (e[1] + 3 e[2]) = 3 (e[1] + 3 e[2]) = 3 ** (e[1] + 3 e[2])`
but
`a * (e[1] + 3 e[2]) = a (e[1] + 3 e[2])`
is different from
`a ** (e[1]+ 3 e[2])= CP[a, e[1]+3 e[2]]= a e[] ** (e[1]+3 e[2])`

- **NPart**
  `NPart[`*cliffor*`, n]` shows only those components of *cliffor* that have degree n.
  For example if we want to select the bivector part of $(a/b)\,e_1 - 2\,e_{12} + \ln z\,e_{23} + a$
  we type:
  `NPart[a/b e[1] - 2 e[1,2] + Log[z] e[2,3] + a e[], 2]`
  and get the answer
  `-2 e[1,2] + e[2,3] Log[z]`

- **CommonFactor, CF**
  `CommonFactor[`*cliffor*`]` extracts the e_I as common factors. It is often useful
  the combination `CommonFactor[CP [ ..., ... ]]` if we want to read the
  outcome of the Clifford product. For instance
  `CommonFactor[e[0,1] + b e[0,1] + c e[2]]`
  gives the result
  `c e[2] + (1 + b) e[0,1]`
  We can also specify which e_I we want as the common factor. This is done with
  command `CommonFactor[`*cliffor*`, list]` where $list = I$.

- **CPower**
  `CPower[`*cliffor*`, n]` is the n-th power (n a positive integer) of *cliffor*. For
  example:
  `CPower[e[] + 2 e[1,2,3], 13]`
  the result is
  `-8839 e[] + 33802 e[1,2,3]`

- **Rev**
  `Rev[`*cliffor*`]` gives the reversion of *cliffor*. For example:
  `Rev[2 e[1,2,19,0]+ Sin[Cos[z]] e[17, 13, 2]]`
  the result is
  `2 e[0,19,2,1] + e[2,13,17] Sin[Cos[z]]`

- **GradeInvol, GI, MI**
  `GradeInvol[`*cliffor*`]` applies the grade or main involution automorphism to *cliffor*. For example

```
GradeInvol[2 e[12,19]+ Cos[z] e[97, 13, 2]]
```
the result is
```
2 e[12,19] - Cos[z] e[97,13,2]
```

- **GRev, GR**
  GradeRev[*cliffor*] gives the Clifford conjugation or grade reversion of *cliffor*.
  For example:
  ```
  GradeRev[2 e[12,19]+ Cos[z] e[97,13,2]]
  ```
  the result is
  ```
  2 e[19,12] - Cos[z] e[2,13,97]
  ```

- **ComplexConjugate, CC; RPart, ReP; IPart, ImP**
  These functions assume that all imaginary dependence is explicitly given by the
  I factor. That is, that all variables or functions are assumed to be real. They
  are most useful in dealing with complex Clifford algebras. For instance:
  ```
  ComplexConjugate[(x0 - I Sin[a]) e[1,2]]
  RPart[(x0 - I Sin[a]) e[1,2]]
  IPart[(x0 - I Sin[a]) e[1,2]]
  ```
  give as results:
  ```
  (x0 + I Sin[a]) e[1,2]]
  x0 e[1,2]
  - Sin[a] e[1,2]
  ```

- **Changei**
  Changei[*cliffor*, *i*] changes the sign of e_i. It may be useful to implement
  time reversal, for example.
  ```
  Changei[a e[1,2] - 3 e[1,2,3], 3]
  ```
  the result is
  ```
  a e[1,2] + 3 e[1,2,3]
  ```

- **Numeric**
  Numeric[*cliffor*, *n*] gives the numerical value of the *cliffor* coefficients with *n*
  digits. It has the same effect as the usual *Mathematica* built-in **N**[*cliffor*, *n*]
  but it does not transform into real numbers the indices in the basis vectors and
  the outcome is still a *cliffor*.
  ```
  Numeric[Pi e[4] - 1/2 e[2], 3]
  ```
  gives
  ```
  -0.5 e[2] + 3.14 e[4]
  ```
  In the form **Numeric**[*cliffor*], we have six decimal precision digits.

- **ExteriorProduct, EP**
  ExteriorProduct[*cliffor*$_1$, *cliffor*$_2$, ..., *cliffor*$_n$] evaluates the exterior product
  of the *cliffors*. For example
  ```
  ExteriorProduct[a e[1] + b e[2], a e[1] + b e[2]]
  ```
  gives a null outcome.

- **InnerProduct, MP**

  InnerProduct[$cliffor_1$, $cliffor_2$, ..., $cliffor_n$] evaluates that part of the Clifford product that depends on the metric. It is not restricted to give the lowest degree of the Clifford product (corresponding to the maximum possible number of metric contractions). For example

  InnerProduct[a e[1] + b e[2], a e[1] + b e[2]]

  gives

  $$a^2\ e[] + b^2\ e[]$$

  As an interesting application of these functions we have written the auxiliary files **inertia.*** that implement the inertia tensor tables in (Hestenes, 1986, p.450-451). This implementation is intended to strengthen Hestenes' claim that the use of Clifford algebra pays for itself in the domain of classical mechanics.

- **Contract**

  Contract[i_1, ..., i_n] contracts the indices following the Clifford algebra currently defined. It is essentially an auxiliary function that may be useful in teaching manual computation of Clifford products. For example,

  Contract[1,2,1,1,2] gives e[1]

- **SubindexOn[]**

  SubindexOn[] produces a subindexed expression for all Clifford multivector units. After **PreReadOn[]** and **SubindexOn[]**

  1**(3 e[1] + 4 e[2,1,3] - a e[1,0,3,2])

  produces

  $$3\ e_1 - 4\ e_{123} - a\ e_{0123}$$

- **CTeX**

  CTeX[$cliffor$] gives the TEXform of $cliffor$. For example:

  CTeX[(b c f - a c g) e[1,2]]

  gives

  \left(b \, c \, f - a \, c \, g \right) \, {\bf e}_{12}

- **RegularSession; RegularMatrix, RM; GroupTable**

  These are the important functions devoted to the construction of the regular representation. Due to the special use of the zero index, intended to be useful to physicists, we have two kinds of regular representations: In the first one the index zero is neglected and the basis vectors are $e_1 \ldots e_n$. The regular matrix representation has then dimension 2^n and is obtained typing:

  RegularSession[]

  RegularMatrix[a e[1] - 4 e[1,2]]

  Evaluation of the regular matrix for a generic element gives the general form of the regular representation of the Clifford algebra generated by the orthonormal vectors $e_1 \ldots e_n$. If instead we want e_0 to be added as the first basis vector we should type:

  RegularSession0[]

  RegularMatrix[a e[1] - 4 e[1,2] - x e[0,3]]

obtaining now a 2^{n+1} real matrix.

The **RegularMatrix** function is only operative after doing a **RegularSession[]** or **RegularSession0[]**. Then, also, in the variable **GroupTable** we have the table product of 2^n or 2^{n+1} multivector units, arranged in such a way that the 'column elements' are the inverses of the 'row elements' producing a unit diagonal in the table.

The auxiliary files **inreg*.m** have been used to generate the regular matrix representations that implement the Clifford product in a *MatLab* environment for all $C\ell_{p,q}, p + q \leq 6$. For dimensions greater than three, the $\mathbf{e}_0$ vector has been taken, on physical grounds, as the first vector in the algebra. In the Appendix we give an outline of how to use these *MatLab* files, intended for readers without any familiarity with the program.

3. Description of Nabla.m functions

Although this program, designed to deal with *cliffor* fields in a manifold, needs the previous algebraic package **Clifford.m**, it is not necessary to load it previously, because it is automatically loaded with **Nabla.m**. However it is necessary to start with **CliffordAlgebra[...]**.

- **MetricTensor, MT**
  MetricTensor[$\{h_0, h_1, \ldots, h_n\}$, $\{q_0, \ldots, q_n\}$ **]** establishes the metric tensor of the manifold and the coordinates we will use. At present only orthogonal sets of coordinates are considered. The h_i are the well known Lamé's coefficients, and q_i are the names of the coordinates. As before, we always must specify h_0 and q_0. Lamé's coefficients are always positive. The signature of the space has been previously introduced with **DefineCliffordAlgebra**. For instance, bipolar (u, v, z) coordinates for ordinary space have

  $$h_1 = h_2 = \frac{a}{\cosh v - \cos u}, \quad h_3 = 1$$

  so they are introduced by:
  MetricTensor[{1, a/(Cosh[v]-Cos[u]), a/(Cosh[v]-Cos[u]), 1},
  　　　　　　{x0, u, v, z}]
  and can be also used to calculate in a four-dimensional Euclidean space or in a Minkowski space of signature ± 2, according to the previously defined Clifford algebra.

  If we are only interested in algebraic computation we can use **Clifford.m**, but if we want differentiation tools **Nabla.m** is the choice.

  In the following examples it is assumed that bipolar coordinates have been already introduced.

- **Ricci**
  Ricci[i, j, k **]** gives the Ricci rotation coefficient according to the derivation formula $\mathbf{e}_i^{\nabla}(\mathbf{e}_j) = \text{Ricci}[i, j, k]\mathbf{e}_k$. For instance, **Ricci[2,2,1]** gives
  Sin[u]
  ------
  　a

- **ExtCod**

  ExtCod[*cliffor*] gives a list of two elements. The first element is the exterior differential of *cliffor* and the second its codifferential. We stress that the sign of the codifferential is the opposite of the accepted standard in harmonic function theory. That the standard convention, rooted in a peculiar definition of Hodge duality, leads to the wrong sign in the Laplacian (or alternatively to define it as the square of the difference between the exterior differential and the codifferential) has been clearly expressed in (Nelson, 1967, p.100). Once realized that in a metric space the Clifford product structure is more fundamental than Hodge duality, there is no reason to maintain a "wrong" convention.

  `ExtCod[Log[u] e[3] + Sin[u v] e[1,2]]`

  gives

  $$\left\{-\frac{\text{Cos}[u]\ e[1,3]}{a\ u} + \frac{\text{Cosh}[v]\ e[1,3]}{a\ u},\right.$$

  $$\frac{u\ \text{Cos}[u]\ \text{Cos}[u\ v]\ e[1]}{a} - \frac{u\ \text{Cos}[u\ v]\ \text{Cosh}[v]\ e[1]}{a} -$$

  $$\left.\frac{u\ \text{Cos}[u]\ \text{Cos}[u\ v]\ e[2]}{a} + \frac{v\ \text{Cos}[u\ v]\ \text{Cosh}[v]\ e[2]}{a}\right\}$$

- **Codif**

  Codif[*cliffor*] gives the codifferential of *cliffor*. For instance

  `Codif[Log[u] e[3] + Sin[u, v] e[1,2]]`

  gives

  $$\frac{u\ \text{Cos}[u]\ \text{Cos}[u\ v]\ e[1]}{a} - \frac{u\ \text{Cos}[u\ v]\ \text{Cosh}[v]\ e[1]}{a} -$$

  $$\frac{u\ \text{Cos}[u]\ \text{Cos}[u\ v]\ e[2]}{a} + \frac{v\ \text{Cos}[u\ v]\ \text{Cosh}[v]\ e[2]}{a}$$

- **ExtD**

  ExtD[*cliffor*] gives the exterior differential of *cliffor*. For example

  `ExtD[Log[u] e[3] + Sin[u, v] e[1,2]]`

  gives

  $$-\frac{\text{Cos}[u]\ e[1,3]}{a\ u} + \frac{\text{Cosh}[v]\ e[1,3]}{a\ u},$$

  For both `Codif` and `ExtD` there is the possibility of applying them to a previous **ExtCod** outcome; in this case no derivation is realized, but merely the selection of the corresponding part.

- **Nabla**

Nabla[*cliffor*] Calculates the Nabla of *cliffor*. For instance:

Nabla[Log[u] e[3] + Sin[u,v] e[1,2]]

gives

```
   Cos[u] e[1,3]        Cosh[v] e[1,3]
- ------------- + ------------- +
       a u                a u


  u Cos[u] Cos[u v] e[1]     u Cos[u v] Cosh[v] e[1]
  --------------------- - ----------------------- -
           a                          a


  u Cos[u] Cos[u v] e[2]     v Cos[u v] Cosh[v] e[2]
  --------------------- + -----------------------
           a                          a
```

The Laplace operator does not need a primitive: **Nabla[Nabla[...]]** is the Laplace operator for any *cliffor* field. A "Copernican change" is needed to put the Nabla operator in its proper central position. The replacement of **Laplace operator** for **Nabla operator** in Nelson's clarifying analysis (Nelson, 1967, p.100) will literally fulfill what Hamilton (the man who found ◁) wrote in 1859 to Tait (who named it ∇, Nabla and extensively developed it): "*Could* anything be simpler or more satisfactory? Don't you *feel*, as well as think, that we are on a *right track*, and shall be *thanked* hereafter. Never mind when."(Crowe, 1993, p.254)

In this line, as two nontrivial didactic applications of the capabilities of the **Nabla.m** we accompany it with two sets of example files related to Maxwell and Dirac equations. In the first one the **rodvaz.*** files show how a non-standard proposed solution to the Maxwell equations (Rodrigues, 1995) is tested, and how this problem is much more easily tackled if we change the expression of the field from Cartesian to spherical components (not merely a coordinate change in the Cartesian components!). Finally we show how the two invariants of the electromagnetic field are computed by Clifford squaring the field.

In the second one the set of **dirac*.*** files show how the standard matrix form of Dirac's equation can be put into exact correspondence with four different sets of Hestenes-type geometric equations. This can be done irrespectively of the signature chosen for the Minkowski space-time. We have separate files for $Cl_{1,3}$ and $Cl_{3,1}$ in which we test the equality of each real component of the Dirac bispinor equation with a geometric component of the Hestenes-type equation. We refer to our work (Parra, 1992) for a detailed discussion of the possible implications of this multiple equivalence.

Note: *No function in the above packages incorporates an error processing safeguard. Thus, the user must know what he wants to do and how. Therefore it may occasionally happen that functions give very strange results because their parameters are not properly matched, or they had been misused.*

The reason why no error processing routines have been included is because they make the calculations substantially slower.

Appendix: MatLab numerical implementation

In order to use the above mentioned *MatLab* files to perform numerical calculations in Clifford Algebras one can proceed as follows:

1. Copy the files `cp.m`, `rg*.m`, `rev*.m`, `grade*.m`, `grev*.m`, `display*.m` to a working subdirectory of *MatLab* and introduce the new path in the initialization file `matlabrc.m` in the `matlab` root directory.

2. Each *cliffor* factor is put in a single line that lists its 2^n components separated by a blank space. No parentheses are needed at any place. The length of the line may be very large (one can use the Norton editor `ne.com` with word-wrap off).

3. The successive factors in the product from left to right are placed in successive lines of a single file from top to bottom. The name and extension of this file are arbitrary. Let us call it, for example `data3.num`.

4. Still in the *MatLab* dialog-command window, we type `load data3.num`

5. Now `data3` is a matrix whose lines are the Clifford factors and whose columns are the components in strict alphabetical order.

6. `result3=cp(p,q,data3)` stores in the variable `result3` the Clifford product of the elements in `data3` in the algebra $Cl_{p,q}$.

7. `save result.num result3 -ascii -double` saves the variable `result3` in the file `result.num`, with double precision (about 20 digits).

In the file `output1.txt` we have six lines, each representing an element of the Clifford algebra $Cl_{3,0}$, where the last one is the product of the five preceding ones. If the five factors were in a `input1.txt` file this `output1.txt` file is generated by the sequence of commands:

```
load input1.txt
x=cp(3,0,input1)
save output1.txt input1 x -ascii -double
```

We can apply to a single-line element `x` the functions `revn(x)` for reversion or main antiautomorphism, `graden(x)` for the grade or main involution, and `grevn(x)` for the grade reversion or Clifford conjugation, where $n = p + q$ is the dimension of the vector space upon which the Clifford algebra is constructed. For example, `z=rev5(data3(2,:))` gives the 32 components of the reverse of the Clifford element stored in the second row of the array `data3`. We have implemented the functions `displayn(x)` (n=3,4,5,6) in order to display in different lines the components of different order. There are trailing zeros after the last component in each order.

Also, at any time, we have at our disposal the regular matrix representation of a Clifford element. For $p = 2, q = 3$ we can obtain the 32×32 regular matrix of the `z` above as `mrz= rg23(z)`. This may prove very useful to find the spectrum of the Clifford element by means of the *MatLab* primitive `eig`: `spectrum=eig(mrz)`. To plot it in the complex plane, type `compass(spectrum)`.

This matrix representation can also be used to find a general function of the Clifford element by means of the *MatLab* primitives `expm`, `logm`, `sqrtm`, `inv` or, in general, with the `funm` primitive. For example `asinz=funm(mrz,'asin')` will give the $\arcsin(z)$ in the regular representation. Of course, we are only interested in the first column of this matrix, and transposed as a row line. So we end with `result=asinz(:,1)'` as a numeric Clifford variable to be saved in a file. The file `output2.txt` contains in descending order the first element of `input1`, its exponential, its logarithm, its square root and its arcsin.

Although in principle there is no limit to the dimensions, we must remember that the regular matrices are much larger than the irreducible or isomorphic matrix representations. As positive aspects of this representation we list the following:

—   The regular representation properly contains all irreducible representations, be them faithful or not. So it is a basic structure of intrinsic value in itself.

—   Using it all geometric algebras are dealt with on the same footing: all are represented by **real** matrices (no quaternions, no complex numbers, no double fields!). In other words, there is no room left for number mysticisms and the matrix form arises in a most natural way from the multiplicative structure of the **geometric** algebra.

—   The Clifford product does not involve matrix by matrix multiplication but only matrix by vector products.

—   There is no mixing of the geometric components which is a characteristic of the irreducible representations. Consequently the encryption-decoding procedures are not needed.

As *MatLab* possesses no symbolic capabilities, it does not seem able to provide a satisfactory tool for Clifford algebra calculations. However it can be quite useful in particular problems and even in graphic representations when taken in conjunction with a symbolic program such as *Mathematica* . For instance we recall that the hard task of constructing the regular matrix representations that *MatLab* uses as primitives has been performed with the Mathematica package `Clifford.m`.

References

E. Artin: *Geometric Algebra*. N. Y., Interscience Pub.,1957.

M. J. Crowe: *A History of Vector Analysis*. N. Y., Dover, 1993.

D. Hestenes: *New Foundations for Classical Mechanics*. Dordrecht, Kluwer, 1986, 1990.

B. Jancewicz: *Multivectors and Clifford algebra in electrodynamics*. Singapore, World Scientific, 1989.

P. Lounesto, R. Mikkola, V. Vierros: *CLICAL*. Helsinki Univ. of Technology, 1987.

MatLab,*High-Performance Numeric Computation and Visualization Software, Reference Guide*. The Math Works Inc., 1992.

E. Nelson: *Tensor Analysis*, Princeton U. P., 1967.

J. M. Parra: 'On Dirac and Dirac-Darwin-Hestenes equations', pp. 463-477 in A. Micali, R. Boudet, J. Helmstetter (eds.): *Proceedings of the Second International Conference on Clifford Algebras and Their Applications to Physics*. Dordrecht, Kluwer, 1992.

T. Regge: 'Group manifold approach to unified gravity', pp. 933-1006 in B. S. Dewitt, R. Stora R. (eds.), *Relativité, groupes et topologie*. Amsterdam, Elsevier, 1984.

W. Rodrigues, J. Vaz: 'Subluminal and Superluminal solutions in vacuum of the Maxwell equations and the massless Dirac equation'. RP 44/95, IMECC, UNICAMP (Brazil).

S. Wolfram: *Mathematica. A System for Doing Mathematics by Computer*. Addison-Wesley, 1991.

PAULI-ALGEBRA CALCULATIONS IN MAPLE V

W. E. BAYLIS
Department of Physics
University of Windsor
Windsor, Ontario, Canada N9B 3P4
e-mail: baylis@uwindsor.ca

Abstract. Procedures have been implemented in the symbolic-algebra program MAPLE V for calculations in the Clifford Algebra $C\ell_3$ of three-dimensional Euclidean space. The MAPLE language is sufficiently flexible to allow the definition of paravectors as the sum of a vector and a scalar, both of which can be complex. Matrix representations are therefore not required for computations, although they can be easily generated. Many problems in relativistic physics are easily handled by the algebra, which includes procedures for paravector differentiation and the evaluation of functions of paravectors, as well as for finding standard factors of elements.

Key words: The Clifford algebra of $\mathbb{R}^3$ – complex structure – Lorentz transformations

1. Introduction

The Clifford algebra $C\ell_3$ is generated from products of real orthonormal basis vectors e_1, e_2, e_3 in the three-dimensional Euclidean space $\mathbb{R}^3$ that satisfy

$$e_j e_k + e_k e_j = 2\delta_{jk}. \tag{1}$$

The linear space of $C\ell_3$ is spanned by an eight-dimensional real basis

$$\{1, e_1, e_2, e_3, e_2 e_3, e_3 e_1, e_1 e_2, e_1 e_2 e_3\} \tag{2}$$

and contains a natural *complex structure*, given by the identification of the trivector

$$e_1 e_2 e_3 = i \tag{3}$$

as an element of the center of the $C\ell_3$ that squares to -1. With this identity, the basis (2) can be written

$$\{1, e_1, e_2, e_3, ie_1, ie_2, ie_3, i\} . \tag{4}$$

Here we have tacitly chosen a privileged subspace as the real part of the linear space $C\ell_3$, namely the subspace of *paravectors* [1] spanned by

$$\{1, e_1, e_2, e_3\} . \tag{5}$$

This means introduction of a *real structure* in $C\ell_3$. Complexification of the subspace of paravectors gives back the linear space $C\ell_3$:

$$\mathbb{C} \otimes (\mathbb{R} \oplus \mathbb{R}^3) = \mathbb{R} \oplus \mathbb{R}^3 \oplus \overset{2}{\bigwedge} \mathbb{R}^3 \oplus \overset{3}{\bigwedge} \mathbb{R}^3 .$$

---

[1] Paravectors are sums of scalars and vectors.

The Clifford algebra $C\ell_3$ with elements written as complex paravectors is called the Pauli algebra. It is often used in physics in its 2×2 Hermitian-matrix representation where the basis elements $1, e_1, e_2, e_3$ are replaced by the unit matrix and the Pauli spin matrices: $1, \sigma_1, \sigma_2, \sigma_3$. Real elements of the Pauli algebra are then represented by hermitian matrices. The scalar and vector parts of a Pauli element are generally complex. The real part of a Pauli scalar is a scalar in $C\ell_3$ and its imaginary part is a trivector (a pseudoscalar) in $C\ell_3$. Similarly, the real part of a Paul vector is a vector in $C\ell_3$ whereas its imaginary part is a bivector (a pseudovector) in $C\ell_3$, see (Baylis & Jones, 1989).

MAPLE V is a powerful symbolic math package that allows a relatively free and natural format for the paravectors as the sum of a scalar and a vector. Thus MAPLE does not object if we define a paravector representing, for example, a spacetime position

```
> r := t + array([x,y,z]);
```

$$r := t + \begin{bmatrix} x & y & z \end{bmatrix}. \tag{6}$$

So far so good, but MAPLE has to be told how to manipulate such elements. Even if we try something as simple as adding two vectors,

```
> rnew := array([x,y,z])+array([1,0,0]);
```

$$rnew := \begin{bmatrix} x & y & z \end{bmatrix} + \begin{bmatrix} 1 & 0 & 0 \end{bmatrix}, \tag{7}$$

we find the MAPLE does not know how to add components. In the case of vectors (or other arrays), we can use MAPLE's **evalm** (evaluate matrix expression) command:

```
> evalm(rnew);
```

$$\begin{bmatrix} x+1 & y & z \end{bmatrix} \tag{8}$$

A similar evaluation command is evidently needed for Pauli elements.

This article describes a package of twenty procedures for adding, multiplying, conjugating, inverting, taking functions of, and differentiating Pauli elements. The procedure **Pmat** also allows elements to be represented in standard form as 2×2 matrices and **matP** can interpret such matrices as Pauli elements. However, unlike Lounesto's CLICAL, see (Mikkola & Lounesto, 1983), the procedures employ algebraic methods rather than matrix multiplication to evaluate expressions, and because it is based on the MAPLE engine, it can handle symbolic expressions as well as numeric ones. On the other hand, the package is written explicitly for the Pauli algebra form of $C\ell_3$, in which real paravectors can represent spacetime vectors. Commands for boosts and rotations of such vectors are included. Higher-dimensional Clifford algebras are not treated, although some of the approaches used here can be readily extended.

Most of the procedures are quite simple. All of them are available on an accompanying diskette and the majority are reproduced here, but the main thrust of this contribution is to explain the methods and capabilities of the package. The complete source code for an earlier version of the package is also available, together with a MAPLE V.2/V.3 worksheet showing its use, on the diskette distributed with my text, see (Baylis, 1994). The package uses the text form (*.t) form of the procedures rather than their compiled (*.m) form for several reasons:

— The text form is more easily ported between different computer platforms.

— It is less dependent on the version of MAPLE V.

— The text form makes it easier for the user to see what the procedure is doing, how it works, and how it an be adapted for his or her own purposes.

— The text form requires less storage space than the compiled form.

There are a couple of disadvantages, as well, but these are viewed as minor:

— The execution may be slower, but the procedures are so simple that this is not a major consideration.

— A bug in releases 2 and 3 of MAPLE V prevents the result from being printed the first time a *.t procedure is invoked. The simple work-around is to repeat the command or add

```
> '';
```

This is only needed the first time a procedure is called during a MAPLE session, and then, only if it is important to display the result.

To use the package, the user needs to read the table of procedures into the MAPLE session. If the directory containing the ASCII file Pauli.tab is made the working directory of the session, the commands are simply

```
> read 'Pauli.tab';
> with(Palg);
```

The file Pauli.tab defines the paravector basis elements as the unit scalar and unit vectors in a three-dimensional vector space, and it defines a table Palg whose elements are all the Palg procedure names, identified with readlib commands that locate the procedures in the subdirectory Pauli of the current working directory. Pauli.tab has the form

```
e0 := 1 :#paravector basis
e1 := array([1,0,0]):
e2 := array([0,1,0]):
e3 := array([0,0,1]):
#Locate procedures:
Palg := table([
 evalPe = 'readlib('evalPe' ,'pauli/evalPe.t' )',
 evalPa = 'readlib('evalPa' ,'pauli/evalPa.t' )',
 Pbar = 'readlib('Pbar' ,'pauli/Pbar.t' )',
 ...)]:
```

For further information about MAPLE packages and procedures, see on-line help in MAPLE V, for example on procedures, readlib, and with, and standard references, see (Heck, 1993).

2. Internal storage

An arbitrary element ("cliffor") in the Pauli algebra is a complex scalar plus a complex vector, but in symbolic notation this form can be ambiguous. For example, in expression (6), the term t could itself have vector as well as scalar parts. To avoid such problems, the internal storage of elements for manipulation is more tightly structured. The casual user does not need to know the format, but most routines in the package use the procedure evalPe which puts the elements in the form of a *list* of two elements: the first element is the scalar part of the element and the second element is the three-dimensional vector part:

```
evalPe := proc(pin) local n, elem, scal, vec, p;
options 'Copyright 1994 by W. E. Baylis, U. Windsor';
scal: = 0; vec: = array([0,0,0]);
p := expand(pin);
if type(p,'+') then
for n to nops(p)
do elem := evalm(op(n,p));
if type(elem,vector) then vec := evalm(vec+elem)
else scal := scal+elem
fi
od;
else elem := evalm(p);
if type(elem, vector) then vec := elem
else scal := elem
fi
fi;
[scal, evalm(vec)]
end:#basic procedure to evaluate a Pauli cliffor as a list
```

This procedure is sufficient to put most expressions of Pauli elements into the desired form. Products of scalars times sums of paravectors are expanded into a sum of products, and these are then segregated into scalar and vector parts and, with the help of **evalm**, added together and placed in the two-member list. (We do not yet have the tools to evaluate products of cliffors. They will be introduced in the following section.)

Example 1. All examples assume that the **Palg** package has been read in as described above.

```
>  xpr1:=3+array([1,2,3]); xpr2:=6+2*e1+e2;
```

$$xpr1 \ : \ = 3 + \begin{bmatrix} 1 & 2 & 3 \end{bmatrix}$$
$$xpr2 \ : \ = 6 + 2\,e1 + e2$$

```
>  evalPe(2*(xpr1+xpr2)/3);
```

$$[6, \begin{bmatrix} 2 & 2 & 2 \end{bmatrix}]$$

It is important to distinguish in MAPLE between lists and vectors. The vector part of a Pauli cliffor must be introduced as a vector, either through the use of **array** or **vector** or through the defined basis elements e1,e2,e3. Although users may never call **evalPe** directly, it helps simplify many of the routines they do call. For example, the routines **Pscal** and **Pvec** split an element into complex scalar and vector parts by simply extracting the first and second members of the list generated by **evalPe**:

```
Pscal := proc(p) local elem;
elem:=evalPe(p); elem[1]
end:# scalar part of cliffor
```

and

```
Pvec := proc(p) local elem
elem:=evalPe(p); elem[2]
end:# vector part of cliffor}
```

The anti-automorphic involution called *spatial reversal* or *Clifford-conjugation* swaps the sign of the complex vector part. It is denoted by a bar over the element and is calculated in the procedure `Pbar`:

```
Pbar := proc(p) local elem
elem:=evalPe(p); elem[1]-elem[2]
end:# bar, the spatial reverse or clifford conjugate
```

Example 2. Using the expression defined in the previous example, we have
> Pscal(xpr1);

$$3$$

> Pvec(xpr1);

$$[1 \quad 2 \quad 3]$$

> Pbar(xpr1);

$$3 - [1 \quad 2 \quad 3]$$

Another important split is that into real and imaginary parts. The corresponding procedures use MAPLE's facility with complex numbers and its `map` command, together with the assumption that imaginary parts have either been declared as such in an `assume` statement or have been written explicitly with MAPLE's unit imaginary I:

```
Preal := proc(p) local elem;
elem:=evalPe(p);
Re(elem[1]) + map(Re,elem[2])
end:# Real part of cliffor
```

and

```
Pimag := proc(p) local elem;
elem:=evalPe(p);
I*Im(elem[1]) + evalm(I*map(Im,elem[2]))
end:# Imaginary part of element
```

The involution that reverses the sign of imaginary part corresponds to the *reversion* in $C\ell_3$, and is called *Hermitian conjugation* in the Pauli algebra to emphasize its relation to the real structure and complex conjugation [of the linear space $C\ell_3$]. It is denoted with a dagger and is affected by

```
Pdag := proc(p) local elem;
elem:=evalPe(p);
conjugate(elem[1]) + evalm(map(conjugate,elem[2]))
end:# hermitian conjugate (reversal) of cliffor
```

Example 3.

```
> xpr:=2*I+3*array([3*I,5,6]);
```

$$2I + 3\begin{bmatrix} 3I & 5 & 6 \end{bmatrix}$$

```
> Preal(xpr);
```

$$\begin{bmatrix} 0 & 15 & 18 \end{bmatrix}$$

```
> Pimag(xpr);
```

$$2I + \begin{bmatrix} 9I & 0 & 0 \end{bmatrix}$$

```
> Pdag(xpr);
```

$$-2I + \begin{bmatrix} -9I & 15 & 18 \end{bmatrix}$$

The evaluation procedure more likely to be invoked directly by the user than **evalPe** is **evalPa**, which expands any element in the basis $\{e_0, e_1, e_2, e_3\}$:

```
evalPa := proc(p) local elem
elem:=evalPe(p);
elem[1]*'e0'+elem[2][1]*'e1'+elem[2][2]*'e2'+elem[2][3]*'e3'
end:# evaluate Pauli element as a sum: scalar+vector
```

Recall that the paravector basis elements are defined in **Pauli.tab**, the set-up file.

The user may also use the standard 2×2 matrix representation, in which e_1, e_2, e_3 are replaced by the Pauli spin matrices and e_0 is represented by the unit matrix. The procedure **Pmat** puts elements into matrix form:

```
Pmat := proc(p) local e;
e:= evalPe(p);
array([[e[1]+e[2][3],e[2][1]-I*e[2][2]],
[e[2][1]+I*e[2][2],e[1]-e[2][3]]])
end:# find standard matrix representation of p
```

The inverse procedure, to put 2×2 matrices into the algebraic form as the sum of a scalar and a vector, is **matP**:

```
matP  := proc(m:array)
(m[1,1]+m[2,2])/2 + array([
(m[1,2]+m[2,1])/2,(m[1,2]-m[2,1])*I/2,(m[1,1]-m[2,2])/2 ])
end:# convert 2x2 matrix to Pauli cliffor
```

Example 4.

```
> mat:=Pmat(xpr1);
```

$$mat := \begin{bmatrix} 6 & 1 - 2I \\ 1 + 2I & 0 \end{bmatrix}$$

```
> matP(mat);
```

$$3 + \begin{bmatrix} 1 & 2 & 3 \end{bmatrix}$$

3. Products

Fundamental to the package is the procedure defining products of Pauli elements. It is convenient to define an in-line product:

```
'&v':= proc(p,q) local pl,ql,i;
pl:=evalPe(p); ql:=evalPe(q);
pl[1]*ql[1]+sum(pl[2][i]*ql[2][i],i=1..3) +
evalm(pl[1]*ql[2]+ql[1]*pl[2] + I*array(
[ pl[2][2]*ql[2][3] - pl[2][3]*ql[2][2],
pl[2][3]*ql[2][1] - pl[2][1]*ql[2][3],
pl[2][1]*ql[2][2] - pl[2][2]*ql[2][1] ]))
end:# algebraic (''vee'') product
```

The (complex) scalar part of this is needed often enough that it is worthwhile to define a separate procedure that is more efficient than `Pscal(p &v q)`. It is the in-line dot product of paravectors:

```
'&dot':= proc(p,q) local pl,ql,i;
pl:=evalPe(p); ql:=evalPe(q);
pl[1]*ql[1]+sum(pl[2][i]*ql[2][i],i=1..3)
end:# dot (scalar) product of two Pauli cliffors
```

Example 5.

```
> xpr1; xpr2;
```

$$3 + \begin{bmatrix} 1 & 2 & 3 \end{bmatrix}$$
$$6 + 2\,e1 + e2$$

```
> xpr1 &v xpr2;
```

$$22 + \begin{bmatrix} 12 - 3I & 15 + 6I & 18 - 3I \end{bmatrix}$$

```
> xpr1 &dot xpr2;
```

$$22$$

The square norm $p\bar{p}$ of a paravector p is another frequently used product that can be more efficiently calculated by a separate procedure:

```
Pnorm := proc(p) local elem,i;
elem:=evalPe(p);
elem[1]^2-sum(elem[2][i]^2,i=1..3)
end:# square norm
```

A procedure to find the inverse of any element is easily constructed from the **Pnorm** and **Pbar** procedures, but an error must be flagged if no inverse exists:

```
Pinv := proc(p) local Pn;
Pn := Pnorm(p);
if Pn = 0
then ERROR('Cliffor is null. No inverse exists.')
else Pbar(p/Pn)
fi
end:# inverse element
```

Example 6.

> `Pnorm(xpr2);`

$$31$$

> `Pinv(xpr2);`

$$\frac{6}{31} - \begin{bmatrix} \dfrac{2}{31} & \dfrac{1}{31} & 0 \end{bmatrix}$$

4. Sample computations

The above procedures provide the means of computing linear combinations and products of elements in the Pauli algebra. Such capabilities are basic to any computational package for Clifford algebras. Before continuing with more involved or specific procedures, we give simple MAPLE examples that illustrate the use of variables in a few of those introduced thus far and show how to compute the wedge product of two vectors. The following session defines the spacetime position (6), reflects it in the plane perpendicular to e_1, and then shows that the sum of r and its reflection have a vanishing component along the normal e_1 to the plane. Then it computes the bivector formed by the vector part of the algebraic product of two vectors, and it finds the vector **n** normal to the plane spanned by the two vectors (*i.e.*, the *Hodge dual* of the bivector):

Example 7. Reflect a spacetime vector r in a spatial plane:

> `r := t + array([x,y,z]);`

$$r := t + \begin{bmatrix} x & y & z \end{bmatrix}$$

> `r_reflect := e1 &v Pbar(r) &v e1;`

$$r\_reflect := t + \begin{bmatrix} -x & y & z \end{bmatrix}$$

> `(r + r_reflect) &dot e1;`

$$0$$

Consider two vectors,

> `v1 := array([1,2,0]; v2 := array([0,1,2]);`

$$v1 := \begin{bmatrix} 1 & 2 & 0 \end{bmatrix}$$
$$v2 := \begin{bmatrix} 0 & 1 & 2 \end{bmatrix}$$

and find the bivector formed by their wedge product and its dual vector:

```
> bivec:=Pvec(v1 &v v2);
```

$$bivec := [4I \quad -2I \quad I]$$

```
> n := evalPa(bivec/(I*sqrt(Pnorm(bivec))));
```

$$n := \frac{4}{21}\sqrt{21}e1 - \frac{2}{21}\sqrt{21}e2 + \frac{1}{21}\sqrt{21}e3$$

5. Lorentz transformations

Physical Lorentz transformations are compositions of rotations and boosts, and they are simply spin transformations in the Pauli algebra. A spacetime vector p is actively transformed to

$$p \to LpL^{\dagger} \tag{9}$$

by the Lorentz transformation L, which can be written as the product $L = BR$ of a boost element

$$B = \exp(\mathbf{w}/2) = \cosh w/2 + \hat{\mathbf{w}} \sinh w/2 \tag{10}$$

and a rotation element

$$R = \exp(-i\boldsymbol{\theta}/2) = \cos\theta/2 - i\widehat{\boldsymbol{\theta}} \sin\theta/2. \tag{11}$$

Here the vector $\mathbf{w} = w\hat{\mathbf{w}}$ is the boost parameter (the "rapidity"), $\boldsymbol{\theta} = \theta\widehat{\boldsymbol{\theta}}$ is the angle of rotation, θ is its magnitude, $i\widehat{\boldsymbol{\theta}}$ is the rotation plane, and $\widehat{\boldsymbol{\theta}}$ is the axis of rotation.

The procedures for calculating the boost and rotation elements in the **Palg** package are

```
boost := proc(w) local wvec, wmag;
wvec:=evalm(w);
if type(wvec,vector) then
wmag:=sqrt(sum(wvec[i]^2 ,i=1..3));
cosh(wmag/2) + evalm(wvec*sinh(wmag/2)/wmag)
else ERROR('expecting vector argument for boost.')
fi
end:# boost with vector parameter w
```

and

```
rotate := proc (theta) local th,thmag;
th := evalm(theta);
if type(th,vector) then
thmag := sqrt(sum(th[i]**2,i = 1 .. 3));
cos(1/2*thmag)+evalm(-I*th*sin(1/2*thmag)/thmag)
else ERROR('expecting vector argument of rotate.')
fi
end:# rotation operator for given vector angle
```

Example 8. Consider a boost along e_1 and a rotation about e_3, given by the vector parameters

```
> w:=0.4*e1; theta:=e3*Pi/3;
```

$$w = .4\,e1$$
$$\theta = \frac{1}{3}\,e3\,\pi$$

The boost and rotation operators are

```
> B:=boost(w); R:=rotate(theta);
```

$$B := 1.020066756 + [.2013360025 \quad 0 \quad 0]$$

$$R := \frac{1}{2}\sqrt{3} + \left[0 \quad 0 \quad -\frac{1}{2}I \right]$$

6. Cliffor-valued functions

The boost and rotation elements are examples of cliffor-valued functions of cliffors, see (Baylis & Jancewicz, 1995). There is a general procedure, based on eigenidempotent elements and eigenvalues, for finding such functions, see (Baylis & Jones, 1989). The idea is that any Pauli cliffor $p = p^0 + \mathbf{p}$ can be associated with the idempotents $P_\pm = \frac{1}{2}(1 \pm \hat{\mathbf{p}})$, where the unit vector $\hat{\mathbf{p}} := \mathbf{p}/\sqrt{\mathbf{p} \cdot \mathbf{p}}$ may be complex. It is straightforward to show

$$pP_\pm = \epsilon_\pm P_\pm . \tag{12}$$

where $\epsilon_\pm = p^0 \pm \sqrt{\mathbf{p} \cdot \mathbf{p}}$ is the scalar eigenvalue. More generally, any function $f(p)$ can be spectrally decomposed into

$$\begin{aligned} f(p) &= f(\epsilon_+) P_+ + f(\epsilon_-) P_- \\ &= \frac{1}{2}[f(\epsilon_+) + f(\epsilon_-)] + \frac{\hat{\mathbf{p}}}{2}[f(\epsilon_+) - f(\epsilon_-)] . \end{aligned} \tag{13}$$

This works unless $\mathbf{p} \cdot \mathbf{p} = 0$. In that case, $\epsilon_+ = \epsilon_-$ and the limiting form of (13) can be used:

$$f(p) = f(p^0) + \mathbf{p}f'(p^0) \tag{14}$$

where f' is the first derivative of the function. In the Pauli package, the MAPLE procedure that calculates functions of cliffors is **Pfun**:

```
Pfun := proc (f:procedure, p) local pl, pmag, asq, n, fP, fN;
pl := evalPe(p);
if member(f,{'simplify','factor','normal','expand',
'convert','combine','eval','evalf','evalc'}) then
if nargs>2 then asq:=seq(args[n],n=3..nargs);
f(pl[1],asq) + map(f,pl[2],asq)
else f(pl[1])+map(f,pl[2])
fi
else pmag := sqrt(sum(pl[2][i]**2,i = 1 .. 3));
```

```
if pmag<>0 then fP:=f(pl[1]+pmag); fN:=f(pl[1]-pmag);
1/2*fP+1/2*fN+evalm(1/2*pl[2]*(fP-fN)/pmag)
else f(pl[1])+evalm((D(f))(pl[1])*pl[2])
fi
fi
end: # Find cliffor-valued function of cliffor
```

With **Pfun** functions such as *exp* and *sqrt* can be found for cliffors. Note that this procedure can also be used to simplify, factor, expand, etc. the cliffor term by term. Additional arguments, for example to the simplify command, can be appended after the cliffor, as the third argument of the procedure call.

Example 9. The proper velocity of a particle moving along e_3 at 0.6 the speed of light is

```
> u := 5/4 + 3/4 *e3;
```

$$u := \frac{5}{4} + \frac{3}{4} e_3$$

and the corresponding boost is

```
> B := Pfun(sqrt,u);
```

$$B := \frac{3}{4}\sqrt{2} + \begin{bmatrix} 0 & 0 & \frac{1}{4}\sqrt{2} \end{bmatrix}.$$

Rotations can be used to relate spherical and Cartesian coordinates. Starting with a vector of length r oriented along e_3, we perform rotations by the angle θ about e_2 and then by ϕ about e_3 to reorient it with a polar angle θ and an azimuthal angle ϕ. First define the rotation elements:

```
> assume(phi > = 0); assume(theta >= 0); r := 'r':
> R_th := rotate(theta*e2); R_phi := rotate(phi*e3);
```

$$R\_th := \cos\left(\frac{1}{2}\theta^\sim\right) + \begin{bmatrix} 0 & -I\sin\left(\frac{1}{2}\theta^\sim\right) & 0 \end{bmatrix} \tag{15}$$

$$R\_phi := \cos\left(\frac{1}{2}\phi^\sim\right) + \begin{bmatrix} 0 & 0 & -I\sin\left(\frac{1}{2}\phi^\sim\right) \end{bmatrix}. \tag{16}$$

The tildes indicate the presence of an assumed property. Next rotate the vector

```
> v:=r*e3:
```

by both rotations:

```
> rvec:=R_phi &v R_th &v (r*e3) &v Pdag(R_th) &v Pdag(R_phi):
```

The calculated expression involves half angles and needs to be simplified. Try the commands

```
> rvec:=Pfun(evalc,rvec):Pfun(combine,rvec,trig);
```

$$\begin{bmatrix} \frac{1}{2}r\sin\left(\theta^\sim + \phi^\sim\right) - \frac{1}{2}r\sin\left(-\theta^\sim + \phi^\sim\right) \\ \frac{1}{2}r\cos\left(-\theta^\sim + \phi^\sim\right) - \frac{1}{2}r\cos\left(\theta^\sim + \phi^\sim\right) \qquad r\cos\left(\theta^\sim\right) \end{bmatrix} \tag{17}$$

This gives the three Cartesian components. The more common expression is
> `Pfun(expand,'');`

$$[r\sin(\theta^\sim)\cos(\phi^\sim) \quad r\sin(\theta^\sim)\sin(\phi^\sim) \quad r\cos(\theta^\sim)] \ . \tag{18}$$

A cliffor-valued function $p(x)$ can be differentiated to give the gradient with respect to the argument r :

$$\partial p(x) \ , \ \bar{\partial} p(x) \tag{19}$$

where for $x = x^\mu \mathbf{e}_\mu$, the gradient operator is $\partial = \mathbf{e}^\mu \partial/\partial x^\mu = \bar{\mathbf{e}}_\mu \partial/\partial x^\mu$. The MAPLE procedure for $\partial p(x)$ is

```
Pdi := proc(p,x) local pl,xl;
options 'Copyright 1994 by W. E. Baylis, U. Windsor';
pl:= evalPe(p); xl:=evalPe(x);
diff(pl[1],xl[1]) - diff(pl[2][1],xl[2][1]) -
diff(pl[2][2],xl[2][2]) - diff(pl[2][3],xl[2][3]) +
array([  diff(pl[2][1],xl[1]) - diff(pl[1],xl[2][1]) +
I*(diff(pl[2][2],xl[2][3])-diff(pl[2][3],xl[2][2])),
diff(pl[2][2],xl[1]) - diff(pl[1],xl[2][2]) +
I*(diff(pl[2][3],xl[2][1])-diff(pl[2][1],xl[2][3])),
diff(pl[2][3],xl[1]) - diff(pl[1],xl[2][3]) +
I*(diff(pl[2][1],xl[2][2])-diff(pl[2][2],xl[2][1])) ])
end:# Pauli differential op di acting on cliffor-valued function
```

The procedure `Pdib` to calculate $\bar{\partial} p(x)$ differs by only a few signs. By combining the two differentiation procedures, one can find the d'Alembertian:

$$\Box p(x) = \partial\bar{\partial} p(x) = \left[\frac{\partial^2}{(\partial x^0)^2} - \frac{\partial^2}{(\partial x^1)^2} - \frac{\partial^2}{(\partial x^3)^2} - \frac{\partial^2}{(\partial x^3)^2}\right] p(x) \ . \tag{20}$$

Example 10. Consider the Pauli cliffor r^2 where $r = t + \mathbf{r}$:

> `r := t + array([x,y,z]):   F := r &v r;`

$$F := t^2 + x^2 + y^2 + z^2 + [2tx \quad 2ty \quad 2tz] \tag{21}$$

Its gradients ∂F and $\bar{\partial} F$ are
> `Pdi(F,r); Pdib(F,r);`

$$-4t + [0\ 0\ 0]$$
$$8t + [4x\ 4y\ 4z] \tag{22}$$

and its d'Alembertian may be found either from $\partial\bar{\partial} F$ or $\bar{\partial}\partial F$:
> `Pdi(Pdib(F,r),r); Pdib(Pdi(F,r),r);`

$$-4 + [0\ 0\ 0]$$
$$-4 + [0\ 0\ 0] \ . \tag{23}$$

7. Factoring cliffors

Any Pauli cliffor with a nonvanishing norm can be considered the product of a dilation factor, a rotation and a boost. All physical Lorentz transformation elements in the Pauli algebra are unimodular:

$$L\bar{L} = 1. \tag{24}$$

An element with zero norm can be factored into a rotation, a boost, and a projector of the form $\frac{1}{2}(1 + \hat{w})$, where $\hat{w}$ is the boost direction. The procedure **Pfactor** makes such factorizations and gives the dilation factor, the rotation angle as a vector, and the boost parameter. Because of its length, the procedure is not listed here, but it may be read from the accompanying diskette, and its use is illustrated below.

Rotations in three dimensions form the three-parameter orthogonal group O(3), whose parameters can, for example, be expressed as components of the vector rotation angle whose direction gives the axis of rotation. The parameters can also be taken to be the three scalar Euler angles. The **Palg** package allows relations to be established between different parametrizations. The rotation elements in the Pauli algebra are elements of the double covering group SU(2) and use the half angles as parameters. Any product of rotations is another rotation, which can be found algebraically.

Example 11. Consider the product of two 90-degree rotations, the first about e_2 and the second about e_3. The rotation on the right is performed first. The product gives

```
> R32 := rotate(e3*Pi/2) &v rotate(e2*Pi/2);
```

$$R32 := \frac{1}{2} + \left[\frac{1}{2}I \quad -\frac{1}{2}I \quad -\frac{1}{2}I\right] \tag{25}$$

which is factored into dilation, rotation, and boost parts by **Pfactor**, described at the end of section 5:

```
> facts := Pfactor(R32);
```

$$facts := \left[1, \left[\frac{2}{9}\sqrt{3}\pi \quad -\frac{2}{9}\sqrt{3}\pi \quad -\frac{2}{9}\sqrt{3}\pi\right], [0\ 0\ 0]\right]. \tag{26}$$

Note that the dilation factor, the first element of the list **facts**, is unity, and therefore the product **R32** is still unimodular. Furthermore, the boost parameter given in the last element of the list vanishes. Consequently, the product **facts** of two rotations is equivalent to a single rotation of angle

```
> theta:=sqrt(-Pnorm(facts[2]));
```

$$\theta := \frac{2}{3}\pi \tag{27}$$

about the axis

```
> evalPa(facts[2]/theta);
```

$$\frac{1}{3}\sqrt{3}e1 - \frac{1}{3}\sqrt{3}e2 - \frac{1}{3}\sqrt{3}e3. \tag{28}$$

Boosts, on the other hand, do not form a group, and products of boosts are boosts if and only if the boost directions are aligned.

Example 12. Consider the product of boosts in orthogonal directions:

```
> B12 := boost(e1) &v boost(e2);
```

$$B12 := \cosh\left(\frac{1}{2}\right)^2 + \left[\frac{1}{2}\sinh(1) \quad \frac{1}{2}\sinh(1) \quad I\sinh\left(\frac{1}{2}\right)^2\right] \tag{29}$$

whose factors are given numerically by

```
> Digits:=4:Pfactor(Pfun(evalf,B12));
```

$$[1.000, [0 \ 0 \ .4202], [.8230 \ 1.269 \ 0]] \ . \tag{30}$$

Thus, the result is the product of a boost and a non-zero rotation in the plane of the boosts.

8. Conclusions

A MAPLE package `Palg` has been written to manipulate elements of the Clifford algebra $C\ell_3$, with elements interpreted as complex paravectors. MAPLE V is sufficiently flexible to admit paravectors as sums of scalars and vectors. By restricting the procedures to the Pauli algebra and hence to $C\ell_3$, one can satisfy fairly complicated needs while keeping the procedures relatively simple. Both numerical and analytical relations are treated within the same formalism.

References

W. E. Baylis and G. Jones, *J. Phys. A: Math. Gen.* **22**, 1 (1989).

W. E. Baylis, *Theoretical Methods in the Physical Sciences: an introduction to problem solving using MAPLE V*, Birkhäuser, Boston, 1994.

W. E. Baylis and B. Jancewicz, in *Clifford Algebras and Spinor Structures*, eds. by R. Abłamowicz and P. Lounesto, Kluwer Academic, Dordrecht, 1995, pp. 313-324.

A. Heck, *Introduction to MAPLE*, Springer-Verlag, New York, 1993.

R. Mikkola and P. Lounesto, *Internat. J. Math. Educ. Sci. Tech.* **14**, 573 (1983).

THE GENERATIVE PROCESS OF SPACE-TIME AND STRONG INTERACTION QUANTUM NUMBERS OF ORIENTATION

BERND SCHMEIKAL

Biofield Laboratory
Interdisciplinary Centre for Biosynergetics and Holistic Research (ICBHR)
Kundmanngasse 26/8
A-1030 Vienna, Austria
e-mail: schmeika@isis.wu-vien.ac.at

Abstract. An image of physical fields is posed to the reader that does no longer rely on a pregiven concept of spacetime and coordinate systems respectively, wherein equations of motion are formulated a posteriori. But a view is offered where well known properties of spacetime and fields are outcomes of one and the same generative process or grammar. This approach is based theoretically on Clifford algebras and practically supported by the computer algebra system of CLICAL. For instance, it is not postulated that the orientation of space is a priori determined, but that it is the outcome or a process and therefore subject to uncertainty as any quantum field in physical spacetime. Making use of the Clifford algebra $C\ell_{3,1}$ basic spin representations of the octahedral space group $\mathbf{O}_h$ are constructed. Next it can be shown that the quarks can be conceived as quantum states of the central symmetry operators and well known Schönfließ symbols $_1C_2, _2C_2, _3C_2$ of $\mathbf{O}_h$ with the eigenvalues $+1$ or -1. Thus the quarks turn out to represent quantized states of orientation which may be the reason why they are confined. Next the algebraic relations between the orientation symmetry $\vee_h$ and the Gell-Mann matrices are established and finally the Gell-Mann-Nishijima relation is derived from geometry. In this way an inner symmetry is linked to an outer symmetry. Orientation numbers can be defined as parties of oriented plane areas $\mathbf{e}_{i,j}$. Their values are calculated for the case of baryon octet of the nucleons and the quarks. This example makes it clear why quarks are the only particles with orientation quantum numbers ± 1.

Key words: Orientation symmetry, orientation quantization, orientation numbers, quantum geometrodynamics, Clifford algebra, Clifford groups, dihedral group, hyperoctahedral group, octahedral group, strong interaction, Dirac algebra, spinors, finite reflection group.

1. Notation

$\mathbf{D}_{2d}$ is the abstract dihedral group consisting of the mirror reflections σ'_d, σ''_d, rotatory reflections S_4, S_4^3, period-2 rotations $C', C'', C = S_4^2$ and unity (notation by Belger & Ehrenberg, 1981 and Schönfließ, 1886). It is the orientation symmetry of the Euclidean plane. In the octahedral group $\mathbf{O}$ the operators C', C'', C are period-2 rotations $_iC_2 = _iC_4^2$. The half spin representation of $\mathbf{D}_{2d}$ by matrices $D^{(\frac{1}{2})}$ is the double group $_\delta\mathbf{D}_{2d}$ (Inui *et al.*, 1990). Its realizations in the coordinate planes $(\mathbf{e}_2, \mathbf{e}_3)$, $(\mathbf{e}_1, \mathbf{e}_3)$, $(\mathbf{e}_1, \mathbf{e}_2)$ of $\mathbb{R}^3$ are $_1\mathbf{D}_{2d}, _2\mathbf{D}_{2d}, _3\mathbf{D}_{2d}$ and are subgroups of the octahedral groups $\mathbf{O}$ and $\mathbf{O}_h$ respectively (Petraschen & Trifonow, 1969). In some works and lexica the octahedral group $\mathbf{O}$ is denoted G_{24} because it is isomorphic with the symmetric group S_4 and thus has $4! = 24$ elements. Following the early works of Schläfli (1855), we use the notation $SO_{3,4}$. The full octahedral group $\mathbf{O}_h = \mathbf{O} \times \mathbf{C}_i$ is obtained from $\mathbf{O}$ by the grade involution $\mathbf{x}^\wedge = -\mathbf{x}$, denoted $\mathbf{C}_i$ in crystallography. The reflection operators σ'_d, σ''_d of the abstract $\mathbf{D}_{2d}$ and the representative SU(2)-matrices of the reflections $_1\sigma'_d$, $_2\sigma'_d$, $_3\sigma'_d$, $_1\sigma''_d$, $_2\sigma''_d$, $_3\sigma''_d$ must not be confused with the Pauli matrices σ_1, σ_2, σ_3, though there exist algebraic relationships between them.[1]

Although the existence condition for a minimal left ideal vector-subbundle is stronger in $C\ell_{3,1}$ than in $C\ell_{1,3}$ (Thelen, 1992), we shall follow the $C\ell_{3,1}$ approach studied by Greider, Weiderman (1988) and Chisholm (1992) which means a well known change of the basis $\{e_i\}$ of the Dirac algebra Mat$(4, \mathbb{C})$ (Lounesto, 1980). By the same reason, the basis of $C\ell_{3,1}$ is constructed by the aid of the *complex Pauli matrices* σ_1, σ_2, σ_3, although that is not necessary, as the Pauli matrices may just as well be considered to be real. But the complex case is widely accepted due to tradition and habit (Morris & Makhool, 1992). We shall therefore perform the transition with patience.

2. Prologue

There is an increasing uncertainty concerning the scientific value of traditional basic equations of motion in physics. One argument is held against their linearity and another against the uncritical acceptance of both spinor analysis and the complex number field. Regarding the space as a non-local synergistic process, second quantization can merely represent some sort of bad approximation. Appreciating the standard complex bispinor Dirac equation, one is finally led to the observation that it is essentially dispensable, if not wrong, and "Clifford's real geometric algebra a valuable substitute" for it (Parra, 1992). Being attentive at last to the concept of time, it turns out that physical time may be a secondary concept, while the primary is a psychological one (Pöppel, 1989). Yet it is clear from the Clifford algebra approaches that the complexity of the internal structure of strong interaction can be based on the assumption of a Minkowski-space $\mathbb{R}^{3,1}$ or $\mathbb{R}^{1,3}$ respectively. Time, as it appears to us, is an intrinsic property of fields.

3. The view of space-time

First, let us make clear that we should say what is important in simple terms and without hesitation. Traditional science is sometimes based on a wrong understanding of basic concepts. If you would hesitate to put aside what is wrong, you would only carry on making the same old mistakes over and over again. The most serious such mistake is that we separate the creative process that brings about the space-time with its orientation, its measure and outer symmetry from that which brings on the energy levels and the inner symmetries of quantum fields. In a way, we believe that we have overcome the nature of space and time by relativity. But this is not true. Still we begin to formulate physical theories by first setting up frameworks such as the Minkowski space-time $\mathbb{R}^{3,1}$, or the four-dimensional Euclidean space $\mathbb{R}^4$, or the direct sum $\mathbb{R} + \mathbb{R}^3$, or still more complex vector spaces, and within such a frame

---

[1] I am indebted to Professor Pertti Lounesto for his valuable information and material on the subject matter. Let me also thank Professor Urbantke for his interesting hint that the Schönfließ symbols $_i\sigma'_d$, $_i\sigma''_d$ denoting mirror reflections of dihedral groups may be confused with the Pauli matrices σ_i.

we build an action functional or we carry out a second quantization. But we could just as well start off with our knowledge about the field and from there derive the features of spacetime. Field and space-time are essentially the same.

Usually we take the existence of a primary physical concept of a real vector space $\mathbb{R}^n$ for granted. Its orthonormal basis e_1, e_2, ..., e_n, that is, its orientation and measurability we regard as self-evident, and next we build up our field, functional, symmetry and so forth. In practice, when we adjust a coordinate system e_1, e_2, e_3, e_4 at laboratory scales and measure distances in, say, Minkowski space-time, we are effectively integrating over a process of involutions and quantum fluctuations of the field. But as soon as we approach smaller scales, such integration is not self-evident, but misleading. Now, it seems generally accepted that topological fluctuations as quantum gravitational effects destroy both orientation and measure at the Planck scale at the order 10^{-35} m, 10^{-44} sec and 10^{19} GeV. But it is not yet seen that the basic reflections and commutations within the field - known to us in the form of the famous Weyl groups or finite reflection groups – are constitutive for the orientation and scaling of the space-time at the much lower energies of strong and weak interaction. The measurability of space-time is thus directly linked to the action of the $SU(3)$- and $SU(2)$-forces and symmetries. As Chisholm stated in 1993 "there is no distinction between space-time and the internal interaction space". Without the phenomenon of spin there is no measurable space-time.

Next I would like to introduce to the reader some of the concepts and ideas fundamental for an understanding of the generative process of space and its grammar, the grammar being that of a Clifford algebra.

It is often said, in more or less hesitant words, that the spin s would be an intrinsic angular momentum of a particle. But this is not correct. In older books such as Quantum Mechanics Vol. II by Messiah (1970, p. 544) this is claimed without any modesty: "let s be the intrinsic angular momentum (or spin vector) of a particle of spin $\frac{1}{2}$ ...", "since s is the angular momentum" a.s.o. In nowadays primers such as "Field Theory" (Ramond, 1990, pp. 14–17) the relation is established in more precise terms by constructing the generators of a rotation J_i in Minkowski space by the aid of the Pauli spin matrices σ^i. But in its glossary of the popular edition on the "Quarks" (Fritzsch, 1984), the spin is still defined as the "Eigendrehimpuls" of elementary particles. Yet, in the same book series, in his book "QED", Richard Feynman (1988) does not identify the electron spin with an intrinsic angular momentum, but he brings in the notation of a "polarization" of its field. Last but not least, Hagelin (1988) employs the rather dubious formulation: "If one adopts a very classical and particulate view of such states, then one can imagine these particles as physically spinning and therefore possessing an intrinsic angular momentum". But this view is misleading, because the spin does not have any classical analogy. As a matter of fact, it does not represent anything like an intrinsic angular momentum. Rather the subsequent observation $\sigma_i \sigma_j$ of two spin-components results in a Schönfließ rotation $_k C_2$ of period 2, e.g. we have to have $_3 C_2 = e_{12} = \sigma_1 \sigma_2$.

Historically, it has been a somewhat unfortunate event that the derivation of the spin matrices σ_i could be based on the assumption of an angular momentum analogy. But what is much more important is to realize that the Pauli matrices are representatives of the orthonormal basis vectors e_i of $\mathbb{R}^3$. We shall see that we can put

$$\mathbf{e}_i = \sigma_i, \qquad i = 1, 2, 3. \tag{1}$$

Thus the spin matrix σ_3 is a spinor-representative of $\mathbf{e}_3$, but it certainly does not represent a quantized field rotation about the z-axis. However, there exist important relations between the $SU(2)$-representative matrix of such a rotation and the spin matrix σ_3. Let ${}_3C_2$ denote the Schönfließ symbol of a π-rotation about $\mathbf{e}_3$ and j the unit director $\mathbf{e}_{123}$ of the Clifford algebra $C\ell_3$ defined by its representative matrix

$$j = \mathbf{e}_{123} = \mathbf{e}_1\mathbf{e}_2\mathbf{e}_3 = \sigma_1\sigma_2\sigma_3 = \begin{bmatrix} i & 0 \\ 0 & i \end{bmatrix} \quad \text{with} \quad i = \sqrt{-1}. \tag{2}$$

In the $SU(2)$ the rotation ${}_3C_2$ is represented by the bivector $\mathbf{e}_{12} = \sigma_1\sigma_2$, and we can thus verify the equation

$$\sigma_3 = -j\,{}_3C_2. \tag{3}$$

This is not a mathematical trick resulting from the arbitrary use of some special spinor space, but it is a natural and correct approach to fields with a spin. Spin is directly related to spatial orientation.

The principal properties of the Pauli matrices which formerly could be deduced from their definition as $\text{Mat}(2, \mathbb{C})$-matrices are of a special importance where observability of space-time is concerned. Consider the well known equations

$$\sigma_i^2 = 1, \tag{4}$$

$$\sigma_i\sigma_j = \delta_{ij} + i\epsilon_{ijk}\,\sigma_k \tag{5}$$

where δ_{ij} denotes the Kronecker delta and ϵ_{ijk} is the Levi-Civitá symbol, totally antisymmetric in all its indices and with $\epsilon_{123} = 1$.

Equations (4, 5) represent nothing else than the Clifford multiplication rules

$$\mathbf{e}_i^2 = 1, \tag{6}$$

$$\mathbf{e}_i\mathbf{e}_j + \mathbf{e}_j\mathbf{e}_i = 0 \quad \text{for} \quad i \neq j \tag{7}$$

and the analog to relations (5) is

$$\mathbf{e}_i\mathbf{e}_j = \delta_{ij} + j\epsilon_{ijk}\mathbf{e}_k \quad \text{with} \quad j = \mathbf{e}_{123}.$$

The anticommutation of spin operators, therefore, has a definite quantum logical meaning for spatial orientation: given the unit vectors $\mathbf{e}_i$, $\mathbf{e}_j$ are observable quantities, these cannot be observed independent of each other since they do not commute. But the measurement of, say, the first and second basis vector in this sequence brings on the same result as a measurement of the rotation ${}_3C_2 = j\sigma_3$ about the z-axis. The equations (5) define the period-2 rotations in the finite reflection group D_{2d}, we have for instance ${}_2C_2 = j\sigma_2$. The vanishing anticommutator (7) represents the fact that the "second value" of any element in the finite reflection group represented in the $SU(2)$ is the negative of its first value, e.g., ${}_{\delta 3}C_2 = -{}_3C_2$ as we have ${}_{\delta 3}C_2 = \mathbf{e}_{21}$ and ${}_3C_2 = \mathbf{e}_{12}$ so that (7) means ${}_{\delta 3}C_2 + {}_3C_2 = 0$.

From these considerations we learn that the multiplication rules of the Clifford algebra $C\ell_3$ and respectively the SU(2) spinor algebra of the Pauli matrices are closely related to the dihedral (double)group $_\delta D_{2d}$, the spatial congruence group of the plane line-cross of a coordinate system or equally of a square. Up to commutations and involutions of the basis vectors e_i, the operators of the dihedral group D_{2d} leave the orientation of the plane coordinate system invariant. It is therefore that D_{2d} shall be denoted the *"orientation group"* of $\mathbb{R}^2$. The symmetries of D_{2d} transform the basis (e_1, e_2) of $\mathbb{R}^2$ in the following way:

D_{2d}-operator and *Schönfließ*-symbol	basis vectors		signature		
E	e_1	e_2	+	+	
$C_2' = {}_1C_2$	e_1	$-e_2$	+	-	partial
$C_2'' = {}_2C_2$	$-e_1$	e_2	-	+	involutions
$C_2 = {}_3C_2 = S_4^2$	$-e_1$	$-e_2$	-	-	grade
σ_d''	e_2	e_1	+	+	involutions
S_4	e_2	$-e_1$	+	-	commuted
$S_4^3 = S_4^{-1B}$	$-e_2$	e_1	-	+	basis
σ_d'	$-e_2$	$-e_1$	-	-	

These operations include all possible commutations and partial involutions of the basis vectors: We use the denotation of a "partial involution" in order to distinguish between the *"grade involution"* of the Clifford algebra, where each e_i is to be replaced by $-e_i$, and a *"partial involution"* which involves only one or more, but less than n basis vectors. Since D_{2d} acts on the plane, there are $2! = 2$ possible commutations and 2^2 possible signatures or involutions, that is, $2!2^2 = 8$ elements. Turning over to spaces $\mathbb{R}^{p,q}$ with $p + q = n$, the full orientation group has to be the hyperoctahedral group H_n which is of the order of $n!2^n$. For the Euclidean space $\mathbb{R}^3$ the hyperoctahedral group H_3 has an order $3!2^3 = 48$ and is isomorphic with the full octahedral group $\mathbf{O}_h$ which equals $\mathbf{O} \times \mathbf{C}_i$ and is obtained from the octahedral group $\mathbf{O}$ by the grade involution $x^\wedge = -x$ also denoted $\mathbf{C}_i$ in crystallography. In some works and lexica the octahedral group $\mathbf{O}$ is denoted G_{24} because it is isomorphic with the symmetric group S_4 and thus has $4! = 24$ elements. Following the early works of Schläfli (1855), we use the notation $SO_{3,4}$. This group acts on the orthonormal basis of $\mathbb{R}^3$ in just the way as the dihedral group D_{2d} acts on the plane, that is, it produces all the possible commutations and partial involutions of the basis vectors. We have to distinguish between the abstract group $\mathbf{O}$, its representation as a subgroup of $SO(3)$ and its representation as a double group in a spinor-space S. The double-group has twice as many elements as the abstract group. Yet, both the abstract and the double-group may be realized in S. But while the tetrahedral operators T with period 3 in $SO(3)$ turn into hexagonal operators $H = -T$ with period 6, in S both groups may contain the same set of tetrahedral operators. The hyperoctahedral group of the Minkowski space-time is a group of the order 384 containing the octahedral group $\mathbf{O}_h$ with the index eight. Each octahedral group $\mathbf{O}$ by its tetrahedral subgroup $\mathbf{T}$ gives rise to one copy of the SU(3). Thus there are eight copies of the color-group, three colors and three anticolors. This is an essential property of the unbroken symmetry of the Minkowski space-time.

4. The generative process of space-time

4.1. THE CASE OF $\mathbb{R}^3$ AND ITS SPIN GROUP SU(2)

The primitive idempotent $f_1 = \frac{1}{2}(1 + e_3)$ in the Clifford algebra $C\ell_3$ gives rise to the minimal left ideal $S = C\ell_3 f_1$ and a linear space $\mathbf{K} = f_1 C\ell_3 f_1$ with the basis

$$f_1 = \frac{1}{2}(1 + e_3), \quad f_i = \frac{1}{2}(e_{12} + e_{123}) \tag{8}$$

with $f_i^2 = -f_1$ and, therefore, $\mathbf{K} \simeq \mathbf{C}$. The mapping $S \times \mathbf{K} \to S, (\Phi, \tau) \to \Phi\tau$ makes S a right-sided $\mathbf{K}$-linear space. Given this linear structure S becomes a spinor space. It has the basis

$$f_1 = \frac{1}{2}(1 + e_3), \quad f_2 = \frac{1}{2}(e_1 + e_{13}). \tag{9}$$

In this basis the unit vectors e_1, e_2, e_3 of $\mathbb{R}^3$ have the following matrix representations (Lounesto, Mikkola & Vierros, 1987, p. 41):

$$\underline{e}_1 = \begin{bmatrix} 0 & 1 \\ 1 & 0 \end{bmatrix}, \quad \underline{e}_2 = \begin{bmatrix} 0 & -i \\ i & 0 \end{bmatrix}, \quad \underline{e}_3 = \begin{bmatrix} 1 & 0 \\ 0 & -1 \end{bmatrix}. \tag{10}$$

These matrices are generators of the spin group SU(2) of $\mathbb{R}^3$ and are known as the Pauli matrices. In the spinor space S the bivectors e_{12}, e_{23}, e_{13} and the director e_{123} of $C\ell_3$ are represented by the following matrices

$$\underline{e}_{12} = \sigma_1\sigma_2 = \begin{bmatrix} i & 0 \\ 0 & -i \end{bmatrix}, \quad \underline{e}_{23} = \sigma_2\sigma_3 = \begin{bmatrix} 0 & i \\ i & 0 \end{bmatrix},$$

$$\underline{e}_{13} = \sigma_1\sigma_3 = \begin{bmatrix} 0 & -1 \\ 1 & 0 \end{bmatrix}, \quad \underline{e}_{123} = \sigma_1\sigma_2\sigma_3 = \begin{bmatrix} i & 0 \\ 0 & i \end{bmatrix}. \tag{11}$$

4.1.1. The double group $_\delta\mathbf{D}_{2d}$ in the spinor space $\mathbf{S}$

Consider the finite reflection group $\mathbf{D}_{2d}$ known to us as the dihedral group or spatial congruence group $\mathbf{D}_4$ of the square. In the spinor-representation it is a double-group, that is, each element of $\mathbf{D}_{2d}$ being a rotation in SO(3) has two values in the SU(2)-group because of the two-fold covering of the SO(3). Its minimal generating basis can be represented by the special unitary matrices of the reflection σ_d' and e.g. the rotatory reflection S_4

$$\sigma_d' = \frac{1}{\sqrt{2}}(e_{13} - e_{23}), \quad S_4 = \frac{1}{\sqrt{2}}(1 + e_{12}). \tag{12}$$

From this minimal basis we derive the other elements of the double group by using the Clifford algebra calculator CLICAL:

$$_\delta\sigma'_d = -\sigma'_d \qquad _\delta S_4 = -S_4 \ldots \text{ second values to } \sigma'_d \text{ and } S_4$$

$$_2C_2 = \sigma'_d S_4 = \mathbf{e}_{13}, \quad _1C_2 = S_4\sigma'_d = -\mathbf{e}_{23}$$

$$\sigma''_d = {}_2C_2 S_4 = \tfrac{1}{\sqrt{2}}(\mathbf{e}_{13} + \mathbf{e}_{23}), \quad _3C_2 = \sigma''_d\sigma'_d = \mathbf{e}_{12} \tag{13}$$

$$S_4^3 = \tfrac{1}{\sqrt{2}}(-1 + \mathbf{e}_{12}), \text{ and the unit matrix 1 together}$$
with the second values of each element g

$$_\delta g = -g, \text{ e.g. } _\delta C''_2 = -C''_2.$$

Using the spinor representation by the Pauli matrices of the basis vectors $\mathbf{e}_i$ of $\mathbb{R}^3$ in the spinor space S (equations 10) we obtain for the central elements of $\mathbf{D}_{2d}$ the special unitary matrices

$$_1\underline{C}_2 = -\sigma_2\sigma_3 = \begin{bmatrix} 0 & -i \\ -i & 0 \end{bmatrix}, \quad _2\underline{C}_2 = \sigma_1\sigma_3 = \begin{bmatrix} 0 & -1 \\ 1 & 0 \end{bmatrix},$$

$$_3\underline{C}_2 = \sigma_1\sigma_2 = \begin{bmatrix} i & 0 \\ 0 & -i \end{bmatrix} \tag{14}$$

as was claimed in section 1 of this chapter.
The equations $\sigma_i\sigma_j = \delta_{ij} + j\epsilon_{ijk}\sigma_k$ now take the new form:

$$_1\underline{C}_2 = -j\sigma_1, \quad _2\underline{C}_2 = -j\sigma_2, \quad _3\underline{C}_2 = j\sigma_3 \qquad \text{(a)}$$
or $\tag{15}$
$$\sigma_1 = j\,_1\underline{C}_2, \quad \sigma_2 = j\,_2\underline{C}_2, \quad \sigma_3 = -j\,_3\underline{C}_2. \qquad \text{(b)}$$

From the equations (14) and (15a) we can learn two things: first that *each Schönfließ period-2 rotation $_iC_2$ results from a successive measurement of two spin operators σ_j, and second that each spin operator is essentially equal to the product of the oriented unit volume $j = \mathbf{e}_{123}$ in $\mathbb{R}^3$ and its corresponding period-2 rotation $_jC_2$.* There is a second way to calculate representative matrices for the dihedral symmetries. This is not based on CLICAL, but on the somewhat arduous calculation of the complex unitary 2×2-matrices of the spin-$\frac{1}{2}$ representation $D^{(\frac{1}{2})}(R(\alpha, \beta, \tau))$ by the Euler angles α, β, and τ of the SO(3)-rotation $R(\alpha, \beta, \tau)$ (Inui *et al.* 1990, p. 128).

$$D^{(\frac{1}{2})}(R(\alpha, \beta, \tau)) = \begin{bmatrix} e^{-i(\alpha+\tau)/2}\cos\frac{\beta}{2} & -e^{-i(\alpha-\tau)/2}\sin\frac{\beta}{2} \\ e^{i(\alpha-\tau)/2}\sin\frac{\beta}{2} & e^{i(\alpha+\tau)/2}\cos\frac{\beta}{2} \end{bmatrix}. \tag{16}$$

This matrix has to be calculated for each element of the dihedral group. Consider for example the period-4 rotation of the plane $\mathbf{e}_1$, $\mathbf{e}_2$ caused by S_4 together with its Euler angles $\alpha = -\frac{1}{2}\pi, \beta = \tau = 0$. We have to have $e^{i\pi/4} = (1/\sqrt{2})(1+i)$ and $e^{-i\pi/4} = (1/\sqrt{2})(1-i)$ and finally obtain the special unitary matrix

$$D^{(\frac{1}{2})}(S_4) = \frac{1}{\sqrt{2}} \begin{bmatrix} 1+i & 0 \\ 0 & 1-i \end{bmatrix}. \tag{17}$$

Note that in SO(3) the operator S_4 has period 4, but period 8 in SU(2).

Comparison with the generator in equations (12) and substituting the expression (11) for $\mathbf{e}_{12}$, we find that $D^{(\frac{1}{2})}(S_4)$ is equal to its Clifford algebra representative $\underline{S}_4$ in the spinor space S. This can be verified of course for all the 16 elements of the double-group $_\delta \mathbf{D}_{2d}$. Usually, in theoretical physics, by the formula (16) we analyze the symmetries of electronic states of molecules in terms of the representation theory of point groups. The representative special unitary matrices of the Schönfließ operations calculated by formula (16) are identical with the matrices obtained from the Clifford algebra relations (13) in the spinor space S. Last but not least we can test the effect of the operations on the basis, e.g.: $_3C_2$ turns $\{\mathbf{e}_1, \mathbf{e}_2, \mathbf{e}_3\}$ into $\{-\mathbf{e}_1, -\mathbf{e}_2, \mathbf{e}_3\}$ as we have

$$_3C_2^{-1}\,\mathbf{e}_1\,_3C_2 = -\mathbf{e}_1, \quad _3C_2^{-1}\,\mathbf{e}_2\,_3C_2 = -\mathbf{e}_2 \quad \text{and} \quad _3C_2^{-1}\,\mathbf{e}_3\,_3C_2 = \mathbf{e}_3. \tag{18}$$

In S the last equation has the following form:

$$\underbrace{\begin{bmatrix} -i & 0 \\ 0 & i \end{bmatrix}}_{_3\underline{C}_2^{-1}} \underbrace{\begin{bmatrix} 1 & 0 \\ 0 & -1 \end{bmatrix}}_{\sigma_3} \underbrace{\begin{bmatrix} i & 0 \\ 0 & -i \end{bmatrix}}_{_3\underline{C}_2} = \underbrace{\begin{bmatrix} 1 & 0 \\ 0 & -1 \end{bmatrix}}_{\sigma_3}$$

Similarly $_1\underline{C}_2$ turns $\{\sigma_1, \sigma_2, \sigma_3\}$ into $\{\sigma_1, -\sigma_2, -\sigma_3\}$ and $_2\underline{C}_2$ turns $\{\sigma_1, \sigma_2, \sigma_3\}$ into $\{-\sigma_1, \sigma_2, -\sigma_3\}$. The rotations $_1\underline{C}_2$ and $_2\underline{C}_2$ thus cause a tilt of the spin-vector σ_3. It is important to notice that even the Pauli matrices possess "second values" in S, that is, $\{-\sigma_1, -\sigma_2, -\sigma_3\}$ are also valid representatives of the unit vectors $\{\mathbf{e}_1, \mathbf{e}_2, \mathbf{e}_3\}$. This means that $_1\underline{C}_2$ flips $\mathbf{e}_3$ between its first and second value thereby causing a tilt over of spin and thus a partial involution in the basis of $\mathbb{R}^3$.

From these considerations it should be clear that the spin is not an "intrinsic angular momentum" but an orientation of the quantum field.

5. The orientation of $\mathbb{R}^{3,1}$ and strong interaction

5.1. THE OCTAHEDRAL ORIENTATION SYMMETRY O AND THE COLOR ROTATION $_1\mathbf{C}_3$

Consider the Minkowski space-time $\mathbb{R}^{3,1}$ and its Clifford algebra $C\ell_{3,1} \simeq \mathrm{Mat}(4, \mathbb{R})$ with a maximal set of mutually annihilating primitive idempotents f_1, f_2, f_3, f_4 which sum up to unity:

$$\begin{aligned} f_1 &= \tfrac{1}{2}(1+\mathbf{e}_1)\tfrac{1}{2}(1+\mathbf{e}_{34}), & f_2 &= \tfrac{1}{2}(1-\mathbf{e}_1)\tfrac{1}{2}(1+\mathbf{e}_{34}), \\ f_3 &= \tfrac{1}{2}(1-\mathbf{e}_1)\tfrac{1}{2}(1-\mathbf{e}_{34}), & f_4 &= \tfrac{1}{2}(1+\mathbf{e}_1)\tfrac{1}{2}(1-\mathbf{e}_{34}). \end{aligned} \tag{19}$$

Consider the finite reflection group $\mathbf{O}$ which is isomorphic to the symmetric group S_4 and to the Weyl group $W(A_3)$ with the root system $A_3 = \{e_1 - e_2, e_2 - e_3, e_3 - e_4\}$. This group $\mathbf{O}$ representing the orientation group of $\mathbb{R}^3$, gives rise to the appearance of the color symmetry SU(3). It can be generated in the Clifford algebra $C\ell_{3,1}$ by the following six reflections (reflections to family 1):

$$
\begin{aligned}
{}_1s_1 &= f_1 + f_2 + e_{12}(f_4 - f_3), & {}_1s_2 &= f_1 + f_2 + e_{12}(f_3 - f_4), \\
{}_2s_1 &= f_1 + f_3 + e_{13}(f_2 - f_4), & {}_2s_2 &= f_1 + f_3 + e_{13}(f_4 - f_2), \\
{}_3s_1 &= f_1 + f_4 + e_{23}(f_3 - f_2), & {}_3s_2 &= f_1 + f_4 + e_{23}(f_2 - f_3).
\end{aligned}
\tag{20}
$$

With the help of the Clifford algebra calculator CLICAL we are able to verify the following equations

$$
{}_is_1^2 = {}_is_2^2 = 1.
\tag{21}
$$

Next we have to verify the multiplication table of $\mathbf{O}$, that is, we first realize that each central element ${}_iC_2$ has to equal the product ${}_is_1 \cdot {}_is_2$:

$$
{}_1C_2 = {}_1s_1\,{}_1s_2 = e_{34}, \quad {}_2C_2 = {}_2s_1\,{}_2s_2 = e_{134}, \quad {}_3C_2 = {}_3s_1\,{}_3s_2 = e_1.
\tag{22}
$$

Tetrahedral period-3 operators can be defined by the following products:

$$
\begin{aligned}
{}_1C_3 &= {}_1s_1\,{}_3s_1, & {}_1C_3^{-1} &= {}_1C_3^2, \\
{}_2C_3 &= {}_1s_1\,{}_2s_2, & {}_2C_3^{-1} &= {}_2C_3^2, \\
{}_3C_3 &= {}_3s_1\,{}_2s_2, & {}_3C_3^{-1} &= {}_3C_3^2, \\
{}_4C_3 &= {}_2s_1\,{}_3s_2, & {}_4C_3^{-1} &= {}_4C_3^2.
\end{aligned}
\tag{23}
$$

Those are rather bulky expressions; ${}_1C_3$ for instance has the form

$$
\begin{aligned}
{}_1C_3 = {}&\tfrac{1}{4} + \tfrac{1}{4}(e_1 - e_2 + e_3 + e_4) + \\
&+\tfrac{1}{4}(e_{12} + e_{13} + e_{14} + e_{23} - e_{24} + e_{34}) + \\
&+\tfrac{1}{4}(e_{123} + e_{124} + e_{134} + e_{234}) - \tfrac{1}{4}e_{1234}
\end{aligned}
\tag{24}
$$

and its inverse is

$$
\begin{aligned}
{}_1C_3^{-1} = {}&\tfrac{1}{4} + \tfrac{1}{4}(e_1 - e_2 + e_3 - e_4) + \\
&+\tfrac{1}{4}(-e_{12} - e_{13} + e_{14} - e_{23} - e_{24} + e_{34}) + \\
&+\tfrac{1}{4}(e_{123} + e_{124} + e_{134} + e_{234}) + \tfrac{1}{4}e_{1234}.
\end{aligned}
\tag{25}
$$

The four operators ${}_jC_3$ (with $j = 1, 2, 3, 4$), their inverses ${}_jC_3^{-1}$, the three central elements ${}_iC_2$ and the unity form the tetrahedral subgroup of order 12 of the octahedral group $\mathbf{O}$. Note that each expression of the form ${}_jH_6 = -{}_jC_3$ is a hexagonal operator of period 6. As we shall see, it is because of the period-3 operators that the color symmetry has to appear. Those can be visualized as 120°-rotations about the 4 spatial diagonals of a cube which cyclically transform 3 corners into one another. This is characteristic for the color symmetry. Identifying the corner 1 with a lepton

f_1, and f_2, f_3, f_4 with three colored quarks of family 1, the tetrahedral operator $_1C_3$ performs a color-rotation:

$$\begin{aligned}
\text{lepton:} \quad &_1C_3\, f_1\, _1C_3^{-1} = f_1, \\
\text{quarks:} \quad &_1C_3\, f_2\, _1C_3^{-1} = f_3, \; _1C_3\, f_3\, _1C_3^{-1} = f_4, \; _1C_3\, f_4\, _1C_3^{-1} = f_2.
\end{aligned} \tag{26}$$

The idempotents f_2 to f_4 were first interpreted as quarks of one family and three colors by Greider and Weiderman. Chisholm and Farwell (1992) extended this image by interpreting f_1 as the corresponding lepton. The three operators of period 4 and their inverses are given by the products

$$\begin{aligned}
_1C_4 &= {}_2C_3\, _2s_1 = \tfrac{1}{2}(\mathbf{e}_1 - \mathbf{e}_{12} + \mathbf{e}_{134} + \mathbf{e}_{1234}), & _1C_4^{-1} &= {}_1C_4^3, \\
_2C_4 &= {}_4C_3\, _3s_1 = \tfrac{1}{2}(\mathbf{e}_1 - \mathbf{e}_4 - \mathbf{e}_{13} + \mathbf{e}_{34}), & _2C_4^{-1} &= {}_2C_4^3, \\
_3C_4 &= {}_3C_3^2\, _1s_1 = \tfrac{1}{2}(-\mathbf{e}_{23} + \mathbf{e}_{34} + \mathbf{e}_{123} + \mathbf{e}_{134}), & _1C_4^{-1} &= {}_3C_4^3.
\end{aligned} \tag{27}$$

So we have a total of 6 reflections $_is_j$, 3 operators $_iC_2$ which represent the center of $\mathbf{O}$, 8 operators of period 3, 6 operators of period 4 and unity, that is, 24 elements. This is the correct order since $\mathbf{O}$ is isomorphic with the symmetric group S_4. The multiplication table of the group $\mathbf{O}$ can easily be verified with CLICAL.

5.2. REPRESENTATION OF O IN THE DIRAC ALGEBRA

In the Dirac algebra of complex 4×4-matrices $\mathrm{Mat}(4, \mathbb{C})$ the unit vectors of the Clifford algebra $C\ell_{3,1}$ can be represented as follows:

$$\mathbf{e}_1 = \begin{bmatrix} 1 & 0 & 0 & 0 \\ 0 & -1 & 0 & 0 \\ 0 & 0 & -1 & 0 \\ 0 & 0 & 0 & 1 \end{bmatrix}, \quad
\mathbf{e}_2 = \begin{bmatrix} 0 & -i & 0 & 0 \\ i & 0 & 0 & 0 \\ 0 & 0 & 0 & i \\ 0 & 0 & -i & 0 \end{bmatrix},$$

$$\mathbf{e}_3 = \begin{bmatrix} 0 & 0 & -i & 0 \\ 0 & 0 & 0 & -i \\ i & 0 & 0 & 0 \\ 0 & i & 0 & 0 \end{bmatrix}, \quad
\mathbf{e}_4 = \begin{bmatrix} 0 & 0 & i & 0 \\ 0 & 0 & 0 & i \\ i & 0 & 0 & 0 \\ 0 & i & 0 & 0 \end{bmatrix}. \tag{28}$$

Calculating the bivectors $\mathbf{e}_{ij}$ and from those the Dirac matrices of the reflections $_1s_1$ and $_3s_1$ (eq. 20) we obtain the color operator

$$_1C_3 = {}_1s_1\, _3s_1 = \begin{bmatrix} 1 & 0 & 0 & 0 \\ 0 & 1 & 0 & 0 \\ 0 & 0 & 0 & -i \\ 0 & 0 & i & 0 \end{bmatrix}\begin{bmatrix} 1 & 0 & 0 & 0 \\ 0 & 0 & 1 & 0 \\ 0 & 1 & 0 & 0 \\ 0 & 0 & 0 & 1 \end{bmatrix} = \begin{bmatrix} 1 & 0 & 0 & 0 \\ 0 & 0 & 1 & 0 \\ 0 & 0 & 0 & -i \\ 0 & i & 0 & 0 \end{bmatrix},$$

$$_1C_3^{-1} = \begin{bmatrix} 1 & 0 & 0 & 0 \\ 0 & 0 & 0 & -i \\ 0 & 1 & 0 & 0 \\ 0 & 0 & i & 0 \end{bmatrix}. \tag{29}$$

Chisholm and Farwell (1992) have demonstrated how a tetrahedral $\text{Mat}(4,\mathbb{C})$-operator of this sort can be interpreted as a charge operator within this family, that is, we should have

$$Q = (_1C_3)^{\frac{3}{2}} \tag{30}$$

with its $C\ell_{3,1}$ term:

$$Q = \frac{1}{3}(e_1 - e_2 + e_3 + e_{14} - e_{24} + e_{34} + e_{124} + e_{134} + e_{234}).$$

We shall find out that the final particle charge operator can be based on a much simpler expression by taking reference to the Gell-Mann-Nishijima relation.

Clearly, the whole set of 24 symmetry operators of the orientation group **O** can be generated by a minimal basis of two complex 4×4-matrices. Those can be chosen to be $_1s_1, _2C_4$. In all these matrices the idempotent f_1 gives the only non zero contribution to the first rows and columns. It is therefore that a crossing out of the first rows and columns of the representative matrices of **O** does not affect the multiplication rules. So we obtain a representation of **O** in terms of complex unitary 3×3-matrices.

5.3. Representation of O in the SU(3) and the Gell-Mann matrices

It is due to this particular algebraic form in the Clifford algebra $C\ell_{3,1}$ (equations 20) of the generating reflections of the octahedral groups that the group SU(3) gains a special importance for quantum chromodynamics (QCD). Consider the first family of quarks corresponding with reflections in eq. (20). If we cross out the first rows and columns of the representative $\text{Mat}(4,\mathbb{C})$-matrices, we obtain a representation of **O** in the SU(3). The calculations are straight forward. For the first family (eq. 20) we obtain the following special unitary 3×3-matrices of reflections:

$$
_1s_1 \begin{bmatrix} 1 & 0 & 0 \\ 0 & 0 & -i \\ 0 & i & 0 \end{bmatrix}, \quad
_2s_1 \begin{bmatrix} 0 & 0 & -i \\ 0 & 1 & 0 \\ i & 0 & 0 \end{bmatrix}, \quad
_3s_1 \begin{bmatrix} 0 & 1 & 0 \\ 1 & 0 & 0 \\ 0 & 0 & 1 \end{bmatrix},
$$

$$
_1s_2 \begin{bmatrix} 1 & 0 & 0 \\ 0 & 0 & i \\ 0 & -i & 0 \end{bmatrix}, \quad
_2s_2 \begin{bmatrix} 0 & 0 & i \\ 0 & 1 & 0 \\ -i & 0 & 0 \end{bmatrix}, \quad
_3s_2 \begin{bmatrix} 0 & -1 & 0 \\ -1 & 0 & 0 \\ 0 & 0 & 1 \end{bmatrix}.
\tag{31}
$$

The period-4 rotations are:

$$
_1C_4 \begin{bmatrix} -1 & 0 & 0 \\ 0 & 0 & i \\ 0 & i & 0 \end{bmatrix}, \quad
_2C_4 \begin{bmatrix} 0 & 0 & -i \\ 0 & -1 & 0 \\ -i & 0 & 0 \end{bmatrix}, \quad
_3C_4 \begin{bmatrix} 0 & -1 & 0 \\ 1 & 0 & 0 \\ 0 & 0 & -1 \end{bmatrix},
$$

$$
_1C_4^3 \begin{bmatrix} -1 & 0 & 0 \\ 0 & 0 & -i \\ 0 & -i & 0 \end{bmatrix}, \quad
_2C_4^3 \begin{bmatrix} 0 & 0 & i \\ 0 & -1 & 0 \\ i & 0 & 0 \end{bmatrix}, \quad
_3C_4^3 \begin{bmatrix} 0 & 1 & 0 \\ -1 & 0 & 0 \\ 0 & 0 & -1 \end{bmatrix}.
$$

The central operators are all diagonal:

$$
\begin{matrix}
{}_1C_2 & {}_2C_2 & {}_3C_2 & 1
\end{matrix}
$$

$$
\begin{bmatrix} 1 & 0 & 0 \\ 0 & -1 & 0 \\ 0 & 0 & -1 \end{bmatrix}, \quad
\begin{bmatrix} -1 & 0 & 0 \\ 0 & 1 & 0 \\ 0 & 0 & -1 \end{bmatrix}, \quad
\begin{bmatrix} -1 & 0 & 0 \\ 0 & -1 & 0 \\ 0 & 0 & 1 \end{bmatrix}, \quad
\begin{bmatrix} 1 & 0 & 0 \\ 0 & 1 & 0 \\ 0 & 0 & 1 \end{bmatrix}.
$$

It is easy to see that there is an important relationship between the invariant central elements of the octahedral group and the only two linearly independent diagonal generators of the SU(3). That is, isospin T_z and hypercharge Y are determined by the equations

$$
T_z = \frac{1}{4}({}_1C_2 - {}_2C_2), \tag{32}
$$

$$
Y = -\frac{1}{6} - {}_3C_2. \tag{33}
$$

The Gell-Mann matrices can be derived from the relations

$$
\begin{aligned}
\lambda_1 &= \tfrac{1}{2}({}_3s_1 - {}_3s_2), & \lambda_2 &= \tfrac{1}{2}i({}_3C_4 - {}_3C_4^{-1}), \\
\lambda_3 &= \tfrac{1}{2}({}_1C_2 - {}_2C_2), & \lambda_4 &= \tfrac{1}{2}i({}_2C_4 - {}_2C_4^{-1}), \\
\lambda_5 &= \tfrac{1}{2}({}_2s_1 - {}_2s_2), & \lambda_6 &= -\tfrac{1}{2}i({}_1C_4 - {}_1C_4^{-1}), \\
\lambda_7 &= \tfrac{1}{2}({}_1s_1 - {}_1s_2), & \lambda_8 &= -\tfrac{1}{\sqrt{12}}(1 + 3\,{}_3C_2).
\end{aligned} \tag{34}
$$

This is a representation of the generators of the SU(3) by partial involutions and commutations in the basis of $C\ell_{3,1}$. Of a special importance are the derivation of the isospin-, hypercharge- and charge operators in terms of the invariant elements $1, {}_1C_2, {}_2C_2, {}_3C_2$ of the octahedral group. This center is isomorphic to $Z_2 \times Z_2$.

5.4. QUANTUM NUMBERS OF ORIENTATION AND TRANSPOSITION OF THE GELL-MANN-NISHIJIMA RELATION – DERIVATION OF THE CHARGE OPERATOR

The diagonal operators ${}_iC_2$ and unity form the center of the orientation group **O** and shall be denoted the *center of orientation*. They can be represented in terms of the isospin- and hypercharge operators T_z and Y. Note that in the Clifford algebra $C\ell_{3,1}$ the following identity has to hold:

$$
f_1 = \frac{1}{4}(1 + {}_1C_2 + {}_2C_2 + {}_3C_2). \tag{35}
$$

Therefore, the expression $\Gamma_2 = 1 + {}_1C_2 + {}_2C_2 + {}_3C_2$ is invariant under color transformation ${}_3C_1$:

$$
{}_3C_1\,\Gamma_2\,{}_3C_1^{-1} = \Gamma_2. \tag{36}
$$

Be aware that the four operators $1, {}_1C_2, {}_2C_2, {}_3C_2$ form the center of the octahedral group **O** and each of these operators contributes a fraction of electric charge to the field. In the Dirac algebra the term ${}_1C_2 + {}_2C_2 + {}_3C_2 = 4f_1 - 1$ is the matrix

$$
\begin{bmatrix} 3 & 0 & 0 & 0 \\ 0 & -1 & 0 & 0 \\ 0 & 0 & -1 & 0 \\ 0 & 0 & 0 & -1 \end{bmatrix}.
$$

One third of this matrix is equal to Chisholm's and Farwell's "particle charge operator C" for one positive lepton and three quarks; that is, Chisholm's C is given by Chisholm and Farwell (1992, p. 32)

$$C = \frac{1}{3}({}_1C_2 + {}_2C_2 + {}_3C_2).$$

Thus, in this image, each rotation operator ${}_iC_2$ contributes 1/3 of an electron charge to the field while its space-time is oriented. In the representation by complex unitary 3×3-matrices we observe the following decomposition of -1 which might remind us of some sort of physical "zero-sum game" (with ${}_iC_2$ from equations 31):

$$
\begin{array}{rccccc}
{}_1C_2 & = & -\frac{1}{3} & + & 2T_z & + & Y \\
{}_2C_2 & = & -\frac{1}{3} & - & 2T_z & + & Y \\
{}_3C_2 & = & -\frac{1}{3} & & & - & 2Y \\
\hline
1 & = & 1 & & & & \\
\hline
0 & = & 0 & & & &
\end{array}
\tag{37}
$$

Finally consider the Gell-Mann-Nishijima relation

$$Q = \frac{1}{2}Y + T_z. \tag{38}$$

Substituting expressions (32, 33) we obtain its geometric form

$$Q = \frac{1}{6} + \frac{1}{2}{}_1C_2. \tag{39}$$

As Y and T_z are known from high energy physics, the quantum numbers corresponding with ${}_1C_2, {}_2C_2, {}_3C_2$ can be calculated. They shall be called the *central orientation numbers* or *quantum numbers of orientation* or briefly *orientations*. Consider the eightfold path diagram of the baryon octet belonging to the nucleons. Calculating *orientations* ${}_1C_2, {}_2C_2, {}_3C_2$ gives the table below.

	T_z	Y	Q	${}_1C_2$	${}_2C_2$	${}_3C_2$	${}_1C_2 + {}_2C_2 + {}_3C_2$	E
n	$-1/2$	1	0	$-1/3$	$5/3$	$-7/3$	-1	1
p	$1/2$	1	1	$5/3$	$-1/3$	$-7/3$	-1	1
Λ^0	0	0	0	$-1/3$	$-1/3$	$-1/3$	-1	1
Σ^0	0	0	0	$-1/3$	$-1/3$	$-1/3$	-1	1
Σ^-	-1	0	-1	$-7/3$	$5/3$	$-1/3$	-1	1
Σ^+	1	0	1	$5/3$	$-7/3$	$-1/3$	-1	1
Ξ^-	$-1/2$	-1	-1	$-7/3$	$-1/3$	$5/3$	-1	1
Ξ^0	$1/2$	-1	0	$-1/3$	$-7/3$	$5/3$	-1	1

The most surprising feature of such tables is the invariance of the sum of operators ${}_1C_2 + {}_2C_2 + {}_3C_2$. The most simple figures are the hadrons Λ^0 and Σ^0 located in the center of the octet which is quite reasonable. Now it would be interesting to find out about the geometrodynamics difference between quarks and particles composed by quarks. In which sense can they be said to be "elementary"? Surprisingly, quarks represent the only states with orientation numbers equal to ± 1 (see the table below).

	T_z	Y	Q	$_1C_2$	$_2C_2$	$_3C_2$	$_1C_2 + {}_2C_2 + {}_3C_2$	E
u red	1/2	1/3	2/3	1	-1	-1	-1	1
d yellow	$-1/2$	1/3	$-1/3$	-1	1	-1	-1	1
s blue	0	$-2/3$	$-1/3$	-1	-1	1	-1	1

So quarks seem to be elementary quanta of orientation. The void is dynamic, and quarks are its essence. Space and orientation cannot be separated, but they constitute a most basic process of nature.

5.5. EIGHT SETS OF QUARKS

Once we have identified the outer symmetry of the space-time with the inner symmetry of strong interaction, it becomes important to determine the number of copies of the SU(3) in the Clifford algebra $C\ell_{3,1}$ in order to identify the number of quark families. By the aid of CLICAL eight such families could be found which are generated by the following set of reflections. Beware that we are using the symbols $_is_j$ ($i = 1, \ldots, 12$ and $j = 1, 2$) instead of the Schönfließ-symbols $_i\sigma'_d$ as we are now completing the set of reflections in Minkowski-space thus going with the index i beyond 3 and realizing that the subscript d actually makes no difference; it is merely a relic of history. What remains, however, is that each $_is_2$ is the inverse of $_is_1$ just as the diagonal reflection $_i\sigma''_d$ was the inverse of the operator $_i\sigma'_d$.

$$
\begin{aligned}
{}_1s_1 &= f_1 + f_2 + \mathbf{e}_{12}(f_4 - f_3), & {}_1s_2 &= f_1 + f_2 + \mathbf{e}_{12}(f_3 - f_4), \\
{}_2s_1 &= f_1 + f_3 + \mathbf{e}_{13}(f_2 - f_4), & {}_2s_2 &= f_1 + f_3 + \mathbf{e}_{13}(f_4 - f_2), \\
{}_3s_1 &= f_1 + f_4 + \mathbf{e}_{23}(f_3 - f_2), & {}_3s_2 &= f_1 + f_4 + \mathbf{e}_{23}(f_2 - f_3), \\
{}_4s_1 &= f_2 + f_3 + \mathbf{e}_{23}(f_1 - f_4), & {}_4s_2 &= f_2 + f_3 + \mathbf{e}_{23}(f_4 - f_1), \\
{}_5s_1 &= f_2 + f_4 + \mathbf{e}_{13}(f_3 - f_1), & {}_5s_2 &= f_2 + f_4 + \mathbf{e}_{13}(f_1 - f_3), \\
{}_6s_1 &= f_3 + f_4 + \mathbf{e}_{12}(f_2 - f_1), & {}_6s_2 &= f_3 + f_4 + \mathbf{e}_{12}(f_1 - f_2).
\end{aligned}
\tag{40}
$$

By these reflections 4 copies of the group **O** are generated and thus 4 tetrahedral subgroups each of which gives rise to one copy of the SU(3). Each copy is linked with one idempotent f_i. The first four families can be generated in the following ways:

— set 1 / f_1: by reflections $_1s_1$ to $_3s_2$,
— set 2 / f_2: by reflections $_1s_1, {}_1s_2$ with $_4s_1$ to $_5s_2$,
— set 3 / f_3: by reflections $_2s_1, {}_2s_2$ with $_4s_1, {}_4s_2, {}_6s_1, {}_6s_2$,
— set 4 / f_4: by reflections $_3s_1, {}_3s_2$ with $_5s_1, {}_5s_2, {}_6s_1, {}_6s_2$.

A complete multiplication table (see below) calculated by CLICAL shows the appearance of only four tetrahedral groups. Their period-3 operators $_jC_3$ are now denoted $_iT_j$ (where $i = 1, \ldots, 8$; $j = 1, 2, 3, 4$) as the index 3 makes no difference, but merely identifies the period 3. These four tetrahedral subgroups of the orientation groups give rise to four copies of the SU(3), and it has been believed for quite a while that they are all we can get. However, the rigor can be improved by considering the isomorphism $Cl_3 \simeq Cl_{3,1}^+$ given by the correspondences $e_1, e_2, e_3 \simeq e_{14}, e_{24}, e_{34}$. First, we construct the idempotents g_i from the f_i

$$
\begin{aligned}
g_1 &= \tfrac{1}{2}(1 + e_{14})\tfrac{1}{2}(1 + e_{34}e_4) = \tfrac{1}{2}(1 - e_3)\tfrac{1}{2}(1 + e_{14}), \\
g_2 &= \tfrac{1}{2}(1 - e_3)\tfrac{1}{2}(1 - e_{14}), \\
g_3 &= \tfrac{1}{2}(1 + e_3)\tfrac{1}{2}(1 + e_{14}), \\
g_4 &= \tfrac{1}{2}(1 + e_3)\tfrac{1}{2}(1 - e_{14}).
\end{aligned}
\tag{41}
$$

Families 5 to 8 are then generated by the reflections

$$
\begin{aligned}
{}_7s_1 &= g_1 + g_2 + e_{12}(g_4 - g_3), & {}_7s_2 &= g_1 + g_2 + e_{12}(g_3 - g_4), \\
{}_8s_1 &= g_1 + g_3 + e_{13}(g_2 - g_4), & {}_8s_2 &= g_1 + g_3 + e_{13}(g_4 - g_2), \\
{}_9s_1 &= g_1 + g_4 + e_{23}(g_3 - g_2), & {}_9s_2 &= g_1 + g_4 + e_{23}(g_2 - g_3), \\
{}_{10}s_1 &= g_2 + g_3 + e_{23}(g_1 - g_4), & {}_{10}s_2 &= g_2 + g_3 + e_{23}(g_4 - g_1), \\
{}_{11}s_1 &= g_2 + g_4 + e_{13}(g_3 - g_1), & {}_{11}s_2 &= g_2 + g_4 + e_{13}(g_1 - g_3), \\
{}_{12}s_1 &= g_3 + g_4 + e_{12}(g_2 - g_1), & {}_{12}s_2 &= g_3 + g_4 + e_{12}(g_1 - g_2).
\end{aligned}
\tag{42}
$$

The multiplication table of these reflections is in a one to one correspondence with the previous one if we turn over from the tetrahedral rotations ${}_1T_j, \ldots, {}_4T_j$ to ${}_5T_j, \ldots, {}_8T_j$.

These four tetrahedral subgroups of the orientation groups give

Clearly, every tetrahedral rotation ${}_iT_j$ satisfying ${}_iT_j^3 = 1$ is turned into a hexagonal operator by merely inverting the signs of its Clifford algebra expression, that is, the term ${}_iH_j = -{}_iT_j$ has period six because we have $(-{}_iT_j)^3 = -1$ and $(-{}_iT_j)^6 = +1$. Each hexagonal operator thus describes transitions between quarks and antiquarks of three colors.

5.6. DECOMPOSITION OF THE PRIMITIVE IDEMPOTENTS BY THE CENTRAL OPERATORS OF ORIENTATION

It is interesting to ask about the role of the idempotents f_i and g_i in their relation to the quantum numbers of orientation. Calculating the rotations ${}_iC_2$ ($i = 1, \ldots, 12$) that build up the centers of the 8 octahedral subgroups of $\mathbf{H}_4$, we find

$$
\begin{aligned}
{}_1C_2 &= e_{34}, & {}_2C_2 &= e_{134}, & {}_3C_2 &= e_1, \\
{}_4C_2 &= -e_1, & {}_5C_2 &= -e_{134}, & {}_6C_2 &= -e_{34}, \\
{}_7C_2 &= -e_3, & {}_8C_2 &= e_{134}, & {}_9C_2 &= e_{14}, \\
{}_{10}C_2 &= -e_{14}, & {}_{11}C_2 &= -e_{134}, & {}_{12}C_2 &= e_3.
\end{aligned}
\tag{43}
$$

From this we obtain the following decomposition of the primitive idempotents of the Clifford algebra $Cl_{3,1}$ in terms of the invariant central operators of the octahedral

	$1s_1$	$1s_2$	$2s_1$	$2s_2$	$3s_1$	$3s_2$	$4s_1$	$4s_2$	$5s_1$	$5s_2$	$6s_1$	$6s_2$
$1s_1$	1	e_{34}	$_1T_1^2$	$_1T_2$	$_1T_1$	$_1T_2^2$	$_2T_1$	$_2T_2^2$	$_2T_2$	$_2T_1^2$	e_{234}	$-e_2$
$1s_2$	e_{34}	1	$_1T_4$	$_1T_3^2$	$_1T_3$	$_1T_4^2$	$_2T_3$	$_2T_4^2$	$_2T_3^2$	$_2T_4$	e_2	$-e_{234}$
$2s_1$	$_1T_1$	$_1T_4^2$	1	e_{134}	$_1T_1^2$	$_1T_4$	$_3T_4$	$_3T_1^2$	e_3	e_{14}	$_3T_1$	$_3T_4^2$
$2s_2$	$_1T_2^2$	$_1T_3$	e_{134}	1	$_1T_3^2$	$_1T_2$	$_3T_2$	$_3T_3^2$	$-e_{14}$	$-e_3$	$_3T_2^2$	$_3T_3$
$3s_1$	$_1T_1^2$	$_1T_3^2$	$_1T_1$	$_1T_3$	1	e_1	e_{124}	$-e_{24}$	$_4T_1$	$_4T_3$	$_4T_1^2$	$_4T_3^2$
$3s_2$	$_1T_2$	$_1T_4$	$_1T_4^2$	$_1T_2^2$	e_1	1	e_{24}	$-e_{124}$	$_4T_4^2$	$_4T_2^2$	$_4T_2$	$_4T_4$
$4s_1$	$_2T_1^2$	$_2T_3^2$	$_3T_4^2$	$_3T_2^2$	e_{124}	e_{24}	1	$-e_1$	$_2T_3$	$_2T_1$	$_3T_2$	$_3T_4$
$4s_2$	$_2T_2$	$_2T_4$	$_3T_1$	$_3T_3$	$-e_{24}$	$-e_{124}$	$-e_1$	1	$_2T_2^2$	$_2T_4^2$	$_3T_1^2$	$_3T_3^2$
$5s_1$	$_2T_2^2$	$_2T_3^2$	e_3	e_{14}	$_4T_1^2$	$_4T_4$	$_2T_3^2$	$_2T_2$	1	$-e_{134}$	$_4T_1$	$_4T_4^2$
$5s_2$	$_2T_1$	$_2T_4^2$	e_{14}	$-e_3$	$_4T_3^2$	$_4T_2$	$_2T_1^2$	$_2T_4$	$-e_{134}$	1	$_4T_2^2$	$_4T_3$
s_1	e_{234}	e_2	$_3T_1^2$	$_3T_2$	$_4T_1$	$_4T_2^2$	$_3T_2^2$	$_3T_1$	$_4T_1^2$	$_4T_2$	1	$-e_{34}$
s_2	$-e_2$	$-e_{234}$	$_3T_4$	$_3T_3^2$	$_4T_3$	$_4T_4^2$	$_3T_4^2$	$_3T_3$	$_4T_4$	$_4T_3^2$	$-e_{34}$	1

orientation groups. We have

$$
\begin{aligned}
f_1 &= \tfrac{1}{4}(1 + {}_1C_2 + {}_2C_2 + {}_3C_2), & f_2 &= \tfrac{1}{4}(1 + {}_1C_2 + {}_4C_2 + {}_5C_2), \\
f_3 &= \tfrac{1}{4}(1 + {}_1C_2 + {}_2C_2 + {}_4C_2), & f_4 &= \tfrac{1}{4}(1 + {}_3C_2 + {}_5C_2 + {}_6C_2), \\
g_1 &= \tfrac{1}{4}(1 + {}_7C_2 + {}_8C_2 + {}_9C_2), & g_2 &= \tfrac{1}{4}(1 + {}_7C_2 + {}_{10}C_2 + {}_{11}C_2), \\
g_3 &= \tfrac{1}{4}(1 + {}_8C_2 + {}_{10}C_2 + {}_{12}C_2), & g_4 &= \tfrac{1}{4}(1 + {}_9C_2 + {}_{11}C_2 + {}_{12}C_2).
\end{aligned}
\tag{44}
$$

Every primitive idempotent is thus decomposed into a sum of the invariant operators of the center of **O**.

5.7. THE ORIENTATION-MATRICES $_iC_2$

It is because of three reasons that the operators $_iC_2$ play an important role in the emergence of both strong interaction and the orientation of space-time. Those are
1. the $_iC_2$ form the centers of the octahedral groups contained in the Clifford algebra of the Minkowski-space $\mathbb{R}^{3,1}$

2. the idempotents f_i of the Clifford algebra $C\ell_{3,1}$ are normalized sums of the $_iC_2$ plus unity, and

3. the invariant operators of the SU(3), being isospin T_z and hypercharge Y, can be decomposed into linear sums of the $_iC_2$ and the unity.

Note that these operators have period 2 in the Clifford algebra $C\ell_{3,1}$. They have the same period also in SO(3). Last but not least, they take the same positions within the group table of the abstract group **O** as the Schönfließ-rotations $_xC_2$, $_yC_2$, $_zC_2$. Yet, they are not rotations at all. Similar to the Pauli-matrices they represent spin-matrices of period 2, but not rotations. This is due to the two-fold covering of the special orthogonal groups by their spin-groups, a typical feature of quantum physics.

Each operator $_iC_2$ causes a partial involution of the basis (28) of $\mathbb{R}^{3,1}$, e.g., we have

$$
\begin{aligned}
_1C_2 &: \{e_1, e_2, e_3, e_4\} \rightarrow \{e_1, e_2, -e_3, -e_4\}, \\
_2C_2 &: \{e_1, e_2, e_3, e_4\} \rightarrow \{e_1, -e_2, e_3, e_4\}, \\
_3C_2 &: \{e_1, e_2, e_3, e_4\} \rightarrow \{e_1, -e_2, -e_3, -e_4\}.
\end{aligned}
\tag{45}
$$

Indeed, that does not resemble the action of period-2 rotations about the e_i. But those are given by the bivectors e_{ij} (with $i, j = 1, 2, 3$) which leave e_4 unaltered. We actually observe the required period-doubling, that is, $(e_{ij})^2 = -1$ and $(e_{ij})^4 = 1$ and the transformative properties of the Schönfließ rotations.

$$
\begin{aligned}
(e_{23})^{-1}\{e_1, e_2, e_3, e_4\}e_{23} &= \{e_1, -e_2, -e_3, e_4\}, \\
(e_{13})^{-1}\{e_1, e_2, e_3, e_4\}e_{13} &= \{-e_1, e_2, -e_3, e_4\}, \\
(e_{12})^{-1}\{e_1, e_2, e_3, e_4\}e_{12} &= \{-e_1, -e_2, e_3, e_4\}.
\end{aligned}
\tag{46}
$$

It is essentially due to the fact that in $C\ell_{3,1}$ we observe octahedral groups with diagonal center operators of period 2 instead of 4 that we obtain quantum numbers of orientation. This situation is similar as in the case of SU(2).

An operator of period 4 such as $_1C_4$ induces the following transformation in the basis of $C\ell_{3,1}$:

$$
_1C_4 : \{e_1, e_2, e_3, e_4\} \rightarrow \{e_{134}, -e_2, e_{24}, e_{23}\}.
\tag{47}
$$

This is unlike the action of a representative period-8 operator of a period-4 rotation in the SO(3,1). Such a period-8 representative matrix of the Dirac algebra rather would have the form $(1/\sqrt{2})(1 + e_{ij})$ with $i, j = 1, 2, 3$.

5.8. The (strong inter)action of the color operator in the Minkowski space-time

Perform a tetrahedral rotation in the Euclidean space $\mathbb{R}^3$ given by the permutation of corners (1 3 2) and transforming the basis as follows

$$
T_3 : \{e_1, e_2, e_3\} \rightarrow \{e_3, -e_1, -e_2\}.
\tag{48}
$$

Consider the temporal basis vector e_4 and the unit director e_{123}. Both have period 4. Carry out an involution of e_4 and map $-e_4$ onto e_{123}. Finally, consider the isomorphism $C\ell_3 \simeq C\ell_{3,1}^+$ given by the correspondences $e_1, e_2, e_3 \simeq e_{14}, e_{24}, e_{34}$.

This is the action of the color operators $_1T_1$ and $_1H_1 = -_1T_1$ respectively. Both map the basis of $\mathbb{R}^{3,1}$ in the same way:

$$
_1T_1, {_1H_1} : \quad
\begin{aligned}
e_1 &\rightarrow e_{34} \\
e_2 &\rightarrow -e_{14} \\
e_3 &\rightarrow -e_{24} \\
e_4 &\rightarrow -e_{123}.
\end{aligned}
\tag{49}
$$

This means that a color rotation interlinks the orientation and the metric of space with that of time.

References

R. Ablamowicz, P. Lounesto, and J. Maks: 1991, 'Conference Report', *Found. of Phys.*, **21** (6), pp. 735–748.

M. Belger and L. Ehrenberg: 1981, *Theorie und Anwendung der Symmetriegruppen*, Thun-Frankfurt am Main.

R. N. Cahn: 1984, *Semi-Simple Lie Algebras and Their Representations*, Berkeley.

J. S. R. Chisholm: 1992, 'Tetrahedral Structure of Idempotents of the Clifford Algebra $C_{3,1}$.' in: A. Micali *et al.*, 1992, *Clifford Algebras and their Applications in Mathematical Physics*, Dordrecht, pp. 27–32.

J. S. R. Chisholm and R. S. Farwell: 1992, 'Unified spin gauge theories of the four fundamental forces' in: A. Micali *et al.*, *Clifford Algebras ...*, *op. cit*, Dordrecht, p. 363–370.

H. S. M. Coxeter and W. O. J. Moser: 1984, *Generators and Relations for Discrete Groups*, Berlin.

R. P. Feynman: 1988, *ED-Die Seltsame Theorie des Lichts und der Materie*, München.

H. Fritzsch: 1984, *Quarks - Urstoff unserer Welt*, München.

M. Gell-Mann: 1991, *Elementary Particles and the Universe*, Cambridge.

K. Greider and T. Weiderman: 1988, 'Generalized Clifford algebras as special cases of standard Clifford Algebras', I'UCD Preprint 16.

J. S. Hagelin: 1988, *Is Consciousness the Unified Field? A Field Theorist's Perspective*, in: *Modern Science and Veidc Science*, MIU Press, Fairfield, Iowa, pp. 29 – 87.

T. Inui, Y. Tanabe and Y. Onodera: 1990, *Group Theory and its Application in Physics*, Berlin.

P. Lounesto: 1986, 'Report on Conference: NATO and SERC Workshop on "Clifford Algebras and Their Applications to Mathematical Physics", 1985', *Found. Phys.*, 16(9), pp. 967–971.

P. Lounesto, R. Mikkola, V. Vierros: 1987, *CLICAL User Manual - Complex Number, Vector Space and Clifford Algebra Calculator for MS-DOS Personal Computers*, Research Report A248, Helsinki Univ. of Technology.

P. Lounesto: 1980, 'Scalar Products of Spinors and an Extension of Brauer–Wall Groups', *Found. Phys.*, 11(9/10), pp. 721–740.

A. Messiah: 1970, *Quantum Mechanics, vol. II*, Amsterdam.

A. Micali, R. Boudet, J. Helmstetter: 1992, *Clifford Algebras and their Applications in Mathematical Physics, Proc. 2nd. Workshop at Montpellier, 1989*, Kluwer, Dordrecht.

A. O. Morris, M. K. Makhool: 1992, 'Real projective representations of real Clifford algebras and reflection groups', in A. Micali *et al.*: 1992, *Clifford Algebras ...*, *op. cit*, pp. 69–82.

J. M. Parra: 1992, 'On Dirac and Dirac-Darwin-Hestenes equations' in: A. Micali *et al.*: 1992, *Clifford Algebras ...*, *op. cit*, Dordrecht, pp. 463–477.

M. I. Petraschen and E. D. Trifonow: 1969, *Anwendungen der Gruppentheorie in der Quantenmechanik*, Leipzig.

E. Pöppel: 1989, 'Erlebte Zeit und die Zeit überhaupt: Ein Versuch der Integration', in: J. Aschoff, J. Assmann, J. P. Blaser *et al.*, (Hrsg.). *Die Zeit - Dauer und Augenblick*, pp. 369–382, München.

P. Ramond: 1990, *Field Theory: A Modern Primer*, Benjamin/Cummings, New York.

B. Schmeikal: 1993, *Logic from Space, Quality and Quantity* 27, Dordrecht, pp. 117–137.

S. Thelen: 1992, 'Algebraic Spin Structures', in: A. Micali *et al.*: 1992, *Clifford Algebras ...*, *op. cit*, Dordrecht, pp. 143–149.

J. A. Wheeler and C. M. Patton: 1977, 'Is physics legislated by cosmogony?' in: R. Duncan, M. Weston-Smith, (eds.), *The Encyclopedia of Ignorance*, Oxford, pp. 19–35.

ON A NEW BASIS FOR A GENERALIZED CLIFFORD ALGEBRA AND ITS APPLICATION TO QUANTUM MECHANICS

A. GRANIK and M. ROSS
Physics Department,
University of the Pacific,
Stockton, CA 95211
e-mail: galois@ix.netcom.com

Abstract. A generalized Clifford algebra is formulated in terms of a simple Euclidean carrier space. Using this space the eigenvalues, shifting operators, and the Casimir operator are easily derived for a few selected quantum mechanical problems.

Key words: Generalized Clifford algebra, decomposition of a basis element, Lie algebra, quantum mechanics.

1. Introduction

An important role played by commuting and anticommuting operators in quantum mechanics lead Weyl (1950) to an idea about an intrinsic relation between commutation rules and a natural interpretation of quantum kinematics. Further development of this idea by Weyl resulted in the introduction of a more general commutation relation incorporating both commuting and anticommuting operators as particular cases. This new relation formed the foundation of an algebra which is called a generalized Clifford algebra (GCA). The name is justifiable if we recall that the anticommutation rule is one of the basic relations of a conventional Clifford algebra.

Later, other researchers [for example, Morinaga and Nono (1952)] studied both purely mathematical and applied aspects of GCA, see Hestenes (1966), Morris (1967), (1968), Raševskiĭ (1955), Riesz (1958), and Yamazaki (1964). However, it is important to remember that Weyl has developed his original idea of GCA for purely applied purposes. Therefore it is not surprising that a large part of the subsequent studies in applied GCA centered around problems in quantum mechanics. In particular, it was found that the concept of a spinor follows directly from GCA as was demonstrated by Raševskiĭ (1955).

In this paper we show how GCA can be used to describe some aspects of quantum mechanics without resorting to a matrix representation. By introducing a new carrier space we obtain some well-known results in what appears as a more natural way. In particular, our approach allows one to derive all the properties of the Lie algebra SU(2) (eigenfunctions, eigenvalues, raising and lowering operators) with an amazing ease. To make our paper more self-contained we would like to review some basic concepts of GCA.

2. A generalized Clifford algebra

According to Morinaga and Nono (1952) [see also Morris (1968)] a generalized Clifford algebra, $C\ell_m[n]$ is defined as a polynomial algebra generated over a field K_n (containing a n-th primitive root of one, ω) by a set of the basic elements or generators $\{e_1, e_2, ..., e_m\}$. These elements obey the following defining relations - the commutation laws

$$e_i e_j = \omega e_j e_i \qquad (i > j,\ i,\ j = 1, 2, \ldots, m) \tag{1}$$
$$e_i^n = e_0$$

where e_0 is the identity element, that is $e_i e_0 = e_0 e_i = e_i$ and $\omega = \exp(2\pi\sqrt{-1}/n)$. From Eq. (1) it immediately follows that an element inverse to e_i is

$$e_i^{-1} = e_i^{n-1}. \tag{2}$$

For the purposes of physical applications we choose a particular GCA, namely $C\ell_2[n]$ generated by two basic elements e_1 and e_2. This choice of GCA is influenced by the fact that $C\ell_2[n]$ is isomorphic to a full matrix algebra of $n \times n$ matrices over K_n (Morris, 1968) and therefore can be used as a tool in a matrix-free treatment of some problems in quantum mechanics.

A general element of $C\ell_2[n]$ is then conventionally represented as follows [for example, Morris (1967)]

$$A = \sum_i \sum_j q_{ij} e_1^i e_2^j = q_{00} e_0 + q_{01} e_2 + \ldots + q_{0,n-1} e_2^{n-1} + \ldots$$
$$+ q_{n-1,n-1} e_1^{n-1} e_2^{n-1} \tag{3}$$

where, generally speaking, q_{ij} is an arbitrary complex number. The general element A is a combination of polivectors of different valences. We denote the zero-valence term by A_0. Because any basic element e_i of the algebra can be viewed simultaneously as a linear operator, a general element A of the generalized Clifford algebra can also be viewed as a linear operator.

For the following discussion we need to find the adjoint of the product of two basic elements e_1 and e_2. According to the definition of the adjoint of a product, we obtain the following expression

$$(e_2 e_1)^\dagger = e_1^\dagger e_2^\dagger = \omega^* e_2^\dagger e_1^\dagger \tag{4}$$

where $\omega^* = \omega^{-1}$ is a complex conjugate of ω, and we use the commutation relation (1). If we assume that e_i's are unitary operators then

$$e_i^\dagger = e_i^{-1},\ i = 1, 2. \tag{5}$$

Therefore in this case Eq. (4) yields

$$(e_2 e_1)^\dagger = \omega^* e_2^{-1} e_1^{-1}. \tag{6}$$

In what follows we consider e_1 and e_2 as unitary operators.

The algebra generated by e_1 is a Cartan subalgebra of the algebra generated by e_1 and e_2. The basic elements of this subalgebra are $e_0, e_1, e_1^2, \ldots, e_1^{n-1}$. According to van der Waerden (1967), we can choose a special new basis $I_0, I_1, \ldots, I_{n-1}$ such that

$$I_i I_j = \delta_{ij} I_i \tag{7}$$

where δ_{ij} is the Kronecker symbol. The identity element e_0 is then easily expressed as the direct sum of the new basic elements I_i $(i = 0, 1, 2, \ldots, n-1)$

$$e_0 = \sum_{i=0}^{n-1} I_i. \tag{8}$$

In turn, we can expand e_1 as

$$e_1 = \sum_{i=0}^{n-1} \beta_i I_i. \tag{9}$$

From (9) and (7) it follows that

$$e_1^k = \sum_{i=0}^{n-1} \beta_i^k I_i. \tag{10}$$

Using (1) we immediately obtain that for each i

$$\beta_i^n = 1. \tag{11}$$

At the first glance one can choose each β_i as the n-th root of -1, that is ω. However if all β_i are the same then e_1 becomes a multiple of the identity element e_0. Therefore each β_i must be a different power of ω. We choose these powers in such a way that they match the index i in expression (10) in the following fashion:

$$\beta_i = \omega^{-i}. \tag{12}$$

It is obvious that now $\sum_{i=0}^{n-1} \beta_i = 0$. Upon substitution of (12) into (10) we obtain

$$e_1^k = \sum_{i=0}^{n-1} \omega^{-ik} I_i. \tag{13}$$

To find the inverse of the expansion (13) we notice that the following identity is satisfied: $\sum_{i=0}^{n-1} \omega^{i(j-k)} = n\delta_{jk}$. Using this identity and (13) we find the expansion of I_k in terms of e_1^i:

$$\frac{1}{n} \sum_{i=0}^{n-1} \omega^{ik} e_1^i = \frac{1}{n} \sum_{i=0}^{n-1} \omega^{ik} \omega^{-ij} I_j = I_k. \tag{14}$$

Comparing (13) and (14) we can see that the relationship between I_i and e_1^k is the same as that of a discrete Fourier transform and its inverse.

Because the basic elements $\mathbf{e}_1$ and $\mathbf{e}_2$ of the old basis are unitary, the new basic elements $\mathbf{I}_i$ are self-adjoint, which follows from (5) and (14):

$$\mathbf{I}_i^\dagger = \frac{1}{n}(w^{ki})^*(\mathbf{e}_1^k)^\dagger = \frac{1}{n}\sum_{k=0}^{n-1}(\omega^i)^{n-k}\mathbf{e}_1^{n-k} = \frac{1}{n}\sum_{j=0}^{n-1}\omega^{ij}\mathbf{e}_1^j = \mathbf{I}_i.$$

Here we use $(\omega^{ki})^* = \omega^{-ki} = \omega^{(n-k)i}$, $(\mathbf{e}_1^k)^\dagger = \mathbf{e}_1^{-k} = \mathbf{e}_1^{n-k}$, and introduce a new summation index $j = n - k$.

Let us consider an element $\mathbf{L}_{kj}$ defined as follows:

$$\mathbf{L}_{kj} = \mathbf{e}_2^k \mathbf{I}_j. \tag{15}$$

Note that $\mathbf{L}_{0j} = \mathbf{I}_j$. It can be easily shown that for $k \neq 0$ the elements $\mathbf{L}_{kj}$ form a set of nilpotent elements of the GCA. If we insert $j = 0$ into (15) then we obtain

$$\Psi_k = \mathbf{L}_{k0} = \mathbf{e}_2^k \mathbf{I}_0. \tag{16}$$

For the subsequent analysis the commutation rule for $\mathbf{I}_i$ and $\mathbf{e}_2^k$ is of crucial importance. This rule is given by the following

Lemma 2.1 If $\mathbf{I}_i$ are the idempotent elements of a generalized Clifford algebra $C\ell_2[n]$, and their direct sum represents the identity element of this algebra then the following identity holds true

$$\mathbf{L}_{mj} = \mathbf{e}_2^m \mathbf{I}_j = \mathbf{I}_{j+m} \mathbf{e}_2^m. \tag{17}$$

Proof: Invoking the basic relation of GCA, Eq. (1), and using (12) we write (13) as follows

$$\mathbf{e}_2^m \mathbf{I}_j = \frac{1}{n}\sum_{k=0}^{n-1}\mathbf{e}_2^m(\omega_j)^k \mathbf{e}_1^k = \frac{1}{n}\sum_{k=0}^{n-1}\omega^{jk}(\mathbf{e}_2^m \mathbf{e}_1^k).$$

On the other hand, according to the basic relation of GCA, Eq. (1), $\mathbf{e}_2^m \mathbf{e}_1^k = \omega^{mk}\mathbf{e}_1^k \mathbf{e}_2^m$. Therefore

$$\mathbf{e}_2^m \mathbf{I}_j = \frac{1}{n}\sum_{k=0}^{n-1}[\omega^{(j+m)k}\mathbf{e}_1^k]\mathbf{e}_2^m = \mathbf{I}_{j+m} \mathbf{e}_2^m. \qquad \blacksquare$$

Now we can rewrite a general element of $C\ell_2[n]$, Eq. (3) as the linear combination of the elements $\mathbf{L}_{ij}$ of the different left ideals [cf., van der Waerden (1967)]: $A = q_{ij}\mathbf{e}_1^i \mathbf{e}_2^j = \omega^{-ki}q_{ij}\mathbf{I}_k \mathbf{e}_2^j \equiv b_{kj}\mathbf{I}_k \mathbf{e}_2^j = b_{kj}\mathbf{e}_2^j \mathbf{I}_{k-j}$ where we employ the summation convention over repeated indices, use equations (13) and (17), and denote $\omega^{-ki}q_{ij} \equiv b_{kj}$. After relabeling $k - j \equiv i$, $b_{k,k-i} \equiv a_{ki}$, we obtain a representation of a general element of GCA in the new basis:

$$A = \frac{1}{n}\sum_{i=0}^{n-1}\sum_{k=0}^{n-1}a_{ki}\mathbf{e}_2^{k-i}\mathbf{I}_i \tag{18}$$

where obviously a_{ij} are complex numbers.

As we have already indicated, $Cl_2[n]$ is isomorphic to a set M of $n \times n$ matrices. To find a matrix representation of (15) we need to find matrices that represent both $\mathbf{I}_j$ $(j = 1, 2, \ldots, n)$ and $\mathbf{e}_2$. These matrices are as follows

$$\mathbf{I}_j = \begin{bmatrix} 0 & 0 & 0 & . & . & 0 \\ 0 & 0 & 0 & . & . & 0 \\ . & . & & . & . & . \\ 0 & 0 & . & 1 & . & 0 \\ . & . & & . & . & . \\ 0 & 0 & 0 & . & . & 0 \end{bmatrix} \quad \mathbf{e}_2 = \begin{bmatrix} 0 & 0 & . & . & . & 1 \\ 1 & 0 & . & . & . & 0 \\ 0 & 1 & 0 & . & . & 0 \\ 0 & 0 & 1 & 0 & . & 0 \\ . & . & . & . & . & . \\ 0 & 0 & 0 & . & 1 & 0 \end{bmatrix}$$

where the first matrix has only non-zero element at the intersection of the j-th column and the j-th row.

It is straightforward now to show that A is represented by a $n \times n$ matrix whose elements are a_{ij}. In fact, we even do not need to invoke the matrix representation, and provide it only for the reference. If we consider the product of two elements $A, B \in M$ then we get an element $C \in M$:

$$\begin{aligned} C = AB &= \sum_{i=0}^{n-1} \sum_{j=0}^{n-1} a_{ij} \mathbf{e}_2^{i-j} \mathbf{I}_j \sum_{k=0}^{n-1} \sum_{m=0}^{n-1} b_{km} \mathbf{e}_2^{k-m} \mathbf{I}_m \\ &= \sum_{m=0}^{n-1} \sum_{i=0}^{n-1} \sum_{j=0}^{n-1} a_{ij} b_{jm} \mathbf{e}_2^{i-m} \mathbf{I}_m \qquad (19) \\ &= \sum_{i=0}^{n-1} \sum_{m=0}^{n-1} c_{im} \mathbf{e}_2^{i-m} \mathbf{I}_m. \end{aligned}$$

Here we use (8) and (15). As could be expected, the numbers c_{im} are obtained from a_{ij} and b_{km} according to the same rule that governs matrix multiplication.

Now with the help of (5) and (17) we obtain from (16) an element adjoint to a general element A

$$A^\dagger = \sum_{i=0}^{n-1} \sum_{j=0}^{n-1} (a_{ij} \mathbf{e}_2^{i-j} \mathbf{I}_j) = \sum_{i=0}^{n-1} \sum_{j=0}^{n-1} a_{ji}^* \mathbf{I}_j \mathbf{e}_2^{j-i} = \sum_{i=0}^{n-1} \sum_{j=0}^{n-1} a_{ij}^* \mathbf{e}_2^{i-j} \mathbf{I}_j \qquad (20)$$

where we use the fact that $\mathbf{I}_j$ is self-adjoint. If A is also self-adjoint then from (18) and (20) follows that $a_{ij}^* = a_{ij}$, so the coefficients a_{ij} are real.

We also define the inner product of two general elements of GCA as follows:

$$<A, B> \equiv \sum_{i=0}^{n-1} \sum_{j=0}^{n-1} b_{ij}^* a_{ji}.$$

Therefore we can immediately see that the inner product $<AB>$ coincides with the trace($B^\dagger A$) if the general elements A and B are replaced by the equivalent matrices.

3. Generalized Clifford algebras and some eigenvalue problems in quantum mechanics

We write a wave function as a linear combination of left ideals of GCA [cf. Eq.(16)]:

$$\Psi = \sum_{i=0}^{n-1} \Psi_i e_2^i \mathbf{I}_0 \tag{21}$$

where Ψ_i is the wave function of the i-th stationary state. In the matrix representation expression (21) is nothing more than a square $n \times n$ matrix whose n-th column coincides with a $1 \times n$ matrix-column describing a wave function with n stationary states $\mathbf{I} = 0, 1, \ldots, n - 1$.

We will look for the operator $\mathbf{J}_+$ (a particular general element of the GCA) such that it transforms the i-th state (i.e., the state $e_2^i \mathbf{I}_0$) into a multiple of the $(i+1)$-st state. Using the expression for a general element (operator) of GCA, Eq.(16) we obtain for an arbitrary component of this operator, $a_{km} e_2^{k-m}$

$$A(e_2^i \mathbf{I}_0) = a_{km} e_2^{k-m} \mathbf{I}_m e_2^i \mathbf{I}_0 = a_{km} e_2^{k-m+i} \mathbf{I}_{m-i} \mathbf{I}_0 = a_{ki} e_2^k \mathbf{I}_0 \tag{22}$$

where we use (17), and $\mathbf{I}_{m-i} \mathbf{I}_0 = \delta_{mi} \mathbf{I}_0$. The right-hand side of (22) can be a multiple of the $(i + 1)$-st state only if $k = i + 1$. The components of the operator $\mathbf{J}_+$, that is $a_{ki} e_2^{k-i} \mathbf{I}_i$, are easily found from this result: they are $a_{i+1,i} e_2 \mathbf{I}_i$. Therefore the general expression for $\mathbf{J}_+$ is

$$\mathbf{J}_+ = \sum_{k=0}^{n-1} a_{k+1,k} e_2 \mathbf{I}_k. \tag{23}$$

If we assume that $a_{k+1,k} = a_{k,k+1} = (a_{k,k+1})^*$ then $\mathbf{J}_+$ is the usual raising operator in quantum mechanics.

We introduce another operator $\mathbf{J}_-$ which is equal to the adjoint of $\mathbf{J}_+$:

$$\mathbf{J}_- = (\mathbf{J}_+)^\dagger = (\sum_{k=0}^{n-1} a_{k+1,k} e_2 \mathbf{I}_k) = (a_{k,k+1})^* \mathbf{I}_k e_2^{-1} \tag{24}$$

where we use the unitary character of e_2 (that is $e_2^\dagger = e_2^{-1}$). Straightforward calculations show that $\mathbf{J}_-$ is a lowering operator: each of its components $(a_{k,k+1})^* \mathbf{I}_k e_2^{-1}$ changes an arbitrary state $e_2^j \mathbf{I}_0$ into $e_2^{j-1} \mathbf{I}_0$.

Let us consider the commutation relations between the operators $\mathbf{J}_+$ and $\mathbf{J}_-$ which we write in the following form

$$[\mathbf{J}_+, \mathbf{J}_-] = \beta \mathbf{J}_3 \tag{25}$$

$$[\mathbf{J}_3, \mathbf{J}] = \pm \mathbf{J}_\pm. \tag{26}$$

From the definitions of $\mathbf{J}_+$ and $\mathbf{J}_-$ it follows that their commutator $\mathbf{J}_3$ is a Hermitian operator. We assume that it can be expanded in a new basis $\mathbf{I}_i$ as follows

$$\mathbf{J}_3^\dagger \equiv \mathbf{J}_3 = \sum_{i=1}^{n} \gamma_i \mathbf{I}_i. \tag{27}$$

Here β is the structure constant, and we consider a case of all γ_i being real. We also introduce an additional assumption that $a_{k,k+1} = a_{k+1,k} = (a_{k,k+1})^*$. Note that for $\beta = 1$ equations (25) and (26) represent the governing relations for SU(2).

Upon substitution of (23), (24) into (25) and (26) we obtain with the help of (17) the following difference equations with constant coefficients

$$(a_{i,i-1})^2 - (a_{i+1,i})^2 = \gamma_i, \quad i = 0, 1, 2, \ldots, n \tag{28}$$

and

$$a_{i+1,i}(\gamma_{i+1} - \gamma_i) = a_{i+1,i}, \quad i = 0, 1, 2, \ldots, n. \tag{29}$$

Here we use $a_{0,-1} = 0$ (corresponding to the absence of a (-1)-st state), and $a_{n+1,n} = 0$ [corresponding to the absence of a $(n+1)$-st state]. Equations (29) allow the trivial solution $b_{i+1,i} = 0$ for any $0 \leq i \leq n$. This means that $\mathbf{J}_\pm = 0$. For $a_{i+1,i} \neq 0$ Equations (29) yield a set of $(n-1)$ difference equations whose solution is

$$\gamma_i = c + i \tag{30}$$

where c is the constant to be determined. Substituting (30) into (29) we obtain

$$(a_{i,i-1})^2 = \frac{i\beta}{2}(1 - 2c - i) + d, \quad i = 0, 1, 2, \ldots, n \tag{31}$$

where d is the constant which can be easily found. In fact, using $a_{0,-1} = 0$ we find from (31) that $d = 0$. Therefore the coefficients $a_{i,i-1}$ are

$$a_{i,i-1} = \sqrt{\frac{i\beta}{2}(1 - 2c - i)}. \tag{32}$$

To determine the constant c we assume that the Hermitian operator $\mathbf{J}_3$ must be traceless [in accordance with the quantum-mechanical operator $\mathbf{J}_3$, see Georgi (1982) or $\mathbf{L}_3$, see Landau (1980)]

$$\sum_{i=0}^{n} \gamma_i = 0. \tag{33}$$

Using (30) in (33) we find the value of the constant c:

$$c = -\frac{n}{2} \equiv -L. \tag{34}$$

If we denote

$$i - L \equiv M \tag{35}$$

then substitution of (34) in (32) yields the eigenvalues of the operators $\mathbf{J}_+$ and $\mathbf{J}_-$

$$\sqrt{\frac{\beta}{2}(L + M)(L - M + 1)}. \tag{36}$$

Using (32) in (28) we obtain the eigenvalues of $\mathbf{J}_3$: $\gamma_i = i - L = M$, $i = 0, 1, \ldots, n$. It is obvious that $-L \leq \gamma_i \leq +L$.

By varying parameter β we can obtain various eigenvalues for $\mathbf{J}_+$ and $\mathbf{J}_-$. In particular, if $\beta = 1$ then equation (36) yields the well-known expression for the eigenvalues of the raising and lowering operators in SU(2), Georgi (1982). If, on the other hand, $\beta = 2$ then (36) yields the same operators in the theory of angular momenta, Landau (1980). The respective Casimir operators follow from (31), and (34) – (36):

$$\mathbf{J}^2 = \frac{\mathbf{J}_+\mathbf{J}_- + \mathbf{J}_-\mathbf{J}_+}{2} + \mathbf{J}_3^2 = \frac{\beta}{2}L(L+1) + M^2(1 - \frac{\beta}{2})$$

$$= \begin{cases} L(L+1), & \beta = 2 \\ \frac{1}{2}[L(L+1) + M^2], & \beta = 1. \end{cases}$$

Another eigenvalue problem is related to the unitary evolution operator U for a universal pure-state quantum computer, Lloyd (1994). Its repetitive action on a computational state, $(|b>)$, [the notation introduced by Dirac (1970)] may lead to two outcomes: 1) either $(|b>)$ never returns to itself for any number of iterations, or 2)

$$U^n|b> = \exp(j\phi)|b>, \tag{37}$$

where $j = \sqrt{-1}$, and n is a positive integer. We concentrate on the second outcome, and rewrite (37) as follows

$$[\exp(j\frac{2\pi - \phi}{n})]^n U^n|b> = 1|b>.$$

On the other hand from the definition of $\mathbf{e}_1$ we get

$$e_1^n|b> = \sum_{i=0}^{n-1} \omega^{in}\mathbf{I}_i|b> \equiv (\sum_{i=0}^{n-1} \omega^i\mathbf{I}_i)^n|b> = 1|b>.$$

Comparing the above two expressions we obtain

$$[\exp(j\frac{2\pi - \phi}{n})U]^n|b> = (\sum_{i=0}^{n-1} \omega^i\mathbf{I}_i)^n|b>. \tag{38}$$

From (38) it immediately follows that

$$U|b> = \sum_{i=1}^{n} \exp(j\frac{\phi - 2\pi}{n})\omega_i\mathbf{I}_i|b> \tag{39}$$

which means that eigenvalues of the unitary operator U are

$$\lambda_i = \exp[\frac{2\pi(i - 1) + \phi}{n}] \tag{40}$$

where $i = 1, 2, \ldots, n$.

Now we can easily find the respective normalized eigenvectors of U. In fact, let us introduce the state which we denote as $|\alpha_i>$:

$$|\alpha_i> \equiv \sqrt{n}\, \mathbf{I}_i\, |b>. \tag{41}$$

From (39) we obtain the expansion of the operator U^k in terms of these states

$$U^k|b> = \frac{1}{\sqrt{n}} \sum_{i=1}^{n} \exp(jk\frac{2\pi(i-1)+\phi}{n})|\alpha_i>.$$

Using (14) we rewrite expression (41)

$$|\alpha_i> = \frac{1}{\sqrt{n}} \sum_{i=0}^{n-1} \omega^{ik} \mathbf{e}_1^k |b>.$$

Taking into account

$$U^k|b> = \exp\left[\frac{jk}{n}(\phi-2\pi)\right] \sum_{i=0}^{n-1} \omega^{ik} \mathbf{I}_i |b> = \exp\left[\frac{jk}{n}(\phi-2\pi)\right] \mathbf{e}_1^k |b>$$

we can see that $\omega_i^k \mathbf{e}_1^k |b> = \exp[\frac{jk}{n}(2\pi-\phi)]U^k|b>$. As a result we express $|\alpha_i>$ as follows, Lloyd (1994)

$$|\alpha_i> = \frac{1}{\sqrt{n}} \sum_{k=0}^{n-1} \exp\left[\frac{jk}{n}2\pi(1+i)-\phi\right] U^k|b>.$$

In conclusion we provide the expressions for the Dirac matrices written in our new basis $\mathbf{I}_k$. Using the results of Ross (1986) we get

$$\gamma_4 = \sum_{k=1}^{n} \frac{\mathbf{I}_k}{2}[\omega^{k+1/2} + \omega^{-(k+1/2)}],$$

$$\gamma_3 = \sum_{k=1}^{n} \frac{\mathbf{I}_k}{2}[(\omega^{k-1} - \omega^{-(k-1)}) - (\omega^k - \omega^{-k})]\mathbf{e}_2^2$$

$$\gamma_2 = \sum_{k=1}^{n} \frac{\mathbf{I}_k}{2}[(\omega^{k+1} + \omega^{-(k+1)})\mathbf{e}_2 - (\omega^k + \omega^{-k})\mathbf{e}_2^{-1}]$$

$$\gamma_1 = \sum_{k=1}^{n} \frac{\mathbf{I}_k}{2}[(\omega^k - \omega^{-k})\mathbf{e}_2 - (\omega^{k+1} - \omega^{-(k+1)})\mathbf{e}_2^{-1}]$$

4. Conclusions

We have shown that the generalized Clifford algebra (GCA) can be formulated with the help of an Euclidean carrier space which simplifies an application of GCA to problems of physical interest. As an example, we have considered formulation of the commutation relations in SU(2) (whose importance for a "natural" description of quantum mechanics was emphasized by Weyl) in terms of this new carrier space. As a result, the commutation relations are readily reduced to two difference equations with constant coefficients. Their solution recovers the eigenvalues of the raising (lowering) operators, and the Casimir operator in a very straightforward manner. As another example, we find the eigenvalues of a cyclic operator that plays an important role in the theory of universal quantum computers. In both of these applications we have achieved our goals without resorting to any particular matrix representation. Because of a simplicity of this approach it seems promising to apply a GCA (based on the proposed Euclidean carrier space) to the study of the higher unitary groups SU(N).

References

P. Dirac: 1970, *The Principles of Quantum Mechanics*, Clarendon Press, London.

D. Hestenes: 1966, *Space-Time Algebra*, Gordon & Breach, New York-London-Paris.

H. Weyl: 1950, *The Theory of Groups and Quantum Mechanics*, Dover, New York.

H. Georgi: 1982, *Lie Algebras in Particle Physics*, Benjamin-Cumming Publishing, Reading, Mass.

L. Landau and E. Lifshitz, *Quantum Mechanics (Non-Relativistic Theory)*, Pergamon Press, Oxford, 1980.

S. Lloyd: 1994, 'Necessary and sufficient conditions for quantum computations', *Journal of Modern Optics*, **41**, pp. 2503–2520.

K. Morinaga and T. Nono: 1952, 'On the linearization of a form of higher degree and its representation', *J. Sci. Hiroshima Univ. Ser. A*, **16**, pp. 13–41.

A. O. Morris: 1967, 'On a generalized Clifford algebra', *Quart. J. Math.* (Oxford), **18**, pp. 7–12.

A. O. Morris: 1968, 'On a generalized Clifford algebra. II', *Quart. J. Math.* (Oxford), **19**, pp. 289–299.

A. Ramakrishnan: 1971, *L-Matrix Theory and the Grammar of Dirac Matrices*, Tata-McGraw Hill, Delhi.

P.K. Raševskiĭ: 1955, 'Teoriya spinorov', *Uspekhi Mat. Nauk*, **10**, pp. 3–110 (in Russian) and 1957, 'The Theory of Spinors', *Transl. A.M.S.*, 6(2), pp. 1-110.

M. Riesz: 1958, *Clifford Numbers and Spinors*, Lecture Series No. 38. The Institute for Fluid Dynamics and Applied Mathematics, Univ. of Maryland. Reprinted by Kluwer, 1993.

M. Ross: 1986, 'Representation-free calculations in relativistic quantum mechanics', in *Clifford Algebras and their Application to Mathematical Physics, Proceedings of the NATO and SERC Workshop*, eds. J. Chisholm and A. Common, Reidel Publishing, pp. 347–352.

B.L. van der Waerden: 1967, *Algebra*, Springer-Verlag, Berlin, Heidelberg, New York.

H. Weyl: 1950, *The Theory of Groups and Quantum Mechanics*, Dover, New York.

K. Yamazaki: 1964, 'On projective representations and ring extensions of finite groups', *J. Fac. Sci. Univ. Tokyo*, **I10**, pp. 147–195.

VECTOR CONTINUED FRACTION ALGORITHMS

D. E. ROBERTS
Department of Mathematics
Napier University
219 Colinton Road
Edinburgh, EH14 1DJ
e-mail: davidr@maths.napier.ac.uk

Abstract. We consider the construction of rational approximations to given power series whose coefficients are vectors. The approximants are in the form of vector-valued continued fractions which may be used to obtain vector Padé approximants using recurrence relations. Algorithms for the determination of the vector elements of these fractions have been established using Clifford algebras. We devise new algorithms based on these which involve operations on vectors and scalars only – a desirable characteristic for computations involving vectors of large dimension. As a consequence, we are able to form new expressions for the numerator and denominator polynomials of these approximants as products of vectors, thus retaining their Clifford nature.

Key words: Rational approximants, vector-valued functions, continued fractions, Viskovatov, modified Euclidean algorithm, Clifford algebra.

1. Introduction

The use of Clifford numbers in the context of vector-valued rational approximants originated in the work of Wynn (1963, 1968) and McCleod (1972), which was motivated by attempts to accelerate the convergence of vector sequences arising from various numerical methods. Interest in these approximants was revived in the eighties by Graves-Morris using an axiomatic treatment not based on Clifford algebras. More recently the author has illustrated some of the advantages of these algebras in the construction of rational approximants to vector-valued functions. The algebraic approach allows a development of the vector theory which follows that of the scalar, so that proofs of theorems and algorithms valid in the scalar case may be carried over to the vector version. For an introduction to the usual theory of Padé approximants and some generalizations the reader is referred to the books by Baker *et al.*, and to the author's 1994 paper for more information on the background to the use of Clifford algebras in the context of vectors.

However, in applications of vector Padé approximants, the dimension of the vectors can be quite large – in some instances of several thousand – thus rendering a matrix representation impractical [N.B. Matrix representations of Cl_n involve dimensions of the order $2^{n/2}$]. Hence, a method of construction is sought which allows the necessary operations to be performed using scalars and vectors only. It was shown by the author in 1992 how this could be accomplished by resorting to (vector) continued fractions and their recurrence relations, provided the vector elements of the continued fraction are known.

In particular, we focus attention on two related algorithms – viz. the Viskovatov and Modified Euclidean – which are used to derive the elements of the *corresponding* continued fraction representations of the given function – see equations (3.5) and (4.1). In 1994 Graves-Morris and the author presented the second algorithm for vectors using Clifford algebras to generalize arguments employed in the scalar case. Viskovatov's algorithm is considered in this paper coupled with a demonstration of how each algorithm may be implemented employing vectors and scalars only, by taking advantage of the algebraic structure of $C\ell_n$.

2. Vector Padé approximants

We first of all recount here various definitions and results for the convenience of the reader, who is referred to Roberts 1990, 1993, and Graves-Morris *et al.* 1994 for more details. Consider a vector-valued function, $\mathbf{f} \colon \mathbb{C} \to \mathbb{C}^n$, which has a MacLaurin series expansion

$$\mathbf{f}(z) = \mathbf{c}_0 + z\mathbf{c}_1 + z^2\mathbf{c}_2 + \ldots, \quad z \in \mathbb{C}, \quad \mathbf{c}_i \in \mathbb{R}^n, \quad i = 0, 1, \ldots \quad (2.1)$$

converging in some neighborhood of the origin. In this paper we restrict attention to *real* vectors, which is the more common situation in practical applications. However, for a discussion of the case of complex vector coefficients the reader is referred to the author's 1995 paper and to Graves-Morris *et al.*, 1994. The right-handed $[l/m]$ vector Padé approximant (VPA) to $\mathbf{f}(z)$, if it exists, is defined by

$$[l/m](z) := p^{[l/m]}(z)[q^{[l/m]}(z)]^{-1}$$

for which

$$[l/m](z) - \mathbf{f}(z) = O(z^{l+m+1}) \quad (2.2)$$

and

$$q^{[l/m]}(0) = \mathbf{e}_0 \quad (2.3)$$

where $p^{[l/m]}(z)$ and $q^{[l/m]}(z)$ are polynomials in $z \in \mathbb{C}$ over $C\ell_n$ of maximum degrees l and m respectively. We are following Baker *et al.* by insisting on condition (2.3). The left-handed version is defined in a similar manner.

Since $q^{[l/m]}(0) \neq 0$ then, on multiplying (2.2) by $q^{[l/m]}(z)$ from the right, we obtain the order condition

$$p^{[l/m]}(z) - \mathbf{f}(z)q^{[l/m]}(z) = O(z^{l+m+1}) \quad (2.4)$$

thus yielding a system of linear equations in the unknown polynomial coefficients.

The resulting approximants share many of the properties enjoyed by the scalar version. In particular we require the following:

Uniqueness: If the $[l/m]$ approximant to $\mathbf{f}(z)$ exists then it is unique. This property implies the identification of the left and right handed approximants.

Duality: If $\mathbf{f}(0) \neq 0$ then, using an obvious notation, in which $\mathbf{g}(z) = [\mathbf{f}(z)]^{-1}$,

$$[l/m]_{\mathbf{f}}(z) \equiv \{[m/l]_{\mathbf{g}}(z)\}^{-1}$$

provided either approximant exists.

To see this, assume that $[m/l]_\mathbf{g}$ exists and is given by

$$p_\mathbf{g}^{[m/l]}(z)[q_\mathbf{g}^{[m/l]}(z)]^{-1}$$

where $p_\mathbf{g}^{[m/l]}(z)$ and $q_\mathbf{g}^{[m/l]}(z)$ are polynomials of maximum degrees m and l respectively. We note that none of $\mathbf{f}(0), q_\mathbf{g}^{[m/l]}(0)$ or $p_\mathbf{g}^{[m/l]}(0)$ vanish, and that $p_\mathbf{g}^{[m/l]}(z)$ is invertible, if constructed, for example, by using either of the continued fractions of this paper. Hence,

$$
\begin{aligned}
\mathbf{f}(z) &- q_\mathbf{g}^{[m/l]}(z)[p_\mathbf{g}^{[m/l]}(z)]^{-1} \\
&= \mathbf{f}(z)\{p_\mathbf{g}^{[m/l]}(z)[q_\mathbf{g}^{[m/l]}(z)]^{-1} - \mathbf{g}(z)\}q_\mathbf{g}^{[m/l]}(z)[p_\mathbf{g}^{[m/l]}(z)]^{-1} \\
&= O(z^{l+m+1})
\end{aligned}
$$

To satisfy the Baker condition we divide each polynomial by $p_\mathbf{g}^{[m/l]}(0)$ from the right. This, together with the uniqueness property, is enough to guarantee that the $[l/m]$ vector Padé approximant to $\mathbf{f}(z)$ is given by $\{[m/l]_\mathbf{g}(z)\}^{-1}$. To complete the proof the above argument may be applied if $[l/m]_\mathbf{f}$ exists.

3. Corresponding vector continued fractions

In this section we consider the problem of expressing the given power series in the form of a continued fraction – as in equation (3.5). A common approach in the scalar context is Viskovatov's algorithm which dates from 1803-6. We develop this method for the non-degenerate case and, in the course of doing so, demonstrate the existence of the inverses necessary for its implementation.

Viskovatov's algorithm as formulated by Baker $et\ al.$ may be adapted for non-commuting elements by constructing the identity

$$
\left.
\begin{aligned}
&\left(\textstyle\sum_{i=0}^{\infty} d_{k,i}z^i\right)\left(\textstyle\sum_{i=0}^{\infty} d_{k+1,i}z^i\right)^{-1} = \\
&d_{k,0}(d_{k+1,0})^{-1} + z[(\textstyle\sum_{i=0}^{\infty} d_{k+1,i}z^i)(\textstyle\sum_{i=0}^{\infty} d_{k+2,i}z^i)^{-1}]^{-1}
\end{aligned}
\right\}
\tag{3.1}
$$

where

$$d_{k+2,i} := d_{k,i+1} - [d_{k,0}(d_{k+1,0})^{-1}]d_{k+1,i+1} \quad \text{for } k,i = 0,1,\ldots. \tag{3.2}$$

The application to the vector-valued power series (2.1) is achieved by setting

$$
\left.
\begin{aligned}
d_{0,i} &:= \mathbf{c}_i, & i = 0,1,\ldots \\
d_{1,0} &:= \mathbf{e}_0, \quad d_{1,i} := 0, & i = 1,2,\ldots
\end{aligned}
\right\}
\tag{3.3}
$$

On defining

$$\boldsymbol{\pi}_k := d_{k,0}(d_{k+1,0})^{-1} \tag{3.4}$$

and using (3.1) repeatedly we obtain a continued fraction expansion of $\mathbf{f}(z)$,

$$\mathbf{f}(z) := \boldsymbol{\pi}_0 + z[\boldsymbol{\pi}_1 + z[\boldsymbol{\pi}_2 + \cdots]^{-1}]^{-1} \tag{3.5}$$

The first few elements are

$$\pi_0 := c_0, \ \pi_1 := c_1^{-1}, \ \pi_2 := -c_1 c_2^{-1} c_1, \ \pi_3 := [c_1 c_2^{-1} c_3 c_2^{-1} c_1 - c_1]^{-1}, \dots$$

which are all vectors in $\mathbb{R}^n$. However, we also require expressions for the $d_{k,i}$, viz.

$$d_{2,i} := c_{i+1}, \ d_{3,i} := -c_1^{-1} c_{i+2}, \ d_{4,i} := c_{i+2} - c_1 c_2^{-1} c_{i+3}, \dots$$

which become increasingly more complicated. In order to develop a version of this algorithm which may be implemented using vectors and scalars only we proceed as follows. Define

$$S_k(z) := \sum_{i=0}^{\infty} d_{k,i} z^i$$

so that (3.2) and (3.4) become

$$S_{k+2}(z) = \frac{1}{z}[S_k(z) - \pi_k S_{k+1}(z)] \tag{3.6}$$

$$\pi_k = S_k(0)[S_{k+1}(0)]^{-1} \tag{3.7}$$

while the identity (3.1) now reads

$$S_k(z)[S_{k+1}(z)]^{-1} = \pi_k + z[S_{k+1}(z)[S_{k+2}(z)]^{-1}]^{-1}]^{-1} \tag{3.8}$$

with

$$S_0(z) := \mathbf{f}(z) \ \text{and} \ S_1(z) := \mathbf{e}_0 \tag{3.9}$$

replacing the initialization (3.3). If we further define

$$\mathbf{V}_{k+1}(z) := S_k(z)\widetilde{S_{k+1}}(z) \ \text{and} \ u_k(z) := S_k(z)\widetilde{S_k}(z) \ \text{for} \ k = 0,1,\dots \tag{3.10}$$

where the tilde denotes the usual reverse anti-automorphism, then it is straight-forward to prove by induction that $\mathbf{V}_{k+1}(z)$ is a real analytic vector function and $u_k(z)$ a real analytic function for $k = 0,1,\dots$ In fact, by considering $S_k(z)\widetilde{S_{k+1}}(z)$ and $S_{k+1}(z)\widetilde{S_{k+1}}(z)$, using (3.6) and (3.9) we may obtain

$$\left. \begin{array}{l} \mathbf{V}_{k+2}(z) = z^{-1}[\mathbf{V}_{k+1}(z) - \pi_k u_{k+1}(z)] \\ u_{k+2}(z) = z^{-2}[u_k(z) - 2\pi_k \cdot \mathbf{V}_{k+1}(z) + (\pi_k \cdot \pi_k)u_{k+1}(z)] \end{array} \right\} \tag{3.11}$$

with the initializations

$$\mathbf{V}_1(z) := \mathbf{f}(z) \ \text{and} \ u_0(z) := \mathbf{f}(z) \cdot \mathbf{f}(z), \ u_1(z) := 1. \tag{3.12}$$

We then have

$$\pi_k = S_k(0)[S_{k+1}(0)]^{-1} = \frac{\mathbf{V}_{k+1}(0)}{u_{k+1}(0)} \tag{3.13}$$

which is a vector in $\mathbb{R}^n$. In this section we assume non-degeneracy i.e. that $\mathbf{V}_k(0)$ is non-null for $k = 1,2,\dots$ Indeed, $[S_{k+1}(z)]^{-1}$ exists as a power series if $S_{k+1}(0) \in \Gamma_n$ which may be proved by induction using the definition of $\mathbf{V}_{k+1}(0)$ and the above assumption. We may then also conclude that $u_{k+1}(0)$ is non-zero which ensures the validity of (3.13).

In summary, the recurrence relations (3.11) with the initializations (3.12) form a version of Viskovatov's algorithm using only vectors and scalars, allowing the determination of the continued fraction elements of (3.5). We now prove that this fraction *corresponds* to the given power series, i.e. that the kth convergent

$$\mathbf{C}_k(z) := \boldsymbol{\pi}_0 + z[\boldsymbol{\pi}_1 + z[\boldsymbol{\pi}_2 + \cdots + z[\boldsymbol{\pi}_k]^{-1} \cdots]^{-1}]^{-1}$$

satisfies the order condition

$$\mathbf{f}(z) - \mathbf{C}_k(z) = O(z^{k+1}). \tag{3.14}$$

From Roberts (1990), we have

$$\mathbf{C}_k(z) = p_k(z)[q_k(z)]^{-1} \tag{3.15}$$

where the polynomials $p_k(z), q_k(z)$, over $C\ell_n$, satisfy the recurrence relations

$$\left. \begin{array}{ll} p_k(z) := p_{k-1}(z)\boldsymbol{\pi}_k + zp_{k-2}(z), & p_{-1}(z) := \mathbf{e}_0, \quad p_0(z) := \boldsymbol{\pi}_0 \\ q_k(z) := q_{k-1}(z)\boldsymbol{\pi}_k + zq_{k-2}(z), & q_{-1}(z) := 0, \quad q_0(z) := \mathbf{e}_0 \end{array} \right\} \tag{3.16}$$

It is straightforward to prove the following by induction

$$\mathbf{f}(z)q_k(z) - p_k(z) = (-1)^k z^{k+1}\widetilde{S_{k+2}}(z). \tag{3.17}$$

Assuming non-degeneracy we observe that (3.14) is then satisfied, since

$$q_k(0) = q_{k-1}(0)\boldsymbol{\pi}_k = \boldsymbol{\pi}_1\boldsymbol{\pi}_2\boldsymbol{\pi}_3 \cdots \boldsymbol{\pi}_k \neq 0$$

using (3.16). We may, in fact, go further and obtain the leading term on the right hand side of the order condition as follows. From (3.13) we have

$$S_{k+2}(0) = [\boldsymbol{\pi}_{k+1}]^{-1}S_{k+1}(0) = [\boldsymbol{\pi}_{k+1}]^{-1}[\boldsymbol{\pi}_k]^{-1} \cdots [\boldsymbol{\pi}_1]^{-1} = [\boldsymbol{\pi}_1\boldsymbol{\pi}_2\boldsymbol{\pi}_3 \cdots \boldsymbol{\pi}_{k+1}]^{-1}$$

since $S_1(0) = \mathbf{e}_0$. Therefore,

$$\mathbf{f}(z)q_k(z) - p_k(z) = (-1)^k z^{k+1}[\boldsymbol{\pi}_{k+1}\boldsymbol{\pi}_k \cdots \boldsymbol{\pi}_2\boldsymbol{\pi}_1]^{-1} + O(z^{k+2})$$

from which we obtain the symmetric result for the order condition (3.14)

$$\mathbf{f}(z) - p_k(z)[q_k(z)]^{-1} = (-1)^k z^{k+1}[q_k(0)\boldsymbol{\pi}_{k+1}\widetilde{q_k}(0)]^{-1} + O(z^{k+2})$$

$$= (-1)^k z^{k+1}[\boldsymbol{\pi}_1]^{-1}[\boldsymbol{\pi}_2]^{-1} \cdots [\boldsymbol{\pi}_k]^{-1}[\boldsymbol{\pi}_{k+1}]^{-1}[\boldsymbol{\pi}_k]^{-1} \cdots [\boldsymbol{\pi}_1]^{-1} + O(z^{k+2}).$$

The successive convergents of (3.5) yield the staircase sequence of approximants $[0/0], [1/0], [1/1], [2/1], \ldots$ of the vector Padé table.

$$[0/0] \quad [1/0] \quad [2/0] \quad \cdots$$
$$[0/1] \quad [1/1] \quad [2/1] \quad \cdots$$
$$[0/2] \quad [1/2] \quad [2/2] \quad \cdots$$
$$\vdots \qquad \vdots \qquad \vdots$$

Fig. 1.   Part of the Vector Padé Table

4. The modified Euclidean algorithm

The Viskovatov algorithm may be adapted to cope with degeneracies i.e. when, $\mathbf{V}_k(0)$ and hence $S_k(0)$ vanish. However, we consider a closely related method which is *reliable* – an algorithm is said to be reliable if, when the approximant does not exist, it detects the degeneracy [from Baker *et al.*]. This is the modified Euclidean algorithm as presented by Graves-Morris and Roberts for the vector case and it generates *diagonal* approximants. Here, we simply state this algorithm, which employs a different definition of $S_k(z)$ for positive k from that of Viskovatov, and refer the reader to the aforementioned paper for a detailed discussion and proof, which broadly uses the ideas of the previous section.

The aim is to generate a continued fraction which corresponds to the power series of $\mathbf{f}(z)$:

$$\mathbf{f}(z) = \boldsymbol{\pi}_0(z) + z^{\mu_1}[\boldsymbol{\pi}_1(z) + z^{\mu_2}[\boldsymbol{\pi}_2(z) + \cdots\cdots]^{-1}]^{-1} \tag{4.1}$$

in which each $\boldsymbol{\pi}_i(z)$ is a vector-valued polynomial of degree ν_i and each μ_i is a positive integer. This is achieved by using (3.9) and repeatedly applying:

$$S_k(z)[S_{k+1}(z)]^{-1} := \boldsymbol{\pi}_k(z) + z^{\mu_{k+1}}[S_{k+1}(z)[S_{k+2}(z)]^{-1}]^{-1},$$

in which the integers μ_k and the polynomials $\boldsymbol{\pi}_k(z)$ are provided by the modified Euclidean algorithm as follows.

We start by defining the quantities:

$$\left.\begin{array}{ll} \boldsymbol{\pi}_0 := S_0(0)S_1(0)^{-1} - \mathbf{f}(0), & \nu_0 := 0 \\ \mu_1 := O(S_0(z) - \boldsymbol{\pi}_0 S_1(z)), & \nu_1 := \mu_1 \end{array}\right\} \tag{4.2}$$

Then the recurrence scheme is implemented:

$$\left.\begin{array}{l} S_{k+1}(z) := z^{-\mu_k}[S_{k-1}(z) - \boldsymbol{\pi}_{k-1}S_k(z)] \\ \boldsymbol{\pi}_k(z) := [S_k(z)S_{k+1}(z)^{-1}]_0^{\nu_k}, \end{array}\right\} \quad k := 1, 2, \ldots \tag{4.3}$$

$$\left.\begin{array}{l} \mu_{k+1} := O(S_k(z) - \boldsymbol{\pi}_k(z)S_{k+1}(z)) \\ \nu_{k+1} := \mu_{k+1} - \nu_k \end{array}\right\} \quad k := 1, 2, \ldots \tag{4.4}$$

employing the Nuttal notation for the Maclaurin section:

$$[\phi(z)]_0^k := \sum_{i=0}^{k} \phi_i z^i.$$

As a consequence of this construction $S_k(0) \in \boldsymbol{\Gamma}_n$ for $k \geq 1$.

We implement this algorithm in such a way as to require scalar and vector functions only, by following the approach outlined in the previous section. Using the definitions (3.10) it may be demonstrated that, as before, $\mathbf{V}_k(z) \in \mathbb{C}^n$ and that $u_k(z) \in \mathbb{C}$. The first of the recurrence relations (4.3) is replaced by

$$\left. \begin{aligned} \mathbf{V}_{k+2}(z) &= z^{-\mu_{k+1}}[\mathbf{V}_{k+1}(z) - \boldsymbol{\pi}_k(z)u_{k+1}(z)] \\ u_{k+2}(z) &= z^{-2\mu_{k+1}}[u_k(z) - 2\boldsymbol{\pi}_k(z) \cdot \mathbf{V}_{k+1}(z) + \boldsymbol{\pi}_k(z) \cdot \boldsymbol{\pi}_k(z)u_{k+1}(z)] \end{aligned} \right\} \quad (4.5)$$

while the second becomes

$$\boldsymbol{\pi}_k(z) := [\mathbf{V}_{k+1}(z)/u_{k+1}(z)]_0^{\nu_k}. \quad (4.6)$$

We note that synthetic division is not necessary to calculate $\boldsymbol{\pi}_k(z)$. For, if we write

$$\begin{aligned} u_{k+1}(z) &= \gamma_0 + \gamma_1 z + \gamma_2 z^2 + \cdots, \\ \mathbf{V}_{k+1}(z) &= \boldsymbol{\beta}_0 + \boldsymbol{\beta}_1 z + \boldsymbol{\beta}_2 z^2 + \cdots, \quad \text{and} \\ \boldsymbol{\pi}_k(z) &= \boldsymbol{\alpha}_0 + \boldsymbol{\alpha}_1 z + \boldsymbol{\alpha}_2 z^2 + \cdots + \boldsymbol{\alpha}_{\nu_k} z^{\nu_k}, \end{aligned}$$

where $\boldsymbol{\alpha}_i, \boldsymbol{\beta}_i \in \mathbb{R}^n$ and $\gamma_i \in \mathbb{R}$ for $i = 0, 1, \ldots$, then we obtain

$$\begin{aligned} (\gamma_0 + \gamma_1 z + \gamma_2 z^2 + \cdots)(\boldsymbol{\alpha}_0 + \boldsymbol{\alpha}_1 z + \boldsymbol{\alpha}_2 z^2 + \cdots + \boldsymbol{\alpha}_{\nu_k} z^{\nu_k}) \\ = \boldsymbol{\beta}_0 + \boldsymbol{\beta}_1 z + \boldsymbol{\beta}_2 z^2 + \cdots + O(z^{\nu_k+1}) \end{aligned}$$

since by construction, $S_{k+1}(0)$, and so γ_0 and $\boldsymbol{\beta}_0$, do not vanish. By comparing coefficients of powers of z, we may derive a triangular set of equations for the coefficients of $\boldsymbol{\pi}_k(z)$, which may be solved by forward substitution to yield

$$\boldsymbol{\alpha}_0 = \frac{\boldsymbol{\beta}_0}{\gamma_0}, \quad \boldsymbol{\alpha}_i = \frac{1}{\gamma_0}[\boldsymbol{\beta}_i - \sum_{j=0}^{i-1} \boldsymbol{\alpha}_j \gamma_{i-j}], \quad i = 1, 2, \ldots, \nu_k.$$

The initializations become

$$\left. \begin{aligned} \mathbf{V}_1(z) &:= \mathbf{f}(z) \quad \text{and} & u_0(z) &:= \mathbf{f}(z) \cdot \mathbf{f}(z), \quad u_1(z) := 1 \\ \boldsymbol{\pi}_0 &:= \mathbf{f}(0) & \nu_0 &:= 0 \\ \mu_1 &:= O[\mathbf{V}_1(z) - \boldsymbol{\pi}_0(z)u_1(z)] & \nu_1 &:= \mu_1 \end{aligned} \right\} \quad (4.7)$$

The successive convergents of (4.1), $\mathbf{C}_k(z)$, are the $[\tau_k/\tau_k]$ vector Padé approximants of $\mathbf{f}(z)$ (correspondence), where $\tau_k := \sum_{i=0}^{k} \nu_i$. In fact, the following order condition is satisfied ($\nu_{k+1} \geq 1$):

$$\mathbf{f}(z) - \mathbf{C}_k(z) = O(z^{2\tau_k+\nu_{k+1}}).$$

If $\nu_{k+1} > 1$ this implies the existence of a non-trivial *Block Structure* – i.e. there are square regions of the vector Padé table inside of which the elements are either identical or do not exist – Graves-Morris *et al.* 1994. This is exactly the situation in the scalar theory, a result that dates from the work of Padé in the 1890's.

In the case of non-degeneracy the blocks are trivial, i.e. of unit size, and we have $\tau_k = k$ since

$$\nu_0 = 0, \quad \nu_1 = \mu_1 = 1, \quad \nu_k = 1, \quad \mu_k = 2 \quad \text{for} \quad k \geq 2$$

so that

$$\mathbf{f}(z) = \boldsymbol{\pi}_0 + z[\boldsymbol{\pi}_1(z) + z^2[\boldsymbol{\pi}_2(z) + z^2[\boldsymbol{\pi}_3(z)\cdots\cdots]^{-1}]^{-1}]^{-1}$$

where each $\boldsymbol{\pi}_k(z)$, $k = 1, 2, \ldots$ is a linear polynomial with real vector coefficients. Hence, in this case, the $[m/m]$ diagonal approximant is given, not by the $2m^{\text{th}}$ convergent as in Viskovatov's approach, but by $\mathbf{C}_m(z)$, which requires fewer applications of the recurrence relations.

5. Construction of approximants

Having determined the elements of a corresponding continued fraction, any entry in the Padé table, if it exists, may now be constructed. For $l > m$, the $[l/m]$ approximant is given by:

$$p^{[l/m]}(z)[q^{[l/m]}(z)]^{-1} = \mathbf{c}_0 + \mathbf{c}_1 z + \cdots + z^{l-m-1}\mathbf{c}_{l-m-1} + z^{l-m}[m/m]_{\mathbf{h}}(z) \quad (5.1)$$

in which $[m/m]_{\mathbf{h}}(z)$ is the diagonal approximant to

$$\mathbf{h}(z) := \mathbf{c}_{l-m} + \mathbf{c}_{l-m+1}z + \cdots + z^{2m}\mathbf{c}_{l+m} + \cdots. \quad (5.2)$$

It is straightforward to show that the Padé order condition (2.2) is satisfied. Furthermore, use of the forward recurrence relations implies that

$$q^{[l/m]}(0) = \boldsymbol{\pi}_1(0)\,\boldsymbol{\pi}_2(0)\cdots\boldsymbol{\pi}_k(0) \in \boldsymbol{\Gamma}_n,$$

for an appropriate value of k, so that it is possible to meet the Baker condition. We point out that only the first $2m + 1$ coefficients of $\mathbf{h}(z)$ are required – i.e. those actually quoted above. The uniqueness property guarantees that (5.1) is *the* $[l/m]$ entry.

In order to construct approximants $[l/m]$ where $l < m$ we consider the inverse series for $\mathbf{f}(z)$, if necessary after extracting an appropriate power of z. To be precise if $\mathbf{f}(z) = z^\lambda \mathbf{f}_1(z)$ with $\mathbf{f}_1(0) \neq \mathbf{0}$, we form the $[m/l]$ approximant to $\mathbf{g}(z) := [\mathbf{f}_1(z)]^{-1}$ using the procedure outlined above. The duality property then allows us to conclude that

$$[l + \lambda/m]_{\mathbf{f}}(z) \equiv z^\lambda\{[m/l]_{\mathbf{g}}(z)\}^{-1}$$

i.e. there is a block of size λ in the top left corner of the Padé table.

Each of the diagonal approximants may be constructed using either backward recurrence relations or a forward version similar to (3.16). Indeed, we point out that, in the spirit of this paper, it is possible to render these relations into a form which does not involve general Clifford elements, but only vectors and scalars – Roberts (1992). This results in objects identical to those of Graves-Morris; *viz.* a numerator vector of polynomials of maximum degree $l+m$ and a scalar denominator polynomial of degree $2m$ rather than l and m, respectively. However, there is another approach to building an approximant which retains, not only the desired degrees, but also the Clifford character of the numerator and denominator polynomials while still only using vectors. If we define

$$\begin{array}{ll}
\mathbf{u}_j(z) := [p_{j-1}(z)]^{-1}p_j(z), & j \geq 0 \\
\mathbf{v}_j(z) := [q_{j-1}(z)]^{-1}q_j(z), & j \geq 1
\end{array} \right\} \quad (5.3)$$

then the recurrence relations (3.16) may be adapted for the Modified Euclidean algorithm to yield, for $l > m$,

$$\mathbf{u}_0(z) = \sum_{i=0}^{l-m} \mathbf{c}_i z^i, \quad \mathbf{u}_1(z) = \boldsymbol{\pi}_1(z) + z^{l-m+\mu_1}[\mathbf{u}_0(z)]^{-1}$$

and

$$\mathbf{u}_j(z) = \boldsymbol{\pi}_j(z) + z^{\mu_j}[\mathbf{u}_{j-1}(z)]^{-1} \quad \text{for } j \geq 2$$

while

$$\mathbf{v}_1(z) = \boldsymbol{\pi}_1(z) \quad \text{and} \quad \mathbf{v}_j(z) = \boldsymbol{\pi}_j(z) + z^{\mu_j}[\mathbf{v}_{j-1}(z)]^{-1} \quad \text{for } j \geq 2$$

It is straightforward to demonstrate, by induction, that all the $\mathbf{u}_j$, $j \geq 0$, and $\mathbf{v}_j$, $j \geq 1$, are vectors in $\mathbb{C}^n$. Hence, we obtain

$$\left. \begin{aligned} p^{[l/m]}(z) &= p_k(z) = \mathbf{u}_0(z)\mathbf{u}_1(z)\cdots\mathbf{u}_k(z) \\ q^{[l/m]}(z) &= q_k(z) = \mathbf{v}_1(z)\mathbf{v}_2(z)\cdots\mathbf{v}_k(z). \end{aligned} \right\} \tag{5.4}$$

This construction, including the determination of the continued fraction elements using the algorithms discussed, lends itself to numerical implementation on a computer, even if the dimension of the vectors is large.

References

G. A. Baker Jr. and P. R. Graves-Morris: 1981, *Padé approximants, Encyclopedia of Mathematics and its Applications*, Vols. **13, 14**, Addison-Wesley.

P. R. Graves-Morris and C. D. Jenkins: 1986, 'Vector-valued rational interpolants III', *Constr. Approx.* **2**, pp. 263 – 289.

P. R. Graves-Morris and D. E. Roberts: 1994, 'From matrix to vector Padé approximants', *J. Comput. Appl. Math.* **51**, pp. 205 – 236.

J. B. McLeod: 1972, 'A note on the ϵ-algorithm', *Computing* **7**, pp. 17 – 24.

H. Padé: 1892, (Thesis) 'Sur la représentation approché d'une fonction pour des fractions rationnelles', *Ann. Sci.École Norm. Sup. Suppl.* [3], **9**, pp. 1 – 93.

D. E. Roberts: 1990, 'Clifford algebras and vector-valued rational forms I', *Proc. Roy. Soc. Lond.* **A 431**, pp. 285 – 300.

D. E. Roberts : 1992, 'Clifford algebras and vector-valued rational forms II', *Numerical Algorithms* **3**, pp. 371 – 382.

D. E. Roberts: 1993, 'Vector-valued rational forms', *Found. Phys.* **23**, pp. 1521 – 1533.

D.E. Roberts: 1995, 'On the algebraic foundations of the vector ϵ-algorithm', in R. Abłamowicz and P. Lounesto (eds.) *Clifford algebras and spinor structures, Mathematics and its Applications* **321**, pp. 343 – 361, Kluwer.

B. Viskovatov: 1803 – 1806, 'De la méthode générale pour réduire toutes sortes des quantités en fractions continues ', *Memoires de L'Academie Impériale des Sciences de St. Petersburg*, **1**, pp. 226 – 247.

P. Wynn: 1963,'Continued fractions whose coefficients obey a non-commutative law of multiplication', *Arch. Ration. Mech. Analysis*, **12**, pp. 273 – 312.

P. Wynn: 1968, 'Vector continued fractions', *Lin. Alg. Applic.* **1**, pp. 357 – 395.

LUCY: A CLIFFORD ALGEBRA APPROACH TO SPINOR CALCULUS

JÖRG SCHRAY, ROBIN W. TUCKER and CHARLES H.-T. WANG
School of Physics and Chemistry
Lancaster University
Lancaster LA1 4YB, UK
e-mail: j.schray@lancaster.ac.uk, r.tucker@lancaster.ac.uk
c.wang@lancaster.ac.uk

Abstract. LUCY is a MAPLE program that exploits the general theory of Clifford algebras to effect calculations involving real or complex spinor algebra and spinor calculus on manifolds in any dimension. It is compatible with both release 2 and release 3 of MAPLE V and incorporates a number of valuable facilities such as multilinearity of the Clifford product and the freedom to adopt arbitrary bases in which to perform calculations. The user can also pass with ease between the purely (real or complex) Clifford algebraic language and the more familiar matrix language. LUCY enables one to explore the structure of spinor covariant derivatives on flat or curved spaces and correlate the various spinor-inner products with the basic involutions of the underlying Clifford algebra. The canonical spinor covariant derivative is based on the Levi-Civita connection and a facility for the computation of connection coefficients has also been included. A self-contained account of the facilities available is provided together with a description of the syntax, illustrative examples for each procedure and a brief survey of the algorithms that are used in the program.

Key words: Clifford algebra, spinor, spinor adjoint, spinor inner product, spinor covariant derivative, Dirac operator, MAPLE, matrix representations.

1. Introduction

Of the various methods used by physicists for describing spinors the approach using the theory of Clifford algebras offers a comprehensive and unifying language. This approach is also ideally suited for describing the action of the (pseudo-)orthogonal groups in any dimension and therefore provides a natural framework for relativistic theories including gravitation. LUCY [1] is a MAPLE program that enables the user to exploit this language at both the teaching and research level. It incorporates a number of valuable facilities such as multilinearity of the Clifford product and the freedom to adopt arbitrary bases in which to perform calculations. The user can also pass with ease between the purely (real or complex) Clifford algebraic language and the more familiar matrix language. LUCY enables one to explore the structure of spinor covariant derivatives on flat or curved spaces and correlate the various spinor-inner products with the basic involutions of the underlying Clifford algebra. The canonical spinor covariant derivative is based on the Levi-Civita connection and a facility for the computation of connection coefficients has also been included. Since there already exist MAPLE packages for exterior algebras, LUCY is restricted to Clifford algebras although to fully appreciate the Clifford approach to differential

---

[1] LUCY code is available on request from: r.tucker@lancaster.ac.uk.

geometry the experienced user may wish to integrate LUCY with the theory of differential forms (Benn and Tucker, 1987). The following is a self-contained account of the facilities available together with a description of the syntax, illustrative examples for each procedure and a brief survey of the algorithms adopted in the program. The program is compatible with both release 2 and release 3 of MAPLE V. The basic theory and notation follows that of (Benn and Tucker, 1987).

2. Clifford algebra

Let $\{e^a\}$ be a basis for an n-dimensional vector space V over $\mathbb{R}$ or $\mathbb{C}$ with a symmetric inner-product (metric) $\mathbf{g}$, such that the metric components in this basis are $g^{ab} = \mathbf{g}(e^a, e^b)$. Suppose the frame indices $\{a\}$ are enumerated in the integer range $u, u + 1, \cdots, w$ with $w = u + n - 1$. The algebraic rules for the associated 2^n-dimensional Clifford algebra $C\ell(V, \mathbf{g})$ may be established with the command:

$$\texttt{Cliffordsetup}(u .. w , \{ (a_1, b_1) = g^{a_1 b_1}, (a_2, b_2) = g^{a_2 b_2}, \cdots \}, e).$$

The indexed names $e[u], e[u + 1], \cdots, e[w]$ are now available to the user and denote the "ambient" basis vectors for any chosen root name e. A symbol in the square brackets [ ] after a *root name for a basis element* corresponds to an "upper" index. In the second argument, only non-zero and either (a_i, b_j) or (b_j, a_i) metric components need to be specified.

Example 1 The following command sets up a 2-dimensional vector space with Lorentzian orthonormal basis $\{e^0, e^1\}$ in which the metric is diag$(-1, 1)$.

```
> Cliffordsetup(0..1,{(0,0)=-1,(1,1)=1},e):
```

Clifford scalars are recognized by LUCY as expressions independent of basis vectors. It recognizes an expression x as a *vector* ($x \in V$) if it is a linear combination of the basis vectors. Clifford multiplication between vectors is carried out with the multi-argument associative linear operator **&v** which first distributes over addition and factors out scalars.[2] In the following text we shall simply use juxtaposition of symbols to denote such Clifford multiplication. Basis vectors are ordered by LUCY using the anticommutation relation:

$$e^a e^b + e^b e^a = 2g^{ab} \tag{1}$$

to obtain a canonical form of any Clifford element (chosen in this program to be sums of products of basis vectors with increasing indices). The unevaluated $e^{a_1} e^{a_2} e^{a_3} \ldots$ ($a_1 < a_2 < a_3 \ldots$) together with the basis vectors and the number 1 form a basis for the Clifford algebra and a linear combination of these basis elements ω is treated as a *Clifford element* ($\omega \in C\ell(\mathbf{g}, V)$). A complete list of the ambient basis vectors and their canonically ordered Clifford products is returned with the command `Cliffordbasis()`. A simplification procedure `simp` is provided in this program taking any expression as argument and returning an equivalent expression with Clifford elements collected into a canonical order and their coefficients simplified.

---

[2] The name of this operator is motivated by the symbol e which is commonly used to denote the operation of Clifford multiplication.

Example 2 Following Example 1, one can perform the following calculations:

```
> (c1*e[0]+c2*e[1]) &v (c3*e[0]+c4*e[1]);
    c1*(-c3+c4*&v(e[0],e[1]))+c2*(-c3*&v(e[0],e[1])+c4)
> simp("");
    (c1*c4-c2*c3)*&v(e[0],e[1])-c1*c3+c2*c4
> Cliffordbasis();
    [e[0],e[1],&v(e[0],e[1])]
```

Two involutions and an anti-involution [3] on $C\ell$ are available in this program. The main anti-involution (reversion involution) ξ on the Clifford algebra is denoted by the linear unary operator **&xi**. It maps a scalar or a vector to itself but reverses the order of elements in a Clifford product. However, one will not see a reversed **&v** product on output, since **&v** canonically reorders factors using the anticommutation relation (1). The main involution (grade involution) η is performed with the linear unary operator **&eta**. Acting on a Clifford product of k vectors it multiplies such a Clifford element by $(-1)^k$. Complex conjugation $*$ is represented by the conjugate linear unary operator **&aster**, which effects the substitution $\sqrt{-1} \rightarrow -\sqrt{-1}$, *i.e.* **&aster** naively regards all names or functions as real except for the symbolic primitive idempotent which will be discussed later.

Example 3 Following Examples 1 and 2, we can compute:

```
> x:=I*c1+c2*e[0]+I*c3*e[1]+c4*e[0] &v e[1]:
> &eta x;
    I*c1-c2*e[0]-I*c3*e[1]+c4*&v(e[0],e[1])
> &xi x;
    I*c1+c2*e[0]+I*c3*e[1]-c4*&v(e[0],e[1])
> &aster x;
    -I*c1+c2*e[0]-I*c3*e[1]+c4*&v(e[0],e[1])
```

LUCY accepts elements that belong to the dual space V^*. They must be expanded in a basis known to the program. If one designates $\{X_a\}$ to be a basis naturally dual to $\{e^a\}$, *i.e.* $e^a(X_b) = \delta^a{}_b$ in terms of the Kronecker symbol, then call **Cliffordsetup** with the chosen root name X labelling the ambient dual basis, as an extra fourth argument:

$$\texttt{Cliffordsetup}(\,u\,..\,w\,,\{\,(a_1,b_1)\,=\,g^{a_1 b_1},(a_2,b_2)\,=\,g^{a_2 b_2},\,\cdots\,\}\,,\,e\,,\,X\,)$$

together with all other arguments as explained previously. This defines the indexed names $X[\{u\}]$, $X[\{u+1\}]$, $\cdots$, $X[\{w\}]$ to be elements of the ambient dual basis. Subsequently a linear combination of dual basis vectors Y is recognized as a *dual vector* ($Y \in V^*$). Clifford scalars are recognized by LUCY as expressions independent of both basis and dual basis vectors. The superscript-subscript convention adopted by LUCY is that each lower frame index is an integer enclosed with

---

[3] A number of alternative notations exist for these operations. We have chosen to use the two maps ξ and η to facilitate readability of the output from LUCY. The following relations offer a translation to other alternative notations: $\omega^\eta = \hat{\omega}$, $\omega^\xi = \tilde{\omega}$ and $\omega^{\xi\eta} = \bar{\omega}$ where $\omega \in C\ell\,(\mathbf{g}, V)$.

[{}], while each upper frame index is an integer enclosed with []. Hence X [{a}], X[b], e[a] and e[{a}] &v e[b] are understood as X_a, X^b, e^a $rmand$ $e_a e^b$ respectively, where $X^a = g^{ab} X_b$ and $e_a = g_{ab} e^b$ ($g^{ac} g_{cb} = \delta^a{}_b$) as usual. Here a, b, c are integers within the frame enumeration range and e, X the root names of the ambient basis and dual basis respectively. (The Einstein summation convention for indices is adopted throughout this text.) The facilities **convert**($expr$, **up**) or **convert**($expr$, **down**) are provided for raising or lowering frame indices on demand. The first command replaces all e_a, X_a appearing in the expression $expr$ with $g_{ab} e^b$, $g_{ab} X^b$ and conversely the second command replaces all e^a, X^a with $g^{ab} e_b$, $g^{ab} X_b$ respectively. Both $\{1, e^a, e^b, \cdots, e^a e^b, \cdots\}$ and $\{1, e_a, e_b, \cdots, e_a e_b, \cdots\}$ provide a basis for $C\ell(V, \mathbf{g})$. Excluding the number 1, the former Clifford basis elements are accessible from **Cliffordbasis(up)** (or equivalently **Cliffordbasis()** introduced in the last section) while the latter set can be obtained with **Cliffordbasis(down)**.

Example 4 As an extension to Example 1, by specifying the dual frame root name X, the following computation can be executed:

```
> Cliffordsetup(0..1,{(0,0)=-1,(1,1)=1},e,X);
> convert(X[{0}],up);
        -X[0]
> convert(",down);
        X[{0}]
> e[{0}] &v e[0];
        1
> cb:=Cliffordbasis(down);
        cb:=[e[{0}], e[{1}], &v(e[{0}],e[{1}])]
> w:=k0+convert([seq((k.i)*cb[i],i=1..nops(cb))],'+');
        w:=k0+k1*e[{0}]+k2*e[{1}]+k3*&v(e[{0}],e[{1}])
> u:=convert(w,up);
        u:=k0-k1*e[0]+k2*e[1]-k3*&v(e[0],e[1])
```

The interior product (contraction) $i_Y z$ between any dual vector Y and Clifford element z can be performed with the linear binary operator **&i** with the call **&i**(Y, z). It returns 0 if the second argument is a scalar. Contraction between naturally dual basis elements produces 1 or 0 according to whether the frame indices are equal or not. **&i** defines a graded derivation on Clifford products and it acts on **&v** terms as follows: Y **&i** **&v**($x, y, z, \cdots$) yields (Y **&i** x) $*$ **&v**($y, z, \cdots$) $-$ x **&v** (Y **&i** **&v** ($y, z, \cdots$)), for $Y \in V^*$ and $x, y, z \in V$. The second term has a similar structure to the original expression; the same procedure is then applied repeatedly until the expansion is completed.

Example 5 Following Example 4, we can calculate:

```
> X[{0}] &i e[0];
        1
> X[{0}] &i e[{0}];
        -1
> (a0*X[{0}] + a1*X[{1}]) &i u ;
```

```
    -a0*k1 - a0*k3*e[1] + a1*k2 + a1*k3*e[0]
```

Rather than using the (contravariant) metric components $g^{ab} = \mathbf{g}(e^a, e^b)$ when specifying the metric, one can equally well use the covariant metric components with the command:

$$\texttt{Cliffordsetup}(u\mathinner{..}w,$$
$$\{(\{a_1\},\{b_1\}) = g_{a_1 b_1}, (\{a_2\},\{b_2\}) = g_{a_2 b_2}, ...\}, e, X)$$

where g_{ab} are defined by $g^{ac} g_{cb} = \delta^a{}_b$ such that $g_{ab} = \mathbf{g}^*(X_a, X_b)$ in which $\mathbf{g}^*$ is the induced metric on V^*.

3. Clifford calculus

The program is also capable of setting up the Clifford algebra of the cotangent space associated with a variable point p of a patch of an n-dimensional manifold M. If $\mathbf{g}$ is the (covariant) metric tensor field on M we identify V with the cotangent space $\mathrm{T}^*_p M$, V^* with the tangent space $\mathrm{T}_p M$ and $\mathbf{g}^*$ with $\mathbf{g}\,rt_p$ at any point p on M. In this sense we will also call elements in V covectors and elements in V^* tangent vectors. Such a setup is initiated by using one of the following two commands:

$$\texttt{Cliffordsetup}(u\mathinner{..}w,$$
$$\{(a_1, b_1) = g^{a_1 b_1}, (a_2, b_2) = g^{a_2 b_2}, ...\}, e, X, \{\langle cobase\ eqns \rangle\})$$

$$\texttt{Cliffordsetup}(u\mathinner{..}w,$$
$$\{(\{a_1\},\{b_1\}) = g_{a_1 b_1}, (\{a_2\},\{b_2\}) = g_{a_2 b_2}, ...\}, e, X, \{\langle cobase\ eqns \rangle\})$$

using the previously defined conventions. The new parameter $\langle cobase\ eqns \rangle$ is a sequence of n equations expressing each cobase vector in terms of "formal" coordinate differentials. If the names x^1, x^2, ... are to be regarded as local coordinates on the patch, then $\{\texttt{d}(x^1), \texttt{d}(x^2), \cdots\}$ syntactically stands for a local coordinate coframe. Each equation in $\langle cobase\ eqns \rangle$ has a cobase vector e^a as the left hand side and a linear combination of the $\texttt{d}(x^i)$, with functions of the coordinates as coefficients, on the right hand side. The metric components may also be functions of the coordinates.

Example 6 The following sets up a spacetime patch with a partially null coframe $\{n^a\}$ local coordinates $\{u, v, x, y\}$, and a plane wave metric $g = n^1 \otimes n^2 + n^2 \otimes n^1 + n^3 \otimes n^3 + n^4 \otimes n^4$ corresponding to the infinitesimal arc length squared $\mathrm{d}s^2 = H(u, x, y)\, \mathrm{d}u\, \mathrm{d}u + 2\, \mathrm{d}u\, \mathrm{d}v + \mathrm{d}x\, \mathrm{d}x + \mathrm{d}y\, \mathrm{d}y$.

```
> alias(H=H_(u,x,y)):
> Cliffordsetup(1..4,{({1},{2})=1,({3},{3})=1,({4},{4})=1},n,Z,
> {n[1]=d(u),n[2]=d(v)+H/2*d(u),n[3]=d(x),n[4]=d(y)}):
> n[2] &v n[1];
        2 - &v(n[1],n[2])
> convert("",down);
        &v(n[{1}],n[{2}])
```

For any tangent vector fields (*i.e.* linear combinations of dual basis vectors from the ambient frame with functions of the coordinates as coefficients), X, Y and Clifford element α (which may be a linear combination of covectors from the ambient coframe with functions of the coordinates as coefficients), the covariant derivatives $\nabla_X Y$ and $\nabla_X \alpha$ with respect to the *Levi-Civita* (pseudo-)Riemannian connection ∇ induced by g, are respectively computed with the commands X &C Y and X &C α, or &C(X, Y) and &C(X, α). The binary operator &C is function-linear in its first argument while it is distributive over addition in the second argument. The Leibnitz rule X &C $(f\Lambda) = X(f)\Lambda + f(X$ &C $\Lambda)$ is activated for any function f of the coordinates and tangent vector or Clifford element Λ. For a Clifford product of the ambient cobase vectors, $\alpha = yz\cdots$, X &C α yields $(X$ &C $y)$ &v z &v $\cdots$ + y &v $(X$ &C $z)$ &v $\cdots$ + $\cdots$ since the connection is compatible with the metric. After this expansion, the remaining covariant derivatives of the ambient cobase or dual basis vectors are automatically computed by LUCY in terms of the connection coefficients $\Gamma^c{}_{ba}$ defined by:

$$\nabla_{X_a} X_b = \Gamma^c{}_{ba} X_c, \qquad \nabla_{X_a} e^c = -\Gamma^c{}_{ba} e^b. \tag{2}$$

The command RIEMCONNEX() returns the table containing these connection coefficients $\Gamma^a{}_{bc}$ with indices $[a, \{b\}, \{c\}]$ (as well as $\Gamma_{abc} \equiv g_{ad}\Gamma^d{}_{bc}$ with indices $[\{a\}, \{b\}, \{c\}]$) for integers a, b, c within the frame enumeration range.

Example 7 Continuing from the setup in Example 6 where the vector field Z_1 is naturally dual to $n^1 = du$, Z_1 represents the direction derivative $\frac{\partial}{\partial u}$. It follows that $\nabla_{Z_1} H = Z_1(H) = \frac{\partial H}{\partial u}$ as verified below:

```
> Z[{1}] &C H;
        diff(H,u)
> Z[{1}] &C Z[{4}];
        1/2*diff(H,y)*Z[{2}]
> Z[2] &C (n[1] &v n[2] &v n[3] &v n[4]);
        1/2*diff(H,x)*&v(n[1],n[4])-1/2*diff(H,y)*&v(n[1],n[3])
> Gamma:=RIEMCONNEX():
> Gamma[2,{4},{1}];
        1/2*diff(H,y)
> Gamma[{1},{2},{3}];
        0
```

4. Spinor modules

Within a Clifford algebra framework spinors may be represented by Clifford elements lying in some minimal left ideal projected by a primitive idempotent (Benn and Tucker, 1987). A Clifford element P is an idempotent if $PP = P$, and is primitive if P cannot be expressed as a sum of idempotents. However the expansion of such an element of $C\ell\, P$ in a Clifford basis is typically a sum of as many terms as the dimension of the Clifford algebra which exceeds the dimension of the spinor space. Such inefficiency can be circumvented by keeping P symbolic in computation. Let s

be the dimension of any minimal left module $C\ell\ P$. We can choose Clifford elements $\alpha_1, \alpha_2, \ldots, \alpha_s$ from the ambient Clifford basis such that $\{\alpha_i P\}$ spans $C\ell\ P$. By specifying the multiplication rules for any ambient Clifford basis element acting on a symbolic P from the left one can convert ωP into $\sum_{i=1}^{s} f_i (\alpha_i P)$ which has less summands, for any $\omega \in C\ell$ and some scalars f_i. In the same fashion one can find elements $\beta_1, \beta_2, \ldots, \beta_s$ such that $\{P\beta_i\}$ spans a right module $PC\ell$ and write $P\omega = \sum_i h_i (P\beta_i)$, for any $\omega \in C\ell$ and some scalars h_i. The Clifford elements $\{\alpha_i\}$ and $\{\beta_i\}$ are called the left and right kernel elements respectively. Although the algorithm used by LUCY can be generalized, in the following implementation *it will be assumed that $C\ell$ is simple and $PC\ell P$ is 1-dimensional* so that one can further compute P {left and right kernels} P to get the rules for converting $P\omega P$ into a real or complex number times P for any $\omega \in C\ell$. This is the case for V even dimensional where the complexified $C\ell\ (V, \mathbf{g})$ is isomorphic to an $s \times s$ complex matrix algebra, i.e. $C\ell\ (V, \mathbf{g}) \simeq \mathrm{Mat}(s, \mathbb{C})$. By equating $\dim(C\ell\ ) = 2^n$ with $\dim(\mathrm{Mat}(s, \mathbb{C})) = s^2$, in this case the dimension of the spinor space is readily seen to be $s = 2^{\frac{n}{2}}$.

In LUCY the command `Spinorsetup` chooses such a symbolic primitive idempotent and returns a list of two lists. The first list contains the chosen left kernels and the second contains the chosen right kernels. 1 is included as the first chosen kernel in both lists. The length of each list should be $2^{\frac{n}{2}}$ if P is primitive. The program informs the user if this condition is violated.

`Spinorsetup` can be called in two forms. Given an explicit form of a primitive idempotent $\langle P\ expr \rangle$ (in terms of the basis elements accessible from either `Cliffordbasis(up)` or `Cliffordbasis(down)`), the command

$$\texttt{Spinorsetup(}\ P = \langle\ P\ expr \rangle\ \texttt{)}$$

defines any symbol P to carry the algebraic property of $\langle P\ expr \rangle$. It is then much more efficient to use P instead of $\langle P\ expr \rangle$ in subsequent computations. The system recognizes a Clifford element ψ as a *spinor* ($\psi \in C\ell\ P$) if ψ is a linear combination of $\&\mathbf{v}(\cdots, P)$ terms. Likewise a Clifford element ϕ is recognized as a *dual spinor* ($\phi \in PC\ell$) if ϕ is a linear combination of $\&\mathbf{v}(P, \cdots)$ terms. P itself may be regarded as either a spinor or a dual spinor. Spinor or dual spinor basis elements are collected and their coefficients simplified by evaluating `simp` on any (dual) spinor. On some occasions one may wish to bring the symbolic P in any expression *expr* back to its explicit form. `expandP(`*expr*`)` is the command for this purpose.

Example 8 Following Examples 6 and 7 one can issue the following commands:

```
> PP:=1/4*(n[1] &v n[2]) &v (1+n[3]):
> PP &v PP - PP;
        0
> Spinorsetup(P=PP);
        [[1,n[2],n[4],&v(n[2],n[4])],[1,n[1],n[4],&v(n[1],n[4])]]
> (H + sin(v)^2*n[2] - n[3] + cos(v)^2*n[2] &v n[3]) &v P;
        H*P+sin(v)^2*&v(n[2],P)-P+cos(v)^2*&v(n[2],P)
> simp(");
```

```
      (H-1)*P+&v(n[2],P)
> w:=H*n[1] + n[1] &v n[2]:
> wP:=w &v P;
      wP:=2*P
> Pw:=P &v w;
      Pw:=H*&v(P,n[1])+2*P
> wPw:=wP &v Pw;
      wPw:=2*H*&v(P,n[1])+4*P
> simp(expandP(wPw));
      H*n[1]-H*&v(n[1],n[3])+&v(n[1],n[2])+&v(n[1],n[2],n[3])
```

If the ambient coframe is orthogonal, i.e. the matrix of metric components used in `Cliffordsetup` is diagonal, one can alternatively call **Spinorsetup** with any chosen symbol P:

$$\text{Spinorsetup}(P)$$

and the system will automatically generate a choice of a primitive idempotent. In this case, the selected explicit Clifford element that P stands for is returned by applying `expandP`(P).

Example 9 The following sets up a local Clifford structure on a 2-dimensional de Sitter space-time with the orthogonal coordinate coframe $\{e^0 = dt, e^1 = dx\}$.

```
> Cliffordsetup(0..1,{({0},{0})=-1,({1},{1})=exp(4*k*t)},e,X,
> {e[0]=d(t),e[1]=d(x)}):
> Spinorsetup(P);
      [[1,e[1]],[1,e[1]]]
> expandP(P);
      1/2+1/2*I*e[0]
```

The differential relations induced by the idempotency of P are also encoded into the program. Thus for any tangent vector field Y, evaluation of Y **&C** P leads to a sum of spinors and dual spinors:

$$\nabla_Y P = \nabla_Y(PP) = \nabla_Y PP + P\nabla_Y P \tag{3}$$

which clearly lies in $C\ell\, P \oplus PC\ell$. (Furthermore, multiplying with P from either the left or the right yields $P\nabla_Y PP = 0$.)

The operators **&eta**, **&aster**, **&xi** and **&i** cause explicit expansion of the symbolic P since the involutions η, $*$ or the anti-involution ξ generally map P to another primitive idempotent and the contraction of P with a dual vector does not even preserve idempotency.

Example 10 Following Example 9, we can differentiate the symbol P:

```
> X[{0}] &C P;
      0
> X[{1}] &C P;
      -I*k*exp(4*k*t)*&v(e[1],P)-I*k*exp(4*k*t)*&v(P,e[1])
```

```
> &eta P;
      1/2-1/2*I*e[0]
> &xi P;
      1/2+1/2*I*e[0]
> &aster P;
      1/2-1/2*I*e[0]
> X[{0}] &i P;
      1/2*I
```

5. Matrix representations

Contact with a concrete matrix representation can be made by constructing a matrix basis $\{\mathbf{m}_{ij}\}$ for $C\ell$. Any Clifford element ω can be expanded as

$$\omega = \sum_{i,j} \gamma_{ij}(\omega)\,\mathbf{m}_{ij} \tag{4}$$

where the Clifford elements $\mathbf{m}_{ij}$ satisfy a "total matrix algebra" (Benn and Tucker, 1987) and serve to define a matrix basis for $C\ell$.

The complex matrix $\boldsymbol{\gamma}(\omega)$ with entries $\{\gamma_{ij}(\omega)\}$ is call the *matrix representation* of ω with respect to $\{\mathbf{m}_{ij}\}$. In particular, the $\boldsymbol{\gamma}^a \equiv \boldsymbol{\gamma}(e^a)$ for the ambient basis of V give rise to the usual γ-matrices when the dimension of V is even. The well known relations $\boldsymbol{\gamma}^a\boldsymbol{\gamma}^b + \boldsymbol{\gamma}^b\boldsymbol{\gamma}^a = 2\,g^{ab}\,\mathbf{1}$ immediately follow from Equations (1) and (4).

The command $\mathtt{Matrixbasis()}$ returns a table of Clifford elements $\mathbf{m}_{ij}$ as entries, and matrix basis labels as indices.

The choice of $\{\mathbf{m}_{ij}\}$ is made by linearly combining the chosen left kernels $\{\alpha_i\}$ to form $\{\alpha'_i\}$ and right kernels $\{\beta_i\}$ to form $\{\beta'_i\}$ so that $\alpha'_1 = \beta'_1 = 1$ and $P\beta'_i\alpha'_j P = \delta_{ij}\,P$. This is equivalent to arranging $\{\,\mathbf{s}_i \equiv \alpha'_i P\,\}$ and $\{\,\mathbf{r}_i \equiv P\beta'_i\,\}$ to be naturally dual bases for spinors ($\in C\ell\,P$) and for dual spinors ($\in PC\ell$) respectively. Clearly the set $\{\,\mathbf{m}_{ij} \equiv \alpha'_i P\beta'_j\,\}$ generate a total matrix algebra:

$$\mathbf{m}_{ij}\mathbf{m}_{kl} = \alpha'_i P\beta'_j \alpha'_k P\beta'_l = \delta_{jk}\,\alpha'_i P\beta'_l = \delta_{jk}\,\mathbf{m}_{il}$$

and since $\alpha'_1 = \beta'_1 = 1$ we have $\mathbf{s}_i = \mathbf{m}_{i1}$, $\mathbf{r}_i = \mathbf{m}_{1i}$ and $\mathbf{m}_{11} = P$.

Converting a Clifford element ω into its matrix representation with respect to the chosen matrix basis can be done with the commands $\mathtt{convert}(\omega, \mathtt{Matrix})$, $\mathtt{convert}(\omega, \mathtt{column})$ or $\mathtt{convert}(\omega, \mathtt{row})$. While the first command returns the matrix $\boldsymbol{\gamma}(\omega)$ determined by (4), the second only returns the first column and the third only returns the first row of $\boldsymbol{\gamma}(\omega)$. Therefore, one of the last two commands is preferable if ω is a spinor or dual spinor.

Conversely, the command $\mathtt{convert}(\boldsymbol{\gamma}, \mathtt{Clifford})$ returns the Clifford element ω that $\boldsymbol{\gamma}$ represents. Here $\boldsymbol{\gamma}$ can be a square ($s \times s$) matrix such that ω will be computed with the symbolic P explicitly expanded. Otherwise $\boldsymbol{\gamma}$ is allowed to be a column ($s \times 1$) matrix representing a spinor, or a row ($1 \times s$) matrix representing a dual spinor. Accordingly ω is returned as a linear combination of $\&\mathbf{v}(\cdots, P)$ or $\&\mathbf{v}(P, \cdots)$ with P symbolic.

The procedure **Matrixbasis()** is invoked automatically when these conversion facilities are called.

Example 11 Following Example 9, examples of conversion between Clifford elements and matrix representations are given:

```
> MB:=Matrixbasis();
  MB := table([
                (1, 1) = P
                (1, 2) = exp(4*k*t)*(P &v e[1])
                (2, 1) = e[1] &v P
                (2, 2) = exp(4*k*t)*&v(e[i], P, e[1])
               ])
> simp(MB[2,2] &v MB[2,2]);
       exp(4*k*t)*&v(e[1], P, e[1])
> simp(MB[2,1] &v MB[1,1] - MB[2,1]);
       0
> G[0]:=convert(e[0],Matrix);
              [ -I   0 ]
      G[0] := [        ]
              [ 0    I ]
> convert(",Clifford);
       e[0]
> G[1]:=convert(e[1],Matrix);
              [ 0    exp(-4*k*t) ]
      G[1] := [                  ]
              [ 1         0      ]
> convert(",Clifford);
       e[1]
> evalm(G[0] &* G[0]);
              [ -1   0 ]
              [        ]
              [  0  -1 ]
> evalm(G[1] &* G[1]);
              [ exp(-4*k*t)        0        ]
              [                             ]
              [      0        exp(-4*k*t)   ]
> evalm(G[0] &* G[1] + G[1] &* G[0]);
              [ 0  0 ]
              [      ]
              [ 0  0 ]
> spinor:=u1*MB[1,1]+u2*MB[2,1]:
> convert(spinor,column);
              [ u1 ]
              [    ]
              [ u2 ]
> convert(",Clifford);
```

```
      u1*P + u2*&v(e[1],P)
> dualspinor:=v1*MB[1,1]+v2*MB[1,2]:
> convert(dualspinor,row);
              [ v1   v2 ]
> convert(",Clifford);
      v1*P + v2*exp(4*k*t)*&v(P,e[1])
```

Example 12 Given a linear or conjugate linear map on spinors $\Lambda : C\ell\, P \to C\ell\, P$, it is often convenient to associate Λ with a transformation matrix $\boldsymbol{\Lambda}$ when working with spinor components. If $\Lambda(\mathbf{s}_i) = \sum_j \mathbf{s}_j\, \Lambda_{ji}$ $(\Lambda_{ij} \in \mathbb{C})$, then for any spinor $\Psi = \sum_i \psi_i\, \mathbf{s}_i$ it follows that

$$\Lambda(\Psi) = \sum_{i,j} \mathbf{s}_j\, \Lambda_{ji}\, \psi_i \qquad i.e. \quad \Lambda : \psi \; \mapsto \; \boldsymbol{\Lambda}\, \psi$$

if Λ is linear, or

$$\Lambda(\Psi) = \sum_{i,j} \mathbf{s}_j \Lambda_{ji}\, \psi_i^* \qquad i.e. \quad \Lambda : \psi \; \mapsto \; \boldsymbol{\Lambda}\, \psi^*$$

if Λ is conjugate linear. Here ψ is a column matrix consisting of spinor components $\{\psi_i\}$. By exploiting the duality of $\{\mathbf{r}_i\}$ and $\{\mathbf{s}_i\}$ introduced in this section, the associated matrix $\boldsymbol{\Lambda}$ is determined by $\Lambda_{ij}\, P = \mathbf{r}_i\Lambda(\mathbf{s}_j)$. The following MAPLE procedure takes such a map as argument and returns the associated transformation matrix:

```
> Ltransmat:=proc(f)
>   local P,M,A,i,j,n;
>   M:=Matrixbasis(); # table of the chosen matrix basis
>   n:=sqrt(nops([indices(M)])); # spinor space dimension
>   P:=M[1,1];        # P is the primitive idempotent
>   A:=array(1..n,1..n);
>   for i to n do
>    for j to n do
>     A[i,j]:=coeff(simp(M[1,i] &v f(M[j,1])),P)
>    od
>   od:
>   eval(A)
> end:
```

Similarly a linear or conjugate linear map from spinors to dual spinors $\Omega : C\ell\, P \to P C\ell$ with $\Omega(\mathbf{s}_i) = \sum_j \Omega_{ij}\, \mathbf{r}_j$ $(\Omega_{ij} \in \mathbb{C})$, gives rise to the relation

$$\Omega(\Psi) = \sum_{i,j} \psi_i\, \Omega_{ij}\, \mathbf{r}_j \qquad i.e. \quad \Omega : \psi \; \mapsto \; \psi^T\, \boldsymbol{\Omega}$$

if Ω is linear, or

$$\Omega(\Psi) = \sum_{i,j} \psi_i^*\, \Omega_{ij}\, \mathbf{r}_j \qquad i.e. \quad \Omega : \psi \; \mapsto \; \psi^{*T}\, \boldsymbol{\Omega}$$

if Ω is conjugate linear. Here ^{T} denotes matrix transposition and so ψ^T is a row matrix. The associated matrix Ω is accordingly determined by $\Omega_{ij} P = \Omega(\mathbf{s}_i)\mathbf{s}_j$. Obtaining the associated transformation matrix for such a given map can be encoded into the following MAPLE procedure:

```
> Rtransmat:=proc(f)
>  local P,M,A,i,j,n;
>   M:=Matrixbasis();
>   n:=sqrt(nops([indices(M)]));
>   P:=M[1,1];
>   A:=array(1..n,1..n);
>   for i to n do
>    for j to n do
>     A[i,j]:=coeff(simp(f(M[i,1]) &v M[j,1]),P)
>    od
>   od:
>    eval(A)
> end:
```

6. Spinor adjoint

The *spinor adjoint* map ‡ with respect to an anti-involution J, ($J = \xi$, $\xi\eta$, ξ^* or $\xi\eta^*$) is defined by (Schray and Wang, 1995)

$$\ddagger : Cl\, P \longrightarrow PCl$$
$$\Psi \longmapsto \Psi^\ddagger = W\Psi^J. \tag{5}$$

for some chosen $W \in P\, Cl\, P^J$. Thus if Ψ is a spinor, $\Psi^\ddagger$ is its adjoint. The crucial defining property of ‡

$$(\omega\Psi)^\ddagger = \Psi^\ddagger \omega^J \tag{6}$$

$\forall\, \omega \in Cl$, $\forall\, \Psi \in Cl\, P$ ensures that the spinor inner product

$$(\,,\,): Cl\, P \times Cl\, P \longrightarrow \mathbb{C}$$
$$\Phi\,,\, \Psi \longmapsto (\Phi,\Psi) \text{ with } (\Phi,\Psi)P = \Phi^\ddagger\Psi \tag{7}$$

is invariant under the action $\Phi \mapsto \alpha\Phi$, $\Psi \mapsto \alpha\Psi$ where $\alpha^J\alpha = 1$. This follows since $(\Phi,\Psi) \to (\alpha\Phi,\alpha\Psi) = (\Phi,\alpha^J\alpha\Psi) = (\Phi,\Psi)$. [4] Given any Clifford element $W \in P\, Cl\, P^J$ that defines a spinor adjoint we have also $W^J \in P\, Cl\, P^J$ which can be used to another spinor adjoint. For $J = \xi$ or $\xi\eta$, $W^J = \pi W$ where the parity $\pi = \pm 1$ depends on the dimension of V. Accordingly the spinor inner product

---

[4] The elements α, satisfying $\alpha^J\alpha = 1$, form a group, which has as a subgroup the group **SPIN**$^+$ of even elements in the group of invertible elements that map vectors $x \in V$ so that $\alpha x \alpha^{-1} \in V$. Further information on the subgroups of the group defined by the condition $\alpha^J\alpha = 1$ can be found on page 271 of (Porteous, 1969, 1981), see also page 42 of (Benn and Tucker, 1987). This group should not be confused with the Clifford group, introduced by Lipschitz.

naturally possesses the (anti-)symmetry $(\Phi, \Psi) = \pi(\Psi, \Phi)$ for any spinors Φ and Ψ. However if $J = \xi^*$ or $\xi\eta^*$ we can choose $W_\pm = \frac{1}{2}(1 \pm J)W \in \frac{1}{2}(1 \pm J)(P\, C\ell\, P^J)$ such that $W_\pm{}^J = \pm W_\pm$. $W_\pm$ can be then used to define ‡ and the corresponding spinor product denoted by $(\,,\,)_\pm$ has the Hermitian (anti-)symmetry: $(\Phi, \Psi)^*_\pm = \pm(\Psi, \Phi)_\pm$. The procedure that generates a choice of spinor adjoint map is **Spinoradjoint** with the syntax

$$\text{Spinoradjoint}(\, \rho, \, \pi \,)$$

where for $\rho = 1$ the anti-involution is ξ, for $\rho = 2$ the anti-involution is $\xi\eta$, for $\rho = 3$ the anti-involution is ξ^* and for $\rho = 4$ the anti-involution is $\xi\eta^*$. π denotes the parity of the adjoint map and may be chosen to be ± 1. The second argument will be ignored if ρ is 1 or 2 since only when $\rho = 3$ or 4 (*i.e.* $J = \xi^*$ or $\xi\eta^*$) does π need to be specified. (In the case $J = \xi$ or $\xi\eta$ the parity is fixed by the dimension n of the vector space V.) **Spinoradjoint** will display to the user the anti-involution as well as the parity associated with the output adjoint map and return the appropriate unary operator that maps a spinor to its adjoint (dual spinor).

Example 13 Following Examples 9 and 10, we can define various spinor adjoint maps:

```
> '&ad1':=Spinoradjoint(1);
      (Spinoradjoint) defining spinor adjoint map associated with
      anti-involution   &xi   and parity   1   ...
      &ad1:=psi -> SPINADJ(psi,1,1)

> '&ad2':=Spinoradjoint(2);
      (Spinoradjoint) defining spinor adjoint map associated with
      anti-involution   @(&eta,&xi)   and parity   -1   ...
      &ad2:=psi -> SPINADJ(psi,2,e[1])

> alias(h=h_(t),u1=u1_(t),u2=u2_(t),f1=f1_(t),f2=f2_(t)):
> ad:=Spinoradjoint(4,1); # Dirac adjoint
      (Spinoradjoint) defining spinor adjoint map associated with
      anti-involution   @(&eta,&xi,&aster)   and parity   1   ...
      ad:=psi -> SPINADJ(psi,4,1)

> '&ad':=psi -> simp(h*ad(psi)):

> psi:=(u1*exp(I*f1) + u2*exp(I*f2)*e[1]) &v P: # psi is a spinor
>     # field with complex components u1*exp(I*f1) and u2*exp(I*f2)

> coeff(simp((&ad psi) &v (I*e[0]) &v psi),P);
>     # get time-component of the Dirac current
        h*u1^2 + h*u2^2*exp(-4*k*t)
```

As we have seen in Example 13, **Spinoradjoint** returns a map involving a system reserved procedure SPINADJ. It has the syntax $\text{SPINADJ}(\,\Psi,\, \rho,\, \mu\,)$ for any spinor

$\Psi \in C\ell \ P$, ρ indicating the adjoint involution J as explained and μ the linear combination the right kernels defined by $W = P\mu$. With a symbolic P and $\Psi = \sigma P$ for σ an appropriate linear combination of the left kernels, SPINADJ simply computes $P\mu\sigma^J = W\Psi^J = \Psi^{\ddagger}$ in accordance with Equations (5) and (6) and returns the result.

The assignment $H := \texttt{Spinoradjoint}(\rho, \pi)$ defines H to be the function name of the chosen spinor adjoint map. A more general spinor adjoint map can be achieved by the rescaling

$$H' := \psi \to h * H(\psi) \tag{8}$$

with a scalar factor f (which can be a function of the coordinates if defined). h is allowed to be complex in the cases $\rho = 1$ or 2 ($J = \xi$ or $\xi\eta$). However h must be real if $\rho = 3$ or 4 ($J = \xi^*$ or $\xi\eta^*$) in order to preserve the parity π.

The conventional definition of spinor adjoint in matrix language follows also from (5). For example consider $J = \xi^*$ or $\xi\eta^*$ with parity $\pi = +1$ leading to the Dirac adjoint. The choice of J is normally taken to be ξ^* if the Lorentzian metric has the signature $(+, -, -, \cdots)$, or to be $\xi\eta^*$ if the signature is $(-, +, +, \cdots)$ in order to get a Dirac current with (time-like) positive density. Given a matrix basis $\{\mathbf{m}_{ij}\}$ for $C\ell$, the Hermitian conjugation $\dagger$ maps a Clifford element $\omega = \sum_{i,j} \gamma_{ij}(\omega) \, \mathbf{m}_{ij}$ to $\omega^{\dagger} = \sum_{i,j} \gamma_{ji}^*(\omega) \, \mathbf{m}_{ij}$. A class of Clifford elements C always exists with the properties:

$$C\omega^J = \omega^{\dagger}C \qquad \forall \, \omega \in C\ell \tag{9}$$

$$C^{\dagger} = C \quad (\text{hence } C^J = C). \tag{10}$$

If $\mathbf{m}_{11} = P$ and $W = PCP^J$ is chosen to define $\ddagger$, it gives rise to a Hermitian symmetric spinor inner product. It follows from (9) that for any $\Psi \in C\ell \ P$

$$\Psi^{\ddagger} = PC\Psi^J = P\Psi^{\dagger}C = \Psi^{\dagger}C. \tag{11}$$

Writing $C = \sum_{i,j} C_{ij}\mathbf{m}_{ij}$ and $\Psi = \sum_i \psi_i \, \mathbf{s}_i$ with the spinor basis $\{\mathbf{s}_i \equiv \mathbf{m}_{i1}\}$ one can easily transcribe Equation (11) into its matrix form

$$\ddagger : \psi \mapsto \psi^{*T} C. \tag{12}$$

The matrix versions of Equations (9) and (10) follow as:

$$\epsilon \, C \, \gamma^a = \gamma^{a*T} \, C \tag{13}$$

$$C^{*T} = C \tag{14}$$

where $\epsilon = +1$ or -1 according to $J = \xi^*$ or $\xi\eta^*$. Together with (13) and (14), Equation (12) is just the usual definition of the Dirac adjoint in terms of γ-matrices. It is unnecessary to solve (13) and (14) for C in order to construct a spinor adjoint if a matrix basis has been established. It follows from (11) that $\Phi^{\ddagger}\Psi = \Phi^{\dagger}C\Psi$ for any spinors Φ and Ψ. Using $C = \sum_{i,j} C_{ij}\mathbf{m}_{ij}$, the substitution $\Phi = \mathbf{s}_i$ and $\Psi = \mathbf{s}_j$ yields

$$\mathbf{s}_i^{\ddagger}\mathbf{s}_j = \mathbf{m}_{i1}^{\dagger}C\mathbf{m}_{j1} = \sum_{k,l} C_{kl} \, \mathbf{m}_{1i}\mathbf{m}_{kl}\mathbf{m}_{j1} = C_{ij} \, P$$

and hence the matrix C as well as the Clifford element C is determined.

Example 14 It is instructive to employ the procedure `Rtransmat` given in Example 12 to compute the "spinor metric matrix" C associated with the spinor adjoint map `&ad` defined in Example 13.

```
> Cmat:=Rtransmat('&ad');
              [ h              0      ]
    Cmat := [                        ]
              [ 0    -h*exp(-4*k*t) ]
> C:=convert(Cmat,Clifford);
        1/2*I*(h + h*exp(-4*k*t))*e[0] + h/2 - h/2*exp(-4*k*t)
> &eta(&xi(&aster C)) - C; # C has parity +1
        0
```

7. Spinor conjugation

Following the definition of the spinor adjoint, we introduce in a similar fashion the *spinor conjugation* map ^{c} on spinors with respect to an involution K ($K = \eta$, $*$, or η^* in the following) by (Schray and Wang, 1995)

$$\begin{aligned} ^c : C\ell\, P &\longrightarrow & C\ell\, P \\ \Psi &\longmapsto & \Psi^c = \Psi^K U \end{aligned} \tag{15}$$

for some $U \in P^K C\ell\, P$. This map is defined to satisfy (cf. Equation (6))

$$(\omega\Psi)^c = \omega^K \Psi^c \tag{16}$$

$\forall\, \omega \in C\ell$, $\forall\, \Psi \in C\ell\, P$. A choice of spinor conjugation map can be made with the command

$$F := \texttt{Spinorconjugation}(\, \kappa\,)$$

whose effect is to assign a chosen function name F to the unary operator returned by `Spinorconjugation`, mapping spinors to spinors. For the parameter $\kappa = 1$ the associated involution is η, for $\kappa = 2$ the associated involution is $*$ and for $\kappa = 3$ the associated involution is η^*. The output of `Spinorconjugation` is a MAPLE map $\Psi \to \texttt{SPINCONJ}(\Psi, \kappa, \nu)$ where ν is a linear combination of the left kernels such that the selected U is given by $U = \nu P$. For $\Psi = \sigma P$ the call $F(\Psi)$ returns the spinor conjugation of Ψ. (The system reserved procedure `SPINCONJ` exploits the algorithm $\Psi^c = \sigma^K \nu P$ suggested from Equations (15) and (16).) Since $P^K C\ell\, P$ is 1-dimensional, implying that U is fixed up to a complex factor, the (optional) subsequent command

$$F' := \psi \to f * F(\psi) \tag{17}$$

with an arbitrary scalar factor f leads to another spinor conjugation map named F'. This enables the user to explicitly normalize the spinor conjugation map.

The *charge conjugation* map is a transformation between complex spinors. In a Lorentzian spacetime, complex spinors can carry representations of the electromagnetic U(1) gauge group describing electrically charged fields. The transformed spinor will then describe a field with opposite charge. Such a mapping is induced by ^{c} with $K = *$ if the metric has signature $(-, +, +, \cdots)$ and is induced by ^{c} with $K = \eta^*$ if the signature is $(+, -, -, \cdots)$. To compare with the definition of charge conjugation in matrix language note that if $\{\mathbf{m}_{ij}\}$ is a matrix basis for $C\ell$ then so is $\{\mathbf{m}_{ij}^K\}$. The simplicity of $C\ell$ guarantees the existence of a Clifford element B (unique up to a complex factor) that relates the two matrix bases $\{\mathbf{m}_{ij}\}$ and $\{\mathbf{m}_{ij}{}^K\}$ via a similarity transformation:

$$\mathbf{m}_{ij}{}^K = \mathrm{B}\mathbf{m}_{ij}\mathrm{B}^{-1}. \tag{18}$$

This implies, for $K = *, \eta^*$ and any Clifford element $\omega = \sum_{i,j} \gamma_{ij}(\omega)\,\mathbf{m}_{ij}$, that $\omega^K = \sum_{i,j} \gamma_{ij}^*(\omega)\,\mathbf{m}_{ij}^K = \sum_{i,j} \gamma_{ij}^*(\omega)\,\mathrm{B}\mathbf{m}_{ij}\mathrm{B}^{-1}$. This can be expressed as

$$\omega^K \mathrm{B} = \mathrm{B}\omega^{\sharp} \tag{19}$$

where $\sharp$ is defined by $\omega^{\sharp} = \sum_{i,j} \gamma_{ij}^*(\omega)\,\mathbf{m}_{ij}$ for any $\omega \in C\ell$. If $\mathbf{m}_{11} = P$ and $\{\mathbf{s}_i = \mathbf{m}_{i1}\}$ then $\Psi = \sum_i \psi_i\,\mathbf{s}_i$ for any spinor $\Psi \in C\ell\,P$. If $U = P^K \mathrm{B} P$ is chosen to define the spinor conjugation map it then follows from (19) that

$$\Psi^c = \Psi^K \mathrm{B} P = \mathrm{B}\Psi^{\sharp}. \tag{20}$$

Therefore the matrix equivalent of charge conjugation is

$$^c : \psi \mapsto \boldsymbol{B}\,\psi^* \tag{21}$$

with the square matrix $\boldsymbol{B}$ and column matrix ψ given by $\mathrm{B} = \sum_{i,j} B_{ij}\,\mathbf{m}_{ij}$ and $\Psi = \sum_i \psi_i\,\mathbf{s}_i$. Together with

$$\epsilon\,\boldsymbol{B}\,\boldsymbol{\gamma}^{a*} = \boldsymbol{\gamma}^a\,\boldsymbol{B} \tag{22}$$

($\epsilon = +1$ or -1 according to $K = *$ or η^*) derived from (19), Equation (21) is just the definition of charge conjugation in terms of γ-matrices and spinor components. Given U, (20) can be used to compute the corresponding matrix $\boldsymbol{B}$:

$$\mathbf{m}_{1i}\mathbf{m}_{j1}^c = \mathbf{s}_i{}^{\dagger}\mathbf{s}_j{}^c = B_{ij}\,P.$$

It can be shown that $\mathrm{B}^K \mathrm{B}$ is a real number hence by scaling, one can always arrange $\mathrm{B}^K \mathrm{B} = \pm 1$ corresponding to

$$\boldsymbol{B}^* \boldsymbol{B} = \pm 1$$

and $U^K U = \pm P$. This is the standard normalization condition (achieved in LUCY by scaling the charge conjugation map using (17)) ensuring that the norm of the Dirac current is preserved under charge conjugation.

Example 15 Following Example 14, a choice of the charge conjugation is made and the corresponding "charge conjugation matrix" $\boldsymbol{B}$, with respect to the chosen matrix basis, computed using `Ltransmat` described in Example 12.

```
> cc:=Spinorconjugation(2);
  (Spinorconjugation) defining spinor conjugation map associated with
   involution   &aster   ...
   cc := psi -> SPINCONJ(psi,2,e[1])
> '&c':=psi -> simp(exp(2*k*t)*cc(psi)):
> Bmat:=Ltransmat('&c');
         [    0         exp(-2*k*t) ]
  Bmat := [                         ]
         [ exp(2*k*t)       0       ]

> B:=convert(Bmat,Clifford);
      B := exp(2*k*t)*e[1]
> simp(B &v &aster B); # B &v &aster B is normalized to 1
      1
```

8. Spinor covariant derivative

A *spinor covariant derivative* with respect to any tangent vector field Y, denoted by S_Y, is a type-preserving directional derivative on spinors (Benn and Tucker, 1987), (Tucker, 1985). Of particular interest is the unique S_Y having the properties:

$$S_Y(\omega\Psi) = \nabla_Y\omega\Psi + \omega S_Y\Psi \quad \text{(Leibnitz rule over Clifford product)} \qquad (23)$$

$$Y(\Phi,\Psi) = (S_Y\Phi,\Psi) + (\Phi, S_Y\Psi) \quad \text{(compatibility with spinor adjoint)} \qquad (24)$$

$\forall\ \omega \in C\ell$, $\forall\ \Phi,\Psi \in C\ell\,P$ where ∇ is the *Levi-Civita* connection on Clifford elements. If the inner product above is fixed by the anti-involution $J = \xi$ or $\xi\eta$ these equations serve to determine S_Y uniquely. For the case $J = \xi^*$ or $\xi\eta^*$, one may additionally require compatibility with charge conjugation:

$$(S_Y\Psi)^c = S_Y\Psi^c. \qquad (25)$$

Although the usual argument leading to the existence and the explicit construction of such a S_Y is made with matrix representations and in an orthonormal basis, this is unnecessary (Benn and Tucker, 1987), (Bade and Jehle, 1953). LUCY identifies S_Y with the map:

$$\begin{aligned} S_Y: C\ell\,P &\longrightarrow & C\ell\,P \\ \Psi &\longmapsto & S_Y\Psi = (\nabla_Y\Psi)P + (\mathrm{i}_Y\lambda)\,\Psi \end{aligned} \qquad (26)$$

for some (in general complex) covector field λ. (23) is satisfied immediately. Using the definition of the spinor adjoint (5) and the spinor invariant inner product (7), the compatibility requirement (24) completely determines λ by

$$2\,(\mathrm{i}_Y\lambda)W = P\nabla_Y WP^J \qquad (27)$$

if $J = \xi$ or $\xi\eta$, but only fixes the real part of λ:

$$\mathrm{i}_Y(\lambda + \lambda^*)W = P\nabla_Y WP^J \qquad (28)$$

if $J = \xi^*$ or $\xi\eta^*$. In these cases, the imaginary part of λ can be determined by demanding (25), which yields

$$i_Y(\lambda - \lambda^*)U = -P^K \nabla_Y U P \tag{29}$$

where $K = *$ or η^* and (15) has been used.

To establish contact with the usual local matrix representation of S_Y one considers local dual bases, $\{X_a\}$ for TM and $\{e^a\}$ for T^*M in which the metric components are *constant* e.g. orthogonal bases. It follows that the Levi-Civita connection coefficients $\Gamma_{abc} = g_{ad}\Gamma^d{}_{bc}$ (cf. Equation (2)) are antisymmetric in their first two indices:

$$\Gamma_{abc} + \Gamma_{bac} = 0. \tag{30}$$

With the introduction of the real bivectors

$$\Sigma_a \equiv \frac{1}{8}\Gamma_{bca}[e^b, e^c] \tag{31}$$

where $[\omega, \sigma] \equiv \omega\sigma - \sigma\omega \;\; \forall \; \omega, \sigma \in C\ell$, it is straightforward to verify the relation $\nabla_{X_a} e^c = [\Sigma_a, e^c]$ and more generally

$$\nabla_{X_a} e^I = [\Sigma_a, e^I] \tag{32}$$

using (30). Here I denotes a multi-index and e^I is any element of the Clifford basis $\{1, e^a, e^b, \cdots, e^a e^b, \cdots\}$. Since the g^{ab} are constants we can find a matrix basis $\{\mathbf{m}_{ij}\}$ for $C\ell$ such that the $e^a = \sum_{i,j}\gamma^a_{ij}\mathbf{m}_{ij}$ give rise to a set of *constant* γ-matrices $\{\boldsymbol{\gamma}^a\}$. This implies that the two bases $\{e^I\}$ and $\{\mathbf{m}_{ij}\}$ for $C\ell$ are linearly related with some constant coefficients. Hence from (32)

$$\nabla_{X_a}\mathbf{m}_{ij} = [\Sigma_a, \mathbf{m}_{ij}]. \tag{33}$$

For any Clifford element $\omega = \sum_{i,j}\gamma_{ij}(\omega)\mathbf{m}_{ij}$ it follows that

$$\nabla_{X_a}\omega = \sum_{i,j}X_a(\gamma_{ij}(\omega))\mathbf{m}_{ij} + [\Sigma_a, \omega]. \tag{34}$$

In such a frame and matrix basis, a "canonical spinor adjoint" corresponding to a *constant* spinor metric matrix C satisfying (13) and (14) can be introduced yielding the Clifford element W, also with a constant matrix representation. Σ_a are real bivectors so $\Sigma_a{}^K = \Sigma_a$ for $K = *$ or η^* and $\Sigma_a{}^J = -\Sigma_a$ for $J = \xi, \xi\eta, \xi^*$ or $\xi\eta^*$. However the scalar $\langle \Sigma_a \rangle$ defined by $\langle \Sigma_a \rangle P = P\Sigma_a P$ is not necessarily real since P is complex in general. It then follows from (34) and noticing $W \in PC\ell P^J$ that

$$P\nabla_{X_a}WP^J = P[\Sigma_a, W]P^J$$
$$= P\Sigma_a W + (P\Sigma_a W^J)^J = (\langle \Sigma_a \rangle + \langle \Sigma_a \rangle^J)W. \tag{35}$$

If $J = \xi$ or $\xi\eta$ then $\langle \Sigma_a \rangle^J = \langle \Sigma_a \rangle$ and one can feed (35) into (27) to obtain the relation

$$\lambda = \langle \Sigma_a \rangle e^a. \tag{36}$$

Otherwise if $J = \xi^*$ or $\xi\eta^*$ then $\langle \Sigma_a \rangle^J = \langle \Sigma_a \rangle^*$ implying

$$i_{X_a}(\lambda + \lambda^*) = \langle \Sigma_a \rangle + \langle \Sigma_a \rangle^* \qquad (37)$$

by virtue of (28) and (35). If a "canonical charge conjugation" is introduced with *constant* charge conjugation matrix B satisfying (22) then the corresponding U is also associated with a constant matrix representation. Therefore for $J = \xi^*$ or $\xi\eta^*$ one further calculates that

$$\begin{aligned} P^K \nabla_{X_a} U P &= P^K [\Sigma_a, U] P \\ &= (P\Sigma_a U^K)^K - U\Sigma_a P = -(\langle \Sigma_a \rangle - \langle \Sigma_a \rangle^*) U \end{aligned} \qquad (38)$$

using the property $U \in P^K \, C\ell \, P$, $K = *$ or η^*. Substituting (38) into (29) gives

$$i_{X_a}(\lambda - \lambda^*) = \langle \Sigma_a \rangle - \langle \Sigma_a \rangle^* . \qquad (39)$$

Interestingly, by adding or subtracting (37) and (39) one obtains Equation (36) again. Thus λ can be used to set up a "canonical spinor covariant derivative" with respect to any canonical spinor adjoint and charge conjugation map: First adopt the spinor basis $\{s_i \equiv m_{i1}\}$ such that any spinor $\Psi \in C\ell \, P$ ($P = m_{11}$) can be written as $\Psi = \sum_i \psi_i \, s_i$ and exploit (34) to calculate the spinor covariant derivative of Ψ as follows:

$$\begin{aligned} S_{X_a}\Psi &= (\nabla_{X_a}\Psi)P + (i_{X_a}\lambda)\,\Psi \\ &= \sum_i X_a(\psi_i)\, s_i + [\Sigma_a, \Psi]P + (i_{X_a}\lambda)\,\Psi \\ &= \sum_i X_a(\psi_i)\, s_i + \Sigma_a\Psi - \Psi P\Sigma_a P + (i_{X_a}\lambda)\,\Psi \\ &= \sum_i X_a(\psi_i)\, s_i + \Sigma_a\Psi \end{aligned}$$

where the cancellation in the last step follows from (36). With $[M, N]$ denoting $MN - NM$ for any square matrices M and N, the above equation gives rise to the canonical representation

$$S_{X_a}: \; \psi \mapsto X_a(\psi) + \frac{1}{8}\Gamma_{bca}[\gamma^b, \gamma^c]\,\psi$$

where the column matrices ψ and $X_a(\psi)$ respectively contain the spinor components $\{\psi_i\}$ and the direction derivatives $\{X_a(\psi_i)\}$.

In LUCY the procedure `Spinorderiv` generates the spinor covariant derivative map. *Consider first the situation where a frame has been chosen in which the metric components are constants.* The command

```
Spinorderiv()
```

returns a binary map which maps any tangent vector field Y and spinor Ψ to $S_Y\Psi$ with the canonical S_Y. The corresponding covector field λ is computed from Equation (36) and, if required, is returned by calling `SPINCONNEX()`.

Example 16 The *Dirac operator* with contributions from the mass term μ, the electromagnetic potential A (a real covector field), and the electric charge q is defined (Benn and Tucker, 1987), (Benn and Tucker, 1985) by

$$Dirac = e^a S_{X_a} + iqA + \mu$$

with respect to any coframe $\{e^a\}$ and dual frame $\{X_a\}$. Hence $Dirac\,\Psi = 0$ is the Dirac equation for a complex spinor field Ψ, minimally coupled to a U(1) gauge field. For simplicity the next illustration evaluates the Dirac equation on a 2-dimensional flat spacetime using Cartesian coordinates.

```
> Cliffordsetup(0..1,{(0,0)=-1,(1,1)=1},e,X,{e[0]=d(t),e[1]=d(x)}):
> knls:=Spinorsetup(P)[1]
        knls:=[1, e[1]]
> alias(f1=f1_(t,x),f2=f2_(t,x)):
> comps:=[f1,f2];
> psi:=convert(zip((x,y)->x*y,comps,knls),'+') &v P;
        psi := f1*P + f2*&v(e[1],P)
> convert(psi,column);
            [ f1 ]
            [    ]
            [ f2 ]
> '&S':=Spinorderiv(); # The canonical spinor covariant derivative
        &S := simp@((Y,psi) -> (Y &C psi) &v P)
> DIRAC:=(psi,mu,qA)->                       # Dirac operator
>        (convert([seq(e[i] &v (X[{i}] &S psi),i=0..1)],'+')
>               + I*qA &v psi - mu*psi):
> qA:=Q*x*e[0]; # Interaction with a uniform electric field
> phi:=DIRAC(psi,mu,qA): # The Dirac equation phi=0 leads to two
> convert(phi,column);   # coupled 1st order differential equations
        [ - I*diff(f1,t) + diff(f2,x) + Q*x*f1 - mu*f1 ]
        [                                              ]
        [   I*diff(f2,t) + diff(f1,x) - Q*x*f2 - mu*f2 ]
> xi:=DIRAC(phi,-mu,qA):   # The induced two 2nd order differential
> mxi:=convert(xi,column): # equations decouple if Q=0
> mxi[1,1];
        - diff(f1,t,t) + diff(f1,x,x) - 2*I*Q*x*diff(f1,t)
        + Q^2*x^2*f1 - mu^2*f1 - Q*f2
> mxi[2,1];
        - diff(f2,t,t) + diff(f2,x,x) - 2*I*Q*x*diff(f2,t)
        + Q^2*x^2*f2 - mu^2*f2 + Q*f1
> alias(v1=v1_(t,x),v2=v2_(t,x),u1=u1_(t,x),u2=u2_(t,x)):
> ccomps:=[u1+I*v1,u2+I*v2]: # Make a spinor with complex components
> zeta:=convert(zip((x,y)->x*y,ccomps,knls),'+') &v P:
> convert(zeta,column);
            [ u1 + I*v1 ]
            [           ]
            [ u2 + I*v2 ]
```

```
> '&ad':=Spinoradjoint(4,1): # The Dirac adjoint
(Spinoradjoint) defining spinor adjoint map associated with
        anti-involution   @(&eta,&xi,&aster)   and parity   1   ...
> # A spinor inner product:
> sprod:=(phi,psi)->coeff(simp(&ad(phi) &v psi),P):
> # The Dirac current with positive density:
> curr:=convert([seq(sprod(zeta,-I*e[{i}] &v zeta)*e[i],i=0..1)],'+');
        curr := (u1^2+v1^2+u2^2+v2^2)*e[0] + (2*u1*v2-2*v1*u2)*e[1]
```

Example 17 This is an example yielding a spinor connection with a non-zero covector λ on a 2-dimensional flat spacetime with a uniformly accelerating orthonormal coframe $\{ e^0 = \sigma d\tau, e^1 = d\sigma \}$.

```
> Cliffordsetup(0..1,{(0,0)=-1,(1,1)=1},e,X,
> {e[0]=sigma*d(tau),e[1]=d(sigma)}):
> Spinorsetup(P=1/2*(1+e[0] &v e[1]));
        [[1, e[0]], [1, e[0]]]
> '&S':=Spinorderiv();
        &S := (Y,psi) ->
                simp(((Y &C psi) &v P) - 1/2/sigma*(Y &i e[0])*psi)
> lambda:=SPINCONNEX();
        lambda := - 1/2/sigma*e[0]
```

For a situation in which *the metric components in the ambient frame are not constant* (due either to a choice of coordinates or the presence of curvature), (Dray *et al.*, 1995), a more flexible facility for computing the spinor covariant is available. The binary map S_Y takes any tangent vector field Y and spinor Ψ to $S_Y \Psi$, the unique spinor covariant derivative compatible with a specified spinor inner product and charge conjugation operation. It is returned by calling **Spinorderiv** with the syntax:

$$\text{Spinorderiv}(\rho, h, \pi, \kappa, f).$$

The covector λ used in defining this S_Y is also available by calling

$$\text{SPINCONNEX}(\rho, h, \pi, \kappa, f).$$

Using the same correspondence employed for **Spinoradjoint**, the integer ρ indicates the choice of adjoint anti-involution. h is a real factor which scales the spinor adjoint via Equation (8). The third to fifth arguments (π, κ and f) are not required if $\rho = 1$ or 2 (Equation (27) is sufficient to determine λ). Otherwise the parity $\pi = \pm 1$ and the involution $*$ for $\kappa = 2$ or η^* for $\kappa = 3$ must be chosen in order that λ can be determined using Equations (28) and (29). f is a scalar chosen to normalize the charge conjugation map via Equation (17). Note that both h and f are allowed to be any functions of the coordinates as described in Sections 6 and 7. If no modification to the default spinor adjoint (or charge conjugation) map is intended, the value of h (or f) should be 1.

Example 18 Following from Examples 9 – 15, this example illustrates how to get the Spinor covariant differential operator compatible with the spinor adjoint and charge conjugation defined in the previous examples. A particular solution to a massless Dirac equation is also presented for arbitrary constants a_1, a_2, b_1 and b_2:

```
> '&S':=Spinorderiv(4,h,1,2,exp(2*k*t));
        &S := (X,psi) ->
            simp((X &C psi) &v P + 1/2*diff(h,t)/h*(Y &i e[0])*psi)
> u1:=a1/sqrt(h)*exp(-k*t):
> u2:=a2/sqrt(h)*exp(+k*t):
> f1:=b1:
> f2:=b2:
> psi;
        a1/h^(1/2)*exp(-k*t)*exp(I*b1)*P +
        a2/h^(1/2)*exp(k*t)*exp(I*b2)*&v(e[1],P)
> Q0:=X[{0}] &S psi;
        Q0:= - a1*k*exp(I*(I*k*t+b1))/h^(1/2)*P -
        a2*k*exp(I*(-I*k*t+b2))/h^(1/2)*&v(e[1],P)
> Q1:=X[{1}] &S psi;
        Q1 := I*exp(I*(-I*k*t+b2))*k*a2/h^(1/2)*P -
        I*exp(I*(-3*I*k*t+b1))*k*a1/h^(1/2)*&v(e[1],P)
> simp(e[0] &v Q0 + e[1] &v Q1); # massless Dirac equation satisfied
        0
> phi:=&c psi: # charge conjugate of psi satisfies the same equation
> simp( e[0] &v (X[{0}] &S phi) + e[1] &v (X[{1}] &S phi) );
        0
```

Acknowledgements

JS would like to thank the School of Physics & Chemistry at Lancaster University for kind hospitality. This work was partially supported by a Human Capital and Mobility Program of the European Union (RWT). CW is grateful to the Committee of Vice-Chancellors and Principals for an Overseas Research Studentship, Lancaster University for a Peel Studentship, and the School of Physics & Chemistry at Lancaster University for a School Studentship.

References

W. L. Bade and H. Jehle: 1953, 'An introduction to spinors', *Rev. Mod. Phys.* **25**, pp. 714–728.

I M. Benn and R. W. Tucker: 1985, 'The Dirac equation in exterior form', *Commun. Math. Phys.* **98**, pp. 53–63.

I. M. Benn and R. W. Tucker: 1987, *An Introduction to Spinors and Geometry with Applications in Physics*, Adam Hilger.

T. Dray, C. A. Manogue, J. Schray, R. W. Tucker, C. Wang: 1955, 'The construction of spinor fields on manifolds with smooth degenerate metrics', submitted for publication.

I. R. Porteous: 1969, 1981, *Topological Geometry*, Van Nostrand Reinhold, Cambridge University Press.

J. Schray and C. Wang: 1995, 'An algorithm for constructing bilinear forms on spinors', submitted for publication.

R. W. Tucker: 1985, 'A Clifford calculus for physical field theories' in *Clifford Algebras and their Applications in Mathematical Physics*, eds. J. S. R. Chisholm, A. K. Common, NATO ASI Series (C) **183**, pp. 177–199.

Appendix

A. List of Procedures

Procedure name	Description	Page(s)
&aster	Complex conjugation	123
&C	Levi-Civita covariant differential	126
&eta	Main Clifford algebra involution	123
&i	Interior product (Contraction operator)	124
&v	Clifford algebra product	122
&xi	Main Clifford algebra anti-involution	123
APPLYP	Internal usage	
Cliffordbasis	Ambient Clifford basis excluding unity	122,124
Cliffordsetup	Initialisation procedure for Clifford algebra	122,123,125
convert/Clifford	Converts to a Clifford element	129
convert/column	Converts to a column matrix representation	129
convert/down	Lowers indices	124
convert/Matrix	Converts to a matrix representation	129
convert/row	Converts to a row matrix representation	129
convert/up	Raises indices	124
expandP	Expands symbolic primitive	127
FILTER	Internal usage	
Matrixbasis	Total matrix basis for Clifford algebra	129
METRIC	Internal usage	
RIEMCONNEX	Table of Levi-Civita connection components	126
simp	Simplification procedure	122,127
SPINADJ	Output map from Spinoradjoint	133
SPINCONJ	Output map from Spinorconjugation	135
SPINCONNEX	Covector used in Spinorderiv	139,141
Spinoradjoint	Spinor adjoint	133
Spinorconjugation	Spinor conjugation	135
Spinorderiv	Levi-Civita spinor covariant differential	139,141
Spinorsetup	Initialisation procedure for spinor modules	127,128
type/&v	Internal usage	

COMPUTER ALGEBRA IN SPINOR CALCULATIONS

FRANCO PIAZZESE
Department of Physics, Politecnico.
Corso Duca degli Abruzzi 24
10129 Torino, Italy
e-mail: PIAZZESE@polfis.polito.it

Abstract. In this paper, some relationships between spinor and vector fields are investigated in detail, with the aid of computer algebra (calculations have been performed with MAPLE). In particular, an exact paradigm based on the representations of vector spaces is shown, strictly linking Pauli spinor fields with 3-dimensional rotations and vector fields. As an application, the problem of a classical particle in a central field of force is described in terms of one 2-component spinor field: the spinor equation of motion resembles the wave equation of quantum mechanics.

Key words: representations of vector spaces, rotations, Euler's angles, 2-component spinor fields, quantum mechanics, wave equation, MAPLE.

1. Introduction

Quantum mechanics has been universally recognized as one of the chief achievements of physics of all times, and the physical predictions it offers are verified with a high degree of accuracy. However, as it is well known, it is not easy to understand: a number of difficulties even logical is implicated so it may appear more as a collection of rules useful to get correct physical predictions than a fully-developed theory. In the state of things it can be hoped that in spite of the peculiar features of quantum mechanics comparisons and formal analogies among the various physical theories help in making the quantum theory more clear. Notice that a better understanding would not only give an intellectual satisfaction but it would also provide a more sound basis to future developments of the theory.

The starting motivation of the present research is a rather modest one, but in the way of thinking outlined above. As it is known, spinor fields are usually employed in quantum theories; on the contrary, they are not employed in macroscopic physics where tensor (in particular, scalar and vector) fields are employed. Thus, the following questions naturally arise. What is the reason of a different formalism? Is any physics hidden in the very formalism? The affirmation by Penrose (1983) comes to mind: "we have still not yet seen the full significance of spinors—particularly the 2-components ones—in the basic structure of physical laws".

To answer, understanding of both the tensor and spinor techniques is, obviously, needed, and it is well known that understanding of spinors has been a problem for many years. A partial reason for this is undoubtedly that many calculations with spinors are very cumbersome. Thus, in many cases they have not been fully developed and the results have been more guessed than exactly derived.

But now: (i) Pauli and Dirac spinor fields (i.e., the non-relativistic and relativistic

2-component spinor fields, respectively) and the related calculus have been clearly *geometrically* interpreted (cf. Piazzese (1992), (1993 a)), and (ii) using computer algebra is a common practice—which is both a relief and a safety in performing heavy calculations, although not necessarily a source of new great ideas...—. Thus, it appears that the time is ripe to start investigating the function of spinors in quantum physics.

This paper deals, first of all, with the (rather trivial) preliminary problem of describing (or "representing") a vector space with an isomorphic space, preserving the consistency of the representation when either of spaces undergoes a linear transformation, and an exact definition of "representation" is introduced (cf. Section 2). (Strangely enough, a number of Authors dealing with spinors—in particular with the "ideal" approach—do not seem concerned about this point, which has been criticized by Piazzese (1993 b)). The representations of the real affine 3-dimension space (the usual "ambient" space of classical mechanics) with both space $\mathbb{R}^3$ and the real space of the traceless Hermitian matrices of second order are discussed, in Section 3. With the aid of such representations, descriptions are given of both proper rotations and vector fields in terms of Pauli spinor fields and the results are offered as explicit expressions of the Euler angles (cf. Sections 4 and 5). Finally, in Section 6, such spinor fields are shown at work, in an alternative description of the classical problem of a particle in a central field of force. Apparently, this is more than visualizing the spinors in some way, which has been the aim of some Authors even recently (cf. Ablamowicz, Lounesto, Maks (1990)).

All calculations have been performed with the aid of MAPLE. The results are clear, and, out of the mystique usually joined with spinors, they look trivial.

2. Representations of vector spaces

In dealing with linear transformation groups on vector spaces, two points of view, called the "active" and the "passive" points of view, are standard (cf. Goldstein (1980, page 137)). In the former, only the components are transformed, but not the basis. In the latter, the basis undergoes a transformation inverse of the one on the components. As a result, the active point of view describes an automorphism of the space. On the contrary, in the passive one the space elements are not changed. This Section deals with the problem of "describing" a vector space V with another vector space U, in such a way that the description is preserved when either of spaces undergoes a linear transformation.

When only the passive point of view is taken into account, the foregoing requirement is clearly fulfilled if spaces V and U are isomorphic: a one-to-one correspondence ω exists, such that $U = \omega(V)$. But if we take into account also the active point of view, apparently the simplest possibility preserving the intrinsic feature of the description is requiring that, when an element $\mathbf{v} \in V$ is transformed into another one $\mathbf{v}' \in V$, a corresponding transformation $\mathbf{u} \to \mathbf{u}'$ occurs in U, in such a way that the one-to-one correspondence ω is preserved. In other words, we require that the following diagram commutes for any $\mathbf{v} \in V$

$$
\begin{array}{ccc}
\mathbf{v} & \to & \mathbf{u} \\
\downarrow & & \downarrow \\
\mathbf{v}' & \to & \mathbf{u}'
\end{array}
\tag{1}
$$

where

$$\mathbf{u}' = (\boldsymbol{\omega}(\mathbf{v}))' = \boldsymbol{\omega}(\mathbf{v}'). \tag{2}$$

If and only if this is the case, we call element $\mathbf{u} = \boldsymbol{\omega}(\mathbf{v})$ the "representation" $\boldsymbol{\omega}(\mathbf{v})$ in U of the element $\mathbf{v} \in V$, and space U the "representation" $\boldsymbol{\omega}(V)$ of space V.

We call the property described with (1) and (2) the "covariance property" of the representation. It is *a necessary and sufficient condition* for the *existence* of a representation of a vector space as above defined.

Remark. Since the one-to-one correspondence $\boldsymbol{\omega}$ is clearly invertible, element $\mathbf{v} = \boldsymbol{\omega}^{-1}(\mathbf{u})$ is the representation $\boldsymbol{\omega}^{-1}(\mathbf{u})$ in V of the element $\mathbf{u} \in U$, and space V the representation $\boldsymbol{\omega}(U)$ of space U.

3. Representing affine 3-dimensional spaces

Let V be the Euclidean affine 3-dimensional real space (which is the usual ambient space of classical mechanics). In this space, any pair of points O, P individualizes an affine vector $\vec{x}$, the Cartesian components of which coincide with the Cartesian coordinates x^h of P with respect to an arbitrary orthonormal frame $\{\vec{e}_h\}$ with the origin at O. In this Section, we check that space V may be represented with both the 3-dimensional real vector space $\mathbb{R}^3$, and the 3-dimensional space of the traceless Hermitian matrices of second order (here denoted by M).

To this end, we introduce one pair of one-to-one correspondences, between V and $\mathbb{R}^3$ and V and M, respectively. The former is defined by associating any vector $\vec{x} \in V$ with the following element of $\mathbb{R}^3$

$$
\mathbf{x} = \begin{bmatrix} x^1 \\ x^2 \\ x^3 \end{bmatrix} \equiv x^h \mathbf{E}_h
\tag{3}
$$

where

$$
\mathbf{E}_1 = \begin{bmatrix} 1 \\ 0 \\ 0 \end{bmatrix} ; \ \mathbf{E}_2 = \begin{bmatrix} 0 \\ 1 \\ 0 \end{bmatrix} ; \ \mathbf{E}_3 = \begin{bmatrix} 0 \\ 0 \\ 1 \end{bmatrix}
\tag{4}
$$

is the standard basis of $\mathbb{R}^3$. The latter is defined by associating any vector $\vec{x} \in V$ with the following element of M

$$
\mathbf{X} = \begin{bmatrix} x^3 & x^1 - ix^2 \\ x^1 + ix^2 & -x^3 \end{bmatrix} \equiv x^h \boldsymbol{\sigma}_h
\tag{5}
$$

where

$$
\boldsymbol{\sigma}_1 = \begin{bmatrix} 0 & 1 \\ 1 & 0 \end{bmatrix} ; \ \boldsymbol{\sigma}_2 = \begin{bmatrix} 0 & -i \\ i & 0 \end{bmatrix} ; \ \boldsymbol{\sigma}_3 = \begin{bmatrix} 1 & 0 \\ 0 & -1 \end{bmatrix}
\tag{6}
$$

are the well-known Pauli matrices, obviously making a basis of M. (The Einstein sum convention is understood everywhere in this paper. With the exception of spinor index a —assuming the values $1, 2$—, all the indices range from 1 to 3).

Consider now a proper rotation on V

$$\vec{x}' = \mathcal{O}\vec{x} \tag{7}$$

$\mathcal{O}$ denoting the rotation operator. Rotation (7) induces both an orthogonal transformation of $\mathbb{R}^3$

$$\mathbf{x}' = \mathbf{A}\mathbf{x} \tag{8}$$

being $\mathbf{A} \in \mathrm{SO}(3, \mathbb{R})$, and, owing to the well known homomorphism of $\mathrm{SO}(3, \mathbb{R})$ on to $\mathrm{SU}(2, \mathbb{C})$, an isometry of M (with determinant function as the quadratic form)

$$\mathbf{X}' = \mathbf{Q}\mathbf{X}\mathbf{Q}^* \tag{9}$$

where $\mathbf{Q} \in \mathrm{SU}(2, \mathbb{C})$, and the star $*$ denotes the adjoint matrix (in particular, the complex conjugate of a number), cf., e.g., Goldstein (1980), Section 4-5. Transformations (7), (8) and (9) can be summarized in the following commuting diagram

$$
\begin{array}{ccccc}
\vec{x} & \rightarrow & \mathbf{x} & \rightarrow & \mathbf{X} \\
\downarrow & & \downarrow & & \downarrow \\
\vec{x}' & \rightarrow & \mathbf{x}' & \rightarrow & \mathbf{X}'
\end{array} \tag{10}
$$

Since the covariance property is fulfilled, we conclude that both the one-to-one correspondences (3) and (5) define representations of V.

Remarks. 1) The elements of both the bases (4) and (6) are representations of the basis $\{\vec{e}_h\}$ of V. As it is known, suitably defining the scalar products on the corresponding spaces, both (4) and (6) are orthonormal, as is the basis they represent.

2) In accordance with the passive point of view (cf. Section 2), both representations (3) and (5) are invariant with respect to any change of the bases.

4. Spinor fields and the proper rotations

Some results about the spinor fields given by Piazzese (1993 a) are here summarized. For any non-relativistic 2-component (or Pauli) spinor field $\psi = \psi(P, t)$ the following "fundamental representations" can be defined (up to sign)

$$\psi = \begin{bmatrix} \psi^1 \\ \psi^2 \end{bmatrix} \; ; \; \tilde{\psi} = \begin{bmatrix} -\psi^{2*} \\ \psi^{1*} \end{bmatrix} \; ; \; \psi^* = [\, \psi^{1*} \quad \psi^{2*} \,] \; ; \; \tilde{\psi}^* = [\, -\psi^2 \quad \psi^1 \,] \tag{11}$$

being $\psi^a = \psi^a(P, t)$ the spinor components. (As no confusion arises, the same symbol denotes both the spinor field and the first of its representations (11)). The well-known spinor transformation law, induced by rotation (7), may be written in any of the following ways

$$\psi' = \mathbf{Q}'\psi \; ; \; \tilde{\psi}' = \mathbf{Q}'\tilde{\psi} \; ; \; \psi^{*\prime} = \psi^*\mathbf{Q}'^* \; ; \; \tilde{\psi}^{*\prime} = \tilde{\psi}^*\mathbf{Q}'^* \tag{12}$$

being $\mathbf{Q}' \in \mathrm{SU}(2, \mathbb{C})$. Thus, the following positive invariant norm (only vanishing for $\psi^1 = \psi^2 = 0$) can be introduced

$$\| \boldsymbol{\psi} \|^2 = \boldsymbol{\psi}^* \boldsymbol{\psi} \equiv \tilde{\boldsymbol{\psi}}^* \tilde{\boldsymbol{\psi}} = \psi^1 \psi^{1*} + \psi^2 \psi^{2*} \tag{13}$$

For any non-vanishing spinor field ψ, a "normalized" spinor field $\hat{\psi}$ (i.e., such that $\| \hat{\psi} \|^2 = 1$) can be defined, with the following components

$$\hat{\psi}^a = \psi^a / \| \boldsymbol{\psi} \| \tag{14}$$

The columns of $\mathbf{Q}$ and the rows of $\mathbf{Q}^*$ are the "fundamental representations" of a unique 2-component spinor field $\phi = \phi(P, t)$, as follows

$$\mathbf{Q} = \begin{bmatrix} \phi^1 & -\phi^{2*} \\ \phi^2 & \phi^{1*} \end{bmatrix}; \ \mathbf{Q}^* \equiv \mathbf{Q}^{-1} = \begin{bmatrix} \phi^{1*} & \phi^{2*} \\ -\phi^2 & \phi^1 \end{bmatrix} \tag{15}$$

Spinor field ϕ is normalized, i.e. such that

$$\| \phi \|^2 = \phi^1 \phi^{1*} + \phi^2 \phi^{2*} \equiv det\mathbf{Q} = 1 \tag{16}$$

Comparing (8) with (9), one may write, with the aid of (15), the elements of matrix $\mathbf{A}$ as the following spinor functions

$$\begin{aligned}
A^1{}_1 &= \tfrac{1}{2}((\phi^1)^2 - (\phi^2)^2 + (\phi^{1*})^2 - (\phi^{2*})^2) \\
A^2{}_1 &= \tfrac{i}{2}((\phi^1)^2 + (\phi^2)^2 - (\phi^{1*})^2 - (\phi^{2*})^2) \\
A^3{}_1 &= -\phi^1 \phi^2 - \phi^{1*} \phi^{2*} \\
A^1{}_2 &= \tfrac{1}{2i}((\phi^1)^2 - (\phi^2)^2 - (\phi^{1*})^2 + (\phi^{2*})^2) \\
A^2{}_2 &= \tfrac{1}{2}((\phi^1)^2 + (\phi^2)^2 + (\phi^{1*})^2 + (\phi^{2*})^2) \\
A^3{}_2 &= i(\phi^1 \phi^2 - \phi^{1*} \phi^{2*}) \\
A^1{}_3 &= \phi^1 \phi^{2*} + \phi^2 \phi^{1*} \\
A^2{}_3 &= i(\phi^1 \phi^{2*} - \phi^2 \phi^{1*}) \\
A^3{}_3 &= \phi^1 \phi^{1*} - \phi^2 \phi^{2*}
\end{aligned} \tag{17}$$

which may be written shortly, as follows

$$\begin{aligned}
A^i{}_1 &= \delta^{ij} \tfrac{1}{2}(\phi^* \sigma_j \tilde{\phi} + \tilde{\phi}^* \sigma_j \phi) \\
A^i{}_2 &= \delta^{ij} \tfrac{i}{2}(\phi^* \sigma_j \tilde{\phi} - \tilde{\phi}^* \sigma_j \phi) \\
A^i{}_3 &= \delta^{ij} \phi^* \sigma_j \phi
\end{aligned} \tag{18}$$

where δ^{ij} denotes the Kronecker symbol ($\delta^{ij} = 1$ for $i = j$, $\delta^{ij} = 0$ for $i \neq j$), as can be easily verified employing (6) and (11), with ϕ for ψ. Setting $\alpha, \gamma, \beta, \delta$ for $\phi^1, \phi^2, -\phi^{2*}, \phi^{1*}$, (17) agrees with Goldstein (1980), formula (4-64).

As it is known, quantities $A^i{}_j$ may be interpreted also as the components—with respect to basis $\{\vec{e}_j\}$—of the vectors $\{\vec{\epsilon}_j\}$ making an orthonormal triad field in V, which is rotated with respect to $\vec{e}_j$, being the rotation described by any of (7), (8), or (9), and, as a consequence, by spinor field ϕ. Thus, we have

$$\epsilon^i_j = A^i{}_j \tag{19}$$

Computer algebra allows an immediate direct check that vectors $\vec{\epsilon}_h$ with the components defined by (19) and (17)—or (18)—fulfill the orthonormal conditions. (Notice that $\vec{e}_j$ and $\vec{\epsilon}_j$ are denoted by $\mathbf{i}, \mathbf{j}, \mathbf{k}$ and $\mathbf{i}', \mathbf{j}', \mathbf{k}'$, respectively, in the Reference by Goldstein, cf. Section 4.1).

Once matrix $\mathbf{Q} \in SU(2, \mathbb{C})$ is known, the components of the corresponding normalized spinor ϕ are known from (15), and conversely. A number of explicit expressions of matrix $\mathbf{Q}$ as functions of various choices of parameters describing the rotation are offered in the literature. We employ the Euler angles ϕ, θ, ψ as defined in the Reference by Goldstein—often quoted in this paper—and assume for $\mathbf{A}$ the matrix describing the inverse rotation, which is helpful with the active point of view as employed in the following. (Thus, our matrix $\mathbf{A}$ coincides with the transpose of matrix $\mathbf{A}$ of the quoted Reference). As a result, we get (in the "x-convention")

$$\phi^1 = e^{-i(\psi+\phi)/2}\cos(\theta/2); \ \phi^2 = -ie^{-i(\psi-\phi)/2}\sin(\theta/2) \tag{20}$$

From (20), setting $\phi + \pi/2$ and $\psi - \pi/2$ for ϕ and ψ, respectively, one gets the following more symmetrical expressions (in the "y-convention")

$$\phi^1 = e^{-i(\psi+\phi)/2}\cos(\theta/2); \ \phi^2 = e^{-i(\psi-\phi)/2}\sin(\theta/2) \tag{21}$$

Notice that, as rotation field generally depends on space P and time t, the Euler angles in (20) and (21) are functions of P and t as well.

Introducing (20) in (17), one easily gets, with the aid of computer algebra

$$\begin{aligned}
A^1{}_1 &= \cos(\psi)\cos(\phi) - \sin(\psi)\cos(\theta)\sin(\phi) \\
A^2{}_1 &= \cos(\psi)\sin(\phi) + \sin(\psi)\cos(\theta)\cos(\phi) \\
A^3{}_1 &= \sin(\psi)\sin(\theta) \\
A^1{}_2 &= -\sin(\psi)\cos(\phi) - \cos(\psi)\cos(\theta)\sin(\phi) \\
A^2{}_2 &= -\sin(\psi)\sin(\phi) + \cos(\psi)\cos(\theta)\cos(\phi) \\
A^3{}_2 &= \cos(\psi)\sin(\theta) \\
A^1{}_3 &= \sin(\theta)\sin(\phi) \\
A^2{}_3 &= -\sin(\theta)\cos(\phi) \\
A^3{}_3 &= \cos(\theta)
\end{aligned} \tag{22}$$

Likewise, introducing (21) in (17)

$$\begin{aligned}
A^1{}_1 &= -\sin(\psi)\sin(\phi) + \cos(\psi)\cos(\theta)\cos(\phi) \\
A^2{}_1 &= \sin(\psi)\cos(\phi) + \cos(\psi)\cos(\theta)\sin(\phi) \\
A^3{}_1 &= -\cos(\psi)\sin(\theta) \\
A^1{}_2 &= -\cos(\psi)\sin(\phi) - \sin(\psi)\cos(\theta)\cos(\phi) \\
A^2{}_2 &= \cos(\psi)\cos(\phi) - \sin(\psi)\cos(\theta)\sin(\phi) \\
A^3{}_2 &= \sin(\psi)\sin(\theta) \\
A^1{}_3 &= \sin(\theta)\cos(\phi) \\
A^2{}_3 &= \sin(\theta)\sin(\phi) \\
A^3{}_3 &= \cos(\theta)
\end{aligned} \tag{23}$$

The above expressions may be compared with those of Goldstein (calculated without recourse to the spinors): (22) agrees with (4-46) and (4-47); on the contrary, (23) does not entirely agree with (B-3y) only because of a misprint in the Reference. (In fact, in the third raw ψ should be replaced by ϕ).

Some of expressions (22) and (23) are employed in the following Sections.

5. Representing 3-dimensional vector fields with Pauli's spinors

When a "reference" vector $\vec{x}_0 \in V$ is given, any vector $\vec{x} \in V$ may be individualized by means of a transformation which is the commuting product of a proper rotation by a dilatation. (A feature of this description is the presence of a gauge invariance, as any vector is invariant with respect to the rotations around its direction). Generalizing (7), one can write

$$\vec{x} = f\mathcal{O}(\vec{x}_0) \tag{24}$$

being f a suitable non-negative real invariant factor. Clearly, the same vector can be also individualized by either of the following representations

$$\mathbf{x} = f\mathbf{A}\mathbf{x}_0 \tag{25}$$

and

$$\mathbf{X} = f\mathbf{Q}\mathbf{X}_0\mathbf{Q}^* \tag{26}$$

in $\mathbb{R}^3$ and M, respectively (cf. Sections 2 and 3). The above equations describe also any real 3-dimensional vector field, if the involved quantities are regarded as fields as well. The following diagram obviously commutes

$$
\begin{array}{ccccc}
\vec{x}_0(P,t) & \rightarrow & \mathbf{x}_0(P,t) & \rightarrow & \mathbf{X}_0(P,t) \\
\downarrow & & \downarrow & & \downarrow \\
\vec{x}(P,t) & \rightarrow & \mathbf{x}(P,t) & \rightarrow & \mathbf{X}(P,t)
\end{array}
\tag{27}
$$

Introducing (18) in (25) we connect the above vector fields with the normalized Pauli spinor fields

$$x^h = \delta^{hj} f\{\frac{1}{2}(\phi^* \sigma_j \tilde{\phi} + \tilde{\phi}^* \sigma_j \phi)x_0^1 + \frac{i}{2}(\phi^* \sigma_j \tilde{\phi} - \tilde{\phi}^* \sigma_j \phi)x_0^2 + \phi^* \sigma_j \phi \; x_0^3\} \tag{28}$$

Both fields ϕ and f may be put together, by defining spinor field ψ, with the components

$$\psi^a = \sqrt{f}\phi^a. \tag{29}$$

Thus, (28) can be rewritten, as follows

$$x^h = \delta^{hj}\{\frac{1}{2}(\psi^* \sigma_j \tilde{\psi} + \tilde{\psi}^* \sigma_j \psi)x_0^1 + \frac{i}{2}(\psi^* \sigma_j \tilde{\psi} - \tilde{\psi}^* \sigma_j \psi)x_0^2 + \psi^* \sigma_j \psi \; x_0^3\} \tag{30}$$

From (30), (29) and (13), one gets

$$\|\psi\|^2 = f = x/x_0 \tag{31}$$

being x, x_0 the modulus of $\vec{x}, \vec{x}_0$, respectively.

Expressions (30) become particularly simple and symmetrical if the reference vector field $\vec{x}_0$ is chosen to be the (constant) unit vector $\vec{e}_3$ of the "z-axis"

$$x^h = \delta^{hk} \psi^* \sigma_k \psi. \tag{32}$$

The above expressions are the 3-dimensional counterparts of some of the well known Dirac bilinear expressions (cf. Berestetskiĭ, Lifshitz, Pitayevskiĭ (1971), Section 28).

Vector components (32) may be rewritten as functions of the Euler angles. With the aid of (29) and either of (20) or (21) one gets in the x-convention

$$x^1 = x\sin(\theta)\sin(\phi); \quad x^2 = -x\sin(\theta)\cos(\phi); \quad x^3 = x\cos(\theta) \tag{33}$$

and in the y-convention

$$x^1 = x\sin(\theta)\cos(\phi); \quad x^2 = x\sin(\theta)\sin(\phi); \quad x^3 = x\cos(\theta) \tag{34}$$

respectively. In the latter case, x, ϕ, θ coincide with the usual spherical coordinates (but in the tangent fiber). (Of course, expressions (33) and (34) could be derived directly, without any recourse to spinor fields!)

It clearly appears that neither the vector components (33) nor (34) involve angle ψ. Thus, the full rotational information contained in spinor field ψ is not used. However angle ψ *is* involved when the same rotation describes also another vector field, with *another* direction (cf. the next Section).

6. An example: Pauli spinor fields at work!

Consider the classical problem of a particle Q with mass m in a central field of force, being O the center (e.g. a planet in the gravitational field of the Sun). This problem is well known as are the solutions, but the interest in it is due to the fact that it can be easily expressed in the language of spinor fields. In this way, spinors are seen at work, in a fully understandable classical context.

First of all we summarize the problem, as follows. Introduce an inertial reference frame with the origin at O and the coordinate axes parallel to the unit vectors $\vec{e}_h$. Assume that the angular momentum $\vec{L}$ of Q with respect to O has the direction of $\vec{\epsilon}_3$, thus we can write

$$\vec{L} \equiv \vec{r} \times m\vec{v} = L\vec{\epsilon}_3 \tag{35}$$

where $\vec{r}, \vec{v}$ denote the position and the velocity of the particle, and L the modulus of $\vec{L}$. The components of $\vec{L}$ are (y-convention, cf. (34))

$$L^1 = L\sin(\theta)\cos(\phi); \quad L^2 = L\sin(\theta)\sin(\phi); \quad L^3 = L\cos(\theta). \tag{36}$$

The conservation of (35) assures that L, θ, ϕ are constants of motion. This implies, in particular, that motion is plane and that the plane of motion passes through the origin O and is normal to $\vec{\epsilon}_3$. As a result, vector $\vec{r}$ can be chosen with the same direction of the unit vector $\vec{\epsilon}_1$, by putting

$$\vec{r} = r\vec{\epsilon}_1 \tag{37}$$

being r the modulus of $\vec{r}$. Since $\vec{\epsilon}_1$ is always in the plane of motion, it is a function only of the angle ψ (ϕ, θ being constant). In other words, the position of particle Q can be described with the polar coordinates r, ψ. Differentiating (37), one gets the velocity

$$\vec{v} = \dot{r}\vec{\epsilon}_1 + r\dot{\psi}\vec{\epsilon}_2 \tag{38}$$

as the sum of both the "radial" and "transverse" velocities. Introducing (37) and (38) in (35), one has

$$L = mr^2\dot{\psi}. \tag{39}$$

As it is well known, from (39) and the energy conservation equation

$$E = \frac{1}{2}m(\dot{r}^2 + r^2\dot{\psi}^2) + V(r) \equiv \frac{1}{2}m\dot{r}^2 + \frac{L^2}{2mr^2} + V(r) \tag{40}$$

the problem is led to the quadratures.

Now, we consider a spinor description of the same problem. In fact, with the aid of (28) one is in a position to describe the dynamical state of the system as functions of the unique normalized spinor field ϕ (and the coordinate r), as follows

$$r^h = r\delta^{hk}\frac{1}{2}(\phi^* \sigma_k \tilde{\phi} + \tilde{\phi}^* \sigma_k \phi), \tag{41}$$

$$v^h = \dot{r}\delta^{hk}\frac{1}{2}(\phi^* \sigma_k \tilde{\phi} + \tilde{\phi}^* \sigma_k \phi) + r\dot{\psi}\delta^{hk}\frac{i}{2}(\phi^* \sigma_k \tilde{\phi} - \tilde{\phi}^* \sigma_k \phi). \tag{42}$$

However, (42) can be also directly obtained by differentiating (41). To this end, spinor field ϕ has to be differentiated and a calculation from (21) gives

$$\frac{\partial \phi}{\partial t} \equiv \dot{\phi} = -\frac{i}{2}(\dot{\psi} + \dot{\phi}\cos(\theta))\phi + \frac{1}{2}(\dot{\theta} + i\dot{\phi}\sin(\theta))e^{-i\psi}\tilde{\phi} \tag{43}$$

which being ϕ, θ constant reduces to

$$\frac{\partial \phi}{\partial t} = -\frac{i}{2}\dot{\psi}\phi. \tag{44}$$

The quantity $\dot{\psi}$ may be easily expressed as a function of the conserved quantities L and E: from (39) and (40) one has

$$\dot{\psi} = \frac{2}{L}(E - \frac{1}{2}m\dot{r}^2 - V(r)). \tag{45}$$

Introducing (45) in (44), one gets the following equation

$$iL\frac{\partial \phi}{\partial t} = (E - \frac{1}{2}m\dot{r}^2 - V(r))\phi \tag{46}$$

which can be regarded as describing the dynamical evolution of the system. The resemblance between (46) and the "wave equation" of quantum mechanics is remarkable (cf., e.g., Landau-Lifshitz (1958), Section 8).

Conversely, assume that equation (46) is given. To extract the information it contains we introduce the general expression (43) of the derivative of the spinor field

$$\frac{L}{2}(\dot{\psi} + \dot{\phi}\cos(\theta))\phi + \frac{iL}{2}(\dot{\theta} + i\dot{\phi}\sin(\theta))e^{-i\psi}\tilde{\phi} = (E - \frac{1}{2}m\dot{r}^2 - V(r))\phi. \tag{47}$$

Since $\{\phi, \tilde{\phi}\}$ is clearly a basis for Pauli spinors, equation (47) can only be fulfilled, if the coefficient of $\tilde{\phi}$ is zero. This implies that $\dot{\theta} = \dot{\phi} = 0$, i.e., both θ and ϕ are constant, being their values given by the initial conditions. As a result, equation (47) reduces to (45). From this equation and (39) the problem is led once again to the quadratures. We conclude that spinor equation (46)—together with (39)—offers a description of the problem which is equivalent to the classical one.

Remark. It can appear singular that a non-normalized spinor field ψ putting the radial coordinate r together with the normalized spinor field ϕ has not been introduced, as it was done in Section 5. Although such a description is quite feasible, it is worth noticing that this approach does not seem so close to quantum mechanics, where the modulus of the wave function gives the probability density to find a particle (see Pauli theory for the spinning electron, cf. e.g. Persico (1950), Section III, 45).

7. Concluding remarks

As the foregoing Sections show, an exact paradigm exists based on the representations of vector spaces, strictly linking Pauli spinor fields with 3-dimensional rotations and vector fields. It appear that, when dealing with spinor fields, if mathematics has to be respected (and in particular, the consistency of the representations, cf. Section 2), the paradigm can not be overridden. This fact, although introducing some constraints in the use of spinor fields, surely helps in the interpretation. Section 6 offers only a first example.

Acknowledgments

This work has been produced under the auspices of the Italian Council for Research, C. N. R. (G. N. F. M.), with the support of the Italian Ministry of Research.

References

R. Ablamowicz, P. Lounesto, J. Maks: 1991, 'Conference report. Second workshop on "Clifford algebras and their applications in mathematical physics", Université des Sciences et Techniques du Languedoc, Montpellier, France, 1989', *Found. Phys.* **21**, pp. 735–748.

V. B. Berestetskiĭ, E. M. Lifshitz, L. P. Pitayevskiĭ: 1971, *Relativistic quantum theory (Part I)*, Pergamon Press, Oxford.

H. Goldstein: 1980, *Classical mechanics* 2nd Ed., Addison-Wesley, Reading.

L. D. Landau, E. M. Lifshitz: 1958, *Quantum mechanics (non-relativistic theory)*, Pergamon Press, Oxford.

R. Penrose: 1983, 'Spinors and torsion in general relativity', *Found. Phys.* **13**, pp. 325–339.

F. Piazzese: 1992, 'On the classical theory of elementary spinors', in M. Goldberg, D. Hershkowitz, H. Schneider (eds.): *Report on "Sixth Haifa Conference on Matrix Theory"*, Haifa, Israel, *1990*, *Lin. Alg. Appl.* **167**, pp. 242–247.

F. Piazzese: 1993 a, '2- and 4-component spinors. What, in fact are they?', *J. Natural Geometry* **3**, pp. 59–79.

F. Piazzese: 1993 b, 'The "ideal" approach to spinors reconsidered', in F. Brackx, R. Delanghe, H. Serras (eds.): *Proceedings of the Third Conference on "Clifford Algebras and their Applications in Mathematical Physics"*, Deinze, Belgium, *1993*, Kluwer, Dordrecht, pp. 325–332.

E. Persico: 1950, *Fundamentals of quantum mechanics*, Prentice-Hall, Englewood Cliffs.

VAHLEN MATRICES FOR NON-DEFINITE METRICS

J. CNOPS
Universiteit Gent,
Galglaan 2, B-9000 Gent, Belgium
e-mail: jc@cage.rug.ac.be

Abstract. The relation between Möbius transformations -conformal mappings mapping generalized spheres to spheres- of a Euclidean or pseudo-Euclidean space, two by two matrices with entries in a Clifford algebra, and covering groups like $O(p+1,q+1)$ are well known. Vahlen gave, already in the beginning of this century, a characterization of matrices describing Möbius transformations for the Euclidean case (without using the background of orthogonal groups). In this paper we give a criterion, in a form quite close to Vahlen's, for the general case.

Key words: Möbius transformations, Lipschitz group.

1. Introduction

In this section we give an overview of the theory of Möbius transformations. Most of the material can be found in a more elaborate form in the excellent paper Fillmore *et al.* (1990). First a few remarks on the Clifford algebras used (for full information on the notions used, see Delanghe *et al.* (1992) or Porteous (1981)). A Clifford algebra $Cl_{p,q}$ can be described using a standard basis $e_1, \ldots, e_n$ of the underlying space $\mathbb{R}^{p,q}$ (where $n = p + q$). These satisfy $e_i^2 = -e_j^2 = 1$ for $1 \leq i \leq p < j \leq n$. To describe the algebra $Cl_{p+1,q+1}$ two elements e_+ and e_- will be added to obtain a basis of $\mathbb{R}^{p+1,q+1}$, such that $e_\pm^2 = \pm 1$. Vectors of the underlying space will be denoted by boldface letters ($\mathbf{x}$, $\mathbf{y}$, $\mathbf{Y}$, etc.). There are three important morphisms (defined in the same way on $Cl_{p,q}$ and $Cl_{p+1,q+1}$):

- The Clifford-conjugation defined by $\bar{\mathbf{x}} = -\mathbf{x}$, $\overline{ab} = \bar{b}\,\bar{a}$.
- The grade involution defined by $\hat{\mathbf{x}} = -\mathbf{x}$, $\widehat{ab} = \hat{a}\hat{b}$.
- Reversion, defined by $\tilde{\mathbf{x}} = \mathbf{x}$, $\widetilde{ab} = \tilde{b}\tilde{a}$.

The Clifford algebra $Cl_{p+1,q+1}$ is isomorphic to the algebra of 2×2 matrices with entries in $Cl_{p,q}$, written as $\mathrm{Mat}(2, Cl_{p,q})$, with the isomorphism

$$a + be_- + ce_+ + de_+e_- \longrightarrow \begin{pmatrix} a+d & -b+c \\ \hat{b}+\hat{c} & \hat{a}-\hat{d} \end{pmatrix},$$

or, in the other direction

$$\begin{pmatrix} a & b \\ c & d \end{pmatrix} \longrightarrow \frac{1}{2}\left((a + \hat{d}) + (a - \hat{d})e_+e_- + (b + \hat{c})e_+ + (-b + \hat{c})e_-\right).$$

For this theorem we refer to Porteous (1981). In the sequel both the matrix notation and the classical notation for an element of $Cl_{p+1,q+1}$ will be used. An explicit calculation gives the formulae for the automorphism and the anti-automorphisms of $Cl_{p+1,q+1}$ in terms of the matrix elements:

$$\widehat{\begin{pmatrix} a & b \\ c & d \end{pmatrix}} = \begin{pmatrix} \hat{a} & -\hat{b} \\ -\hat{c} & \hat{d} \end{pmatrix},$$

$$\overline{\begin{pmatrix} a & b \\ c & d \end{pmatrix}} = \begin{pmatrix} \bar{d} & -\tilde{b} \\ -\tilde{c} & \tilde{a} \end{pmatrix},$$

$$\begin{pmatrix} a & b \\ c & d \end{pmatrix}^{\sim} = \begin{pmatrix} \bar{d} & \bar{b} \\ \bar{c} & \bar{a} \end{pmatrix}.$$

It should be noticed that the matrix form of a vector $\mathbf{Y} \in \mathbb{R}^{p+1,q+1}$ is $\begin{pmatrix} \mathbf{y} & \mu \\ \nu & -\mathbf{y} \end{pmatrix}$ where $\mathbf{y}$ is a vector in $\mathbb{R}^{p,q}$ and μ and ν are scalars.

The *pseudodeterminant* $\lambda(A)$ of the matrix $A = \begin{pmatrix} a & b \\ c & d \end{pmatrix}$ is given by $a\tilde{d} - b\tilde{c}$. Notice that this is the first entry of the product $A\overline{A}$ and so, if $A\overline{A}$ is a scalar, then $\lambda(A) = A\overline{A}$. Moreover in this case $\lambda(BA) = \lambda(A)\lambda(B)$ for arbitrary B, since $BA\overline{(BA)} = BA\overline{A}\,\overline{B} = \lambda(A)B\overline{B}$.

The square of a vector is always a scalar. Hence a vector $\mathbf{x}$ is invertible iff $\mathbf{x}^2 \neq 0$. The group generated by invertible vectors is called the Lipschitz group (or the Clifford group) $\Gamma_{p,q}$. For an element a in $\Gamma_{p,q}$ there exists an orthogonal transformation on $\mathbb{R}^{p,q}$ sending $\mathbf{x}$ to $a\mathbf{x}\hat{a}^{-1}$, somewhat anomalously called the $\mathbf{Pin}(p,q)$ group representation of a.

A unit vector is a vector $\mathbf{x}$ such that $\mathbf{x}^2 = \pm 1$. The group generated by unit vectors is called the $\mathbf{Pin}(p,q)$ group and it is obviously a subgroup of $\Gamma_{p,q}$; the spin group $\mathbf{Spin}(p,q)$ consists of even products of vectors. The $\mathbf{Pin}(p,q)$ group representation gives a double covering of $O(p,q)$.

Similar definitions of course hold in $Cl_{p+1,q+1}$.

For the $\mathbf{Pin}(p,q)$-group representation we need $\hat{A}^{-1}$ for $A \in \Gamma_{p+1,q+1}$. For such A the equality $\hat{A}^{-1} = \pm \tilde{A}$ holds. As we shall work projectively, the sign doesn't matter, and we shall be using $\tilde{A}$.

Möbius transformations. Let $\mathbb{R}^{p,q}$ be a finite-dimensional metric space. Möbius transformations are those conformal mappings of $\mathbb{R}^{p,q}$ which map generalized spheres to generalized spheres. For $p + q > 2$ this second condition is superfluous, as in this case every conformal mapping is a Möbius transformation (see Haantjes (1937)). The group of Möbius transformations will be denoted by $\mathcal{M}(p,q)$ (notice we do not demand that a Möbius transformation be sense preserving). It is well known that the orthogonal group $O(p+1, q+1)$ gives a double covering of $\mathcal{M}(p,q)$, and so the group $\mathbf{Pin}(p+1, q+1)$ (which gives a double covering of $O(p+1, q+1)$), is a good instrument to express Möbius transformations.

A *sphere* in this context is a set of points of the form

$$\{\mathbf{y} : (\mathbf{y} - \mathbf{m}) \cdot (\mathbf{y} - \mathbf{m}) = -r^2\}$$

where r^2 is a real scalar, but not necessarily positive. If $r^2 = 0$, we speak of a null sphere. $\mathbf{m}$ is called the center of the sphere. The equation can also be written as

$$-\mathbf{y}^2 - 2\mathbf{y} \cdot \mathbf{m} + \left(-\mathbf{m}^2 - r^2\right) = 0. \tag{1}$$

This sphere is identified with the ray $\vec{s}$ in $\mathbb{R}^{p+1,q+1}$ through the point $\mathbf{m} + (1/2)(1 + \mathbf{m}^2 + r^2)e_+ + (1/2)(1 - \mathbf{m}^2 - r^2)e_-$. As an element of the Clifford algebra it has matrix notation

$$\begin{pmatrix} \mathbf{m} & -\mathbf{m}^2 - r^2 \\ 0 & -\mathbf{m} \end{pmatrix}.$$

The rays in $\mathbb{R}^{p+1,q+1}$ which do not represent spheres represent generalized spheres. These are hyperplanes, or the null sphere at infinity, which should be added to $\mathbb{R}^{p,q}$ to make all Möbius transformations bijective. The null sphere at infinity is the image under the transformation $\mathbf{x} \to \mathbf{x}^{-1}$ of the null sphere with center $\mathbf{0}$, and so for Euclidean space has one point (one point compactification of the space), while in the non-definite case it is the image of the light cone through zero. Points in $\mathbb{R}^{p,q}$ can be identified with null spheres. For a null sphere $\vec{s}$ is isotropic.

It is not hard to prove that the spheres $\vec{s}$ and $\vec{k}$ intersect orthogonally if and only if $\vec{s}$ and $\vec{k}$ are orthogonal in $C\ell_{p+1,q+1}$. A point lies on the sphere $\vec{s}$ if and only if its associated null sphere intersects $\vec{s}$ orthogonally. More generally, if two spheres intersect, the (non-oriented) angle of intersection in $\mathbb{R}^{p,q}$ is the angle between the rays $\vec{s}$ and $\vec{k}$. Hence the relation between Möbius group and $O(p+1, q+1)$ becomes immediately clear: since a Möbius transformation maps generalized spheres to generalized spheres it induces in a natural way a mapping of rays in $\mathbb{R}^{p+1,q+1}$, which preserves angles between these rays, and so induces an orthogonal mapping (there is a choice between two possibilities) of $\mathbb{R}^{p+1,q+1}$. This procedure can be used in the opposite direction. The orthogonal mapping in $\mathbb{R}^{p+1,q+1}$ τ given by $\tau(\mathbf{Y}) = -\mathbf{Y}$ induces the identity in $\mathbb{R}^{p,q}$. This way $O(p+1, q+1)$ is a double covering of $\mathcal{M}(p,q)$. Notice that

- If $p + q$ is odd, then τ is indirect. Hence, for a given $g \in \mathcal{M}(p,q)$, there are two elements of $O(p+1, q+1)$ over g, exactly one of which is direct, and $SO(p+1, q+1)$ gives a single covering of $\mathcal{M}(p,q)$.

- If $p + q$ is even, then τ is direct, and $SO(p+1, q+1)$ gives a double covering of the subgroup of even Möbius transformations.

Let $A = \begin{pmatrix} a & b \\ c & d \end{pmatrix}$ be an element of $\mathbf{Pin}(p+1, q+1)$. The sphere $\vec{s}$ is then mapped to the sphere $A\vec{s}\hat{A}^{-1}$ (or $A\vec{s}\tilde{A}$), and the point $\mathbf{x}$ to the point

$$g\mathbf{x} = \frac{a\mathbf{x} + b}{c\mathbf{x} + d}$$

(we use the notation $\frac{u}{v}$ instead of uv^{-1} or u/v for aesthetical reasons: the inverse is always supposed to be taken from the right).

2. Characterization of Lipschitz group

In this section a characterization of Lipschitz group elements in terms of matrices is given. This characterization generalizes the characterization of the matrices given by Vahlen, which is valid in the case $pq = 0$. It is given here in a slightly different form from the one in Ahlfors (1986) because we do not use $\Gamma_{p+1,q}$.

Theorem 2.1 *(Vahlen) A matrix* $A = \begin{pmatrix} a & b \\ c & d \end{pmatrix}$ *in* $\mathrm{Mat}(2, C\ell_{0,n})$ *represents a Möbius transformation in* $\mathbb{R}^{0,n}$ *if and only if*
(a) $a, b, c, d \in \Gamma_{0,n} \cup \{0\}$
(b) $a\tilde{b}, c\tilde{d}, \tilde{c}a, \tilde{d}b \in \mathbb{R}^{0,n}$
(c) the pseudodeterminant of A *is a scalar.*

We can compare this with the characterization for the general case, due to Maks (1989) in this form. It was given in Elstrodt *et al.* (1987) for the paravector formalism:

Theorem 2.2 *(Maks) A matrix* $A = \begin{pmatrix} a & b \\ c & d \end{pmatrix}$ *belongs to the Lipschitz group* $\Gamma_{p+1,q+1}$ *if and only if*
(i) $a\tilde{a}, b\tilde{b}, c\tilde{c}, d\tilde{d}$ *are scalars;*
(ii) $b\tilde{d}, a\tilde{c} \in \mathbb{R}^{p,q}$ *;*
(iii) the pseudodeterminant of A *is a non-zero scalar;*
(iv) for each $\mathbf{y} \in \mathbb{R}^{p,q}$, $a\mathbf{y}\tilde{d} + b\mathbf{y}\tilde{c}$ *is a vector;*
(v) $a\tilde{b} = b\tilde{a}, \; c\tilde{d} = d\tilde{c}$.
(vi) for each $\mathbf{y} \in \mathbb{R}^{p,q}$, $a\mathbf{y}\tilde{b} + b\mathbf{y}\tilde{a}$ *and* $c\mathbf{y}\tilde{d} + d\mathbf{y}\tilde{c}$ *are scalars.*

The proof of this is based on the fact that A is in Lipschitz group if and only if $A\overline{A}$ is a scalar and $AY\overline{A} \in \mathbb{R}^{p+1,q+1}$ for each $Y \in \mathbb{R}^{p+1,q+1}$.

For a better characterization we wish to eliminate conditions (iv) and (vi) of the theorem and write condition (v) in a form comparable to the Vahlen condition (b). However for the general case ($p \neq 0 \neq q$) Maks gave an example of a matrix A which is in Lipschitz group but where none of the entries is invertible (and all are non-zero), which proves that condition (a) has to be modified, and which will necessitate a completely different method in order to prove the characterization. Indeed a generalization of Lipschitz group is needed, which will replace $\Gamma_n \cup \{0\}$ in (a). Following Fillmore *et al.* (1990) we define $T(p,q)$ to be the set of all products of vectors in $\mathbb{R}^{p,q}$. Clearly Lipschitz group $\Gamma_{p,q}$ is the group of invertible elements of $T(p,q)$ and for p or q equal to zero $T(p,q)$ equals $\Gamma_{p,q} \cup \{0\}$. With this set we can formulate our theorem:

Theorem 2.3 *Lipschitz group* $\Gamma_{p+1,q+1}$ *is the set of matrices* A *of the form* $\begin{pmatrix} a & b \\ c & d \end{pmatrix}$ *satisfying*
(a) a, b, c, d *are elements of* $T(p,q)$.
(b) $b\tilde{d}, a\tilde{c}, \tilde{a}b, \tilde{c}d \in \mathbb{R}^{p,q}$.
(c) the pseudodeterminant of A, $\lambda(A)$, *is a non-zero scalar.*

Notice that not only condition (a) has been changed, but also that condition (b) is slightly different from the condition in Vahlen's theorem. Indeed, in the case $p = 0$

or $q = 0$ the order in the products (e.g. $a\tilde{b}$ instead of $\tilde{a}b$) is immaterial. This is no longer true in the general case. Take as an example, in $\mathbb{R}^{1,2}$, $a = (1 + e_{12})$ and $b = e_3(1 + e_{12})$. Then $a\tilde{b} = 0$, while $\tilde{a}b = 2(1 - e_{12})e_{13}$, which is clearly not a vector.

In order to prove the result we shall need several lemmas. In the first one we collect some important properties of $T(p, q)$.

Lemma 2.4 *For each non-zero $a \in T(p, q)$ and $\mathbf{x}$, $\mathbf{y}$ arbitrary*
(i) $a\tilde{a} = \tilde{a}a$ is a scalar. It is non-zero if and only if a is invertible.
(ii) $a\mathbf{y}\tilde{a}$ again is in $\mathbb{R}^{p,q}$;
(iii) Either $a\mathbf{z}\tilde{a} = 0$ for all vectors $\mathbf{z}$ or $a = \mathbf{t}\alpha$ for some $\alpha \in \Gamma_{p,q}$ and some vector $\mathbf{t}$;
(iv) $1 + \mathbf{x}\mathbf{y} \in T(p, q)$.
(v) If there exists a vector $\mathbf{z}$ such that $a\mathbf{z}\tilde{a} \neq 0$ then $\mathbf{x}a + a\mathbf{y} \in T(p, q)$.

Proof.
(i) is obvious.

For (ii) it is clearly sufficient to show this when a is a vector, so suppose $a = \mathbf{t}$. Then $\mathbf{t}\mathbf{y}\tilde{\mathbf{t}} = (\mathbf{t} \cdot \mathbf{y})\tilde{\mathbf{t}} - \mathbf{y}(\mathbf{t}\tilde{\mathbf{t}})$.

(iii) is obvious if a is invertible, when the second alternative is true. Suppose now $a\tilde{a} = 0$. Then a can be written as a product $\mathbf{t}_1\mathbf{t}_2 \ldots \mathbf{t}_k$ where we can assume that all the not-invertible vectors are in the beginning. We prove the proposition explicitly for a product of two vectors, the induction step being easy. Let $a = \mathbf{s}\mathbf{t}$ where $\mathbf{s}$ and $\mathbf{t}$ are not invertible. If $\mathbf{s} \cdot \mathbf{t} \neq 0$ then $(\mathbf{s} + \mathbf{t})$ is invertible and $a = \mathbf{s}(\mathbf{s} + \mathbf{t})$, giving the second alternative. If $\mathbf{s} \cdot \mathbf{t} = 0$ then for $\mathbf{z}$ arbitrary, $\mathbf{t}\mathbf{z}\mathbf{t} = 2(\mathbf{t} \cdot \mathbf{z})\mathbf{t}$ while $\mathbf{s}\mathbf{t}\mathbf{s} = 2(\mathbf{s} \cdot \mathbf{t})\mathbf{t} = 0$. The composition gives $a\mathbf{z}\tilde{a} = 0$, the first alternative.

(iv) is obvious when $\mathbf{x}$ is invertible ($1 + \mathbf{x}\mathbf{y} = \mathbf{x}(\mathbf{x}^{-1} + \mathbf{y})$) or $\mathbf{y}$ is invertible. If both are isotropic, there are two possibilities:
(a) $\mathbf{x} \cdot \mathbf{y} \neq 0$. Then $\mathbf{x} + \mathbf{y}$ is invertible, and $1 + \mathbf{x}\mathbf{y} = (\mathbf{x} + \mathbf{y})((\mathbf{x} + \mathbf{y})^{-1} + \mathbf{y})$.
(b) $\mathbf{x} \cdot \mathbf{y} = 0$. Then $(1 + \mathbf{x}\mathbf{y})\overline{(1 + \mathbf{x}\mathbf{y})} = 1$, and so $1 + \mathbf{x}\mathbf{y}$ is in $T(p, q)$ if and only if it is in Lipschitz group. What we have to prove is that $(1 + \mathbf{x}\mathbf{y})\mathbf{z}(1 + \mathbf{y}\mathbf{x})$ is a vector for arbitrary $\mathbf{z}$. But

$$(1 + \mathbf{x}\mathbf{y})\mathbf{z}(1 + \mathbf{y}\mathbf{x}) = \mathbf{x}\mathbf{y}\mathbf{z}\mathbf{y}\mathbf{x} + \mathbf{x}\mathbf{y}\mathbf{z} + \mathbf{z}\mathbf{y}\mathbf{x} + \mathbf{z}.$$

$\mathbf{x}\mathbf{y}\mathbf{z}\mathbf{y}\mathbf{x}$ is a vector because $\mathbf{x}\mathbf{y}$ is in $T(p, q)$, and

$$\begin{aligned}
\mathbf{x}\mathbf{y}\mathbf{z} = \mathbf{x}(\mathbf{y} \cdot \mathbf{z}) - \mathbf{x}\mathbf{z}\mathbf{y} &= \mathbf{x}(\mathbf{y} \cdot \mathbf{z}) - (\mathbf{x} \cdot \mathbf{z})\mathbf{y} + \mathbf{z}\mathbf{x}\mathbf{y} \\
&= \mathbf{x}(\mathbf{y} \cdot \mathbf{z}) - \mathbf{y}(\mathbf{x} \cdot \mathbf{z}) + \mathbf{z}(\mathbf{x} \cdot \mathbf{y}) - \mathbf{z}\mathbf{y}\mathbf{x},
\end{aligned}$$

which proves that $\mathbf{x}\mathbf{y}\mathbf{z} + \mathbf{z}\mathbf{y}\mathbf{x}$ is a vector, and so $1 + \mathbf{x}\mathbf{y}$ is in the Lipschitz group.
(v) is proved using (iii). Suppose first $a = \mathbf{t}$ for some vector (isotropic or not) then

$$\mathbf{x}\mathbf{t} + \mathbf{t}\mathbf{y} = (\mathbf{x} - \mathbf{y})\mathbf{t} + 2\mathbf{t} \cdot \mathbf{y}$$

and according to (iv) this is in $T(p, q)$. Next put a in its standard form $\mathbf{t}\alpha$, where α is a product of invertible vectors. Putting $\mathbf{y}_1 = \alpha\mathbf{y}\alpha^{-1}$ gives

$$\mathbf{x}a + a\mathbf{y} = \mathbf{x}\mathbf{t}\alpha + \mathbf{t}\alpha\mathbf{y} = (\mathbf{x}\mathbf{t} + \mathbf{t}\mathbf{y}_1)\alpha,$$

both factors of which are in $T(p,q)$.   ∎

The proof of (iv) can be found in Fillmore *et al.* (1990). We have given it here for the sake of completeness.

We now define G to be the set of matrices satisfying the conditions (a)-(c) of the theorem, and so the theorem itself can be rephrased as

$$\Gamma_{p+1,q+1} = G.$$

Lemma 2.5 G *is closed for the Clifford-conjugation* $^{-}$.

Proof.
It has to be proved that $\overline{A}$ is in G if A is. Calculating $A\overline{A}$ gives

$$\begin{pmatrix} a & b \\ c & d \end{pmatrix} \begin{pmatrix} \tilde{d} & -\tilde{b} \\ -\tilde{c} & \tilde{a} \end{pmatrix} = \begin{pmatrix} a\tilde{d} - b\tilde{c} & -a\tilde{b} + b\tilde{a} \\ c\tilde{d} - d\tilde{c} & -c\tilde{b} + d\tilde{a} \end{pmatrix} = \begin{pmatrix} \lambda(A) & 0 \\ 0 & \lambda(A) \end{pmatrix}$$

since the second and third entry are zero. Indeed, multiplying e.g. the second entry at right with $\lambda(A)$ gives $-a\tilde{b}a\tilde{d} + a\tilde{b}b\tilde{c} + b\tilde{a}a\tilde{d} - b\tilde{a}b\tilde{c} = -(a\tilde{a})b\tilde{d} + (b\tilde{b})a\tilde{c} + (a\tilde{a})b\tilde{d} - (b\tilde{b})a\tilde{c} = 0$ since $\tilde{a}b = \tilde{b}a$ because it is a vector. Moreover obviously $\lambda(A) = \tilde{\lambda}(A)$. Hence $\lambda(A)^{-1}\overline{A}$ is the inverse of A. Clearly $\overline{A}$ satisfies (a) and (b). To prove (c) we calculate $\overline{A}A$ which must also give $\lambda(A)$ since A and $\overline{A}$ commute:

$$\begin{aligned} \overline{A}A &= \begin{pmatrix} \tilde{d} & -\tilde{b} \\ -\tilde{c} & \tilde{a} \end{pmatrix} \begin{pmatrix} a & b \\ c & d \end{pmatrix} \\ &= \begin{pmatrix} \tilde{d}a - \tilde{b}c & \tilde{d}b - \tilde{b}d \\ -\tilde{c}a + \tilde{a}c & -\tilde{c}b + \tilde{a}d \end{pmatrix} \\ &= \begin{pmatrix} \lambda(A) & 0 \\ 0 & \lambda(A) \end{pmatrix} \end{aligned}$$

so $\lambda(\overline{A}) = \lambda(A)$.   ∎

Lemma 2.6 *(i) Each entry of a matrix in G can be written as* $\mathbf{t}\alpha$*, where α is in Lipschitz group $\Gamma_{p,q}$.*
(ii) There exist vectors $\mathbf{u}$*,* $\mathbf{v}$*,* $\mathbf{w}$*,* $\mathbf{s}$ *such that in each case at least one of the following alternatives hold:*
  - *$a = c\mathbf{u}$ or $c = a\mathbf{u}$.*
  - *$a = \mathbf{v}b$ or $b = \mathbf{v}a$.*
  - *$b = d\mathbf{w}$ or $d = b\mathbf{w}$.*
  - *$c = \mathbf{s}d$ or $d = \mathbf{s}c$.*

Proof.
We prove the first part explicitly for a, the second part for a and c, separating different possibilities:
  - c is invertible. Then $a = \mathbf{t}c$ with $\mathbf{t} = a\tilde{c}/(\tilde{c}c)$ and we can put $\mathbf{u} = c^{-1}\mathbf{t}c$, and both (i) and (ii) are proved.

- a is invertible. Then (i) is clearly satisfied and we can put $\mathbf{u} = \tilde{a}(c\tilde{a})a/(a\tilde{a})^2$.
- both are not invertible. Then $\lambda(A)c = (a\tilde{d} - b\tilde{c})c = a(\tilde{d}c)$ and $\tilde{\lambda}(A)a = (d\tilde{a} - c\tilde{b})a = c(\tilde{b}a)$ (proving (ii)). Since they cannot both be zero (the pseudodeterminant is not zero) they are both non-zero. So $a(\tilde{d}c) = (a\tilde{c})d$ is not zero and so $a\tilde{c}(d\tilde{a} - c\tilde{b}) = a(\tilde{d}c)\tilde{a}$ is not. According to property (iii) of lemma 2.4, a can be written as $t\alpha$. ∎

This allows us to construct a decomposition of a matrix $A \in G$.

Assume first that none of the elements a, b, c or d is invertible. Then none of them is zero (indeed, assume e.g. $a = 0$. Then, since $\lambda(A) = -b\tilde{c}$ is a scalar different from zero, both b and c are invertible). Then $a = t\alpha$, $b\lambda(\overline{A}) = (b\tilde{d})a$, $\lambda(A)c = a(\tilde{d}c)$ and $\lambda(A)d = -b(\tilde{c}d)$. Putting $\mathbf{v} = \lambda(A)^{-1}(b\tilde{d})$, $\mathbf{u} = \lambda(A)^{-1}\alpha(\tilde{d}c)\alpha^{-1}$ gives the decomposition

$$A = \begin{pmatrix} \mathbf{t} & \mathbf{vt} \\ \mathbf{tu} & -\mathbf{vtu} \end{pmatrix} \begin{pmatrix} \alpha & 0 \\ 0 & \alpha \end{pmatrix}.$$

Multiplying A with $\mathbf{V} = \begin{pmatrix} \mathbf{v} & 1 \\ 1 & -\mathbf{v} \end{pmatrix}$ (notice that $\mathbf{V}^2 = 1$ since $\mathbf{v}^2 = 0$) gives

$$A = \mathbf{V} \begin{pmatrix} \mathbf{vt} + \mathbf{tu} & -\mathbf{vtu} \\ \mathbf{t} - \mathbf{vtu} & \mathbf{vt} \end{pmatrix} \begin{pmatrix} \alpha & 0 \\ 0 & \alpha \end{pmatrix}.$$

The second matrix has an invertible first entry, since

$$(\mathbf{vt} + \mathbf{tu})(\mathbf{tv} + \mathbf{ut}) = \mathbf{vtut} + \mathbf{tutv}$$

which is up to sign the pseudodeterminant of $\begin{pmatrix} \mathbf{t} & \mathbf{vt} \\ \mathbf{tu} & -\mathbf{vtu} \end{pmatrix}$, and hence non-zero. Moreover, the matrix is in G. Indeed, the first, second and fourth entry are clearly in $T(p,q)$. The condition $a\tilde{c} \in \mathbb{R}^{p,q}$ is clearly satisfied, and since a is invertible we have that $c = (c\tilde{a})a/(a\tilde{a})$, which is in $T(p,q)$, and so condition (a) is satisfied. Checking that the other properties hold is elementary.

Assume next that one of the elements e.g. $a = \alpha$ is invertible and hence in Lipschitz group. Then we can multiply all entries at right with α^{-1} and assume that $a = 1$. Hence b and c must be vectors, say $\mathbf{v}$ and $\mathbf{u}$. The expression for the pseudodeterminant becomes $\mu = \tilde{d} - b\tilde{c}$ and we obtain the decomposition

$$A = \begin{pmatrix} 1 & \mathbf{v} \\ \mathbf{u} & \mu + \mathbf{uv} \end{pmatrix} \begin{pmatrix} \alpha & 0 \\ 0 & \alpha \end{pmatrix},$$

where again both terms are in G.

Lemma 2.7 $G \subset \Gamma_{p+1,q+1}$.

Proof.
It is straightforward to check that all factors in the decomposition above satisfy the conditions of Theorem 2.2. ∎

Lemma 2.8 *Let A be an element of G and $\mathbf{Y}$ be an invertible vector in $\mathbb{R}^{p+1,q+1}$. Then $\mathbf{Y}A \in G$.*

Proof.

Let $\mathbf{Y}$ have the form $\begin{pmatrix} \mathbf{y} & \mu \\ \nu & -\mathbf{y} \end{pmatrix}$ and let $A = \begin{pmatrix} a & b \\ c & d \end{pmatrix}$. This gives the product

$$\begin{pmatrix} \mathbf{y}a + \mu c & \mathbf{y}b + \mu d \\ \nu a - \mathbf{y}c & \nu b - \mathbf{y}d \end{pmatrix}.$$

First we prove that all the entries are in $T(p,q)$. For the first entry we have two possibilities. If c is invertible then $a = \mathbf{x}c$ and the first entry becomes $(\mathbf{y}\mathbf{x}+\mu)c$. If c is not invertible then the first entry has the form $\mathbf{y}a + az$ with $\mathbf{z} = \mu \tilde{d}c/(ad - b\tilde{c})$, which is in $T(p,q)$ by property (v) of lemma 2.4. For the other entries a similar procedure is followed. That the other conditions hold follows easily by explicit calculation, the most difficult part is proving that $(\mathbf{y}a + \mu c)\check{}(\mathbf{y}b + \mu d) \in \mathbb{R}^{p,q}$. Explicitly we get $\tilde{a}\mathbf{y}^2 b + \mu^2 \tilde{c}d + \mu(\tilde{a}\mathbf{y}d + \tilde{c}\mathbf{y}b)$. The first two terms are vectors. Since A is in G, it is in Lipschitz group, and so is $\overline{A}$, and $\tilde{d}\mathbf{y}a + \tilde{b}\mathbf{y}c$ is a vector by property (iv) of theorem 2.2. Taking the $\tilde{}$-adjoint of this gives the expression between brackets. ∎

Proof of the theorem. We already know that $G \subset \Gamma_{p+1,q+1}$. Since G contains the unit matrix and is closed under left multiplication by invertible vectors, it contains the set of all products of invertible vectors, i.e. $\Gamma_{p+1,q+1}$. ∎

Remark.

None of the three requirements (a)-(c) can be derived from the other two. Indeed, counter-examples are

$$\begin{pmatrix} 1 + \mathbf{e}_1 & 0 \\ 0 & 1 - \mathbf{e}_1 \end{pmatrix}, \quad \begin{pmatrix} 1 & \mathbf{e}_1 \mathbf{e}_2 \\ 0 & 1 \end{pmatrix}, \quad \begin{pmatrix} 1 & 0 \\ 0 & \mathbf{e}_1 \mathbf{e}_2 \mathbf{e}_3 \end{pmatrix}$$

which violate (a), (b) and (c) respectively, but not the other two conditions, and which are clearly not in Lipschitz group, because all three have both an odd and an even part which is different from zero.

Remark.

The condition (b) can be split into two conditions

$$\text{(b')} \quad a\tilde{b}, c\tilde{d} \in \mathbb{R}^{p,q}$$
$$\text{(b'')} \quad \tilde{c}a, \tilde{d}b \in \mathbb{R}^{p,q}.$$

In the non-definite case it is possible to find matrices satisfying (a), (c) and one of these conditions, but not the other. As an example in $C\ell_{1,2}$, one has the matrix

$$\begin{pmatrix} 1 + \mathbf{e}_{12} & (1 - \mathbf{e}_{12})\mathbf{e}_3 \\ (1 + \mathbf{e}_{12})\mathbf{e}_3 & 1 - \mathbf{e}_{12} \end{pmatrix}$$

which satisfies (a), (b'') and (c), but not (b').

Corollary 2.9 *Let g be an element of $C\ell_{p,q+1}$, $g = g_0 + g_1 \mathbf{e}_-$, $g_0, g_1 \in C\ell_{p,q}$. Then g is in Lipschitz group if and only if*

(i) $g_0, g_1 \in \Gamma_{p,q}$.
(ii) $g_0 \tilde{g}_1 \in \mathbb{R}^{p,q}$
(iii) $g_0 \overline{g_0} - g_1 \overline{g_1} \neq 0$.

Proof.
An element of $C\ell_{p,q+1}$ is in Lipschitz group $\Gamma_{p,q+1}$ if and only if it is in Lipschitz group $\Gamma_{p+1,q+1}$. Since it has the matrix representation $\begin{pmatrix} g_0 & -g_1 \\ \hat{g}_1 & \hat{g}_0 \end{pmatrix}$ the conditions are equivalent with those of the theorem. ∎

Remark.
A similar decomposition holds of course in $C\ell_{p+1,q}$. This is a generalization of a theorem by Porteous.

Classical subgroups of Möbius transformations are defined as follows:

- The group of translations given by

$$T = \left\{ T_{\mathbf{v}} = \begin{pmatrix} 1 & \mathbf{v} \\ 0 & 1 \end{pmatrix} \right\}$$

satisfying $T_{\mathbf{u}} T_{\mathbf{v}} = T_{\mathbf{u}+\mathbf{v}}$.

- The group of special transformations

$$K = \left\{ K_{\mathbf{v}} = \begin{pmatrix} 1 & 0 \\ \mathbf{v} & 1 \end{pmatrix} \right\}$$

satisfying $K_{\mathbf{u}} K_{\mathbf{v}} = K_{\mathbf{u}+\mathbf{v}}$.

- The group of dilatations

$$D = \left\{ D_{\lambda} = \begin{pmatrix} \lambda & 0 \\ 0 & \lambda^{-1} \end{pmatrix}, \ \lambda \text{ is a positive real scalar} \right\}$$

satisfying $D_{\lambda} D_{\mu} = D_{\lambda\mu}$.

- The group of orthogonal transformations

$$R = \left\{ R_{\alpha} = \begin{pmatrix} \alpha & 0 \\ 0 & \hat{\alpha} \end{pmatrix}, \ \alpha \in \Gamma_{p,q} \right\}$$

satisfying $R_{\alpha} R_{\beta} = R_{\alpha\beta}$.

It is easy to see that each element of Lipschitz group having an invertible entry can be written as $E K_{\mathbf{u}} T_{\mathbf{v}} D_{\lambda} R_{\alpha} F$ where E and F are either the identity or $\begin{pmatrix} 0 & 1 \\ 1 & 0 \end{pmatrix}$ in order to put the invertible element on the appropriate place. If A has no invertible entry the decomposition proved above can be used to obtain the form $K_{-\mathbf{v}} T_{\mathbf{v}} E K_{\mathbf{u}} T_{\mathbf{w}} D_{\lambda} R_{\alpha}$ where now in each case $E = \begin{pmatrix} 0 & 1 \\ 1 & 0 \end{pmatrix}$.

3. Coverings of Möbius group

Since the $\mathbf{Pin}(p,q)$ group gives a two-fold covering of the orthogonal group, which in turn gives a two-fold covering of Möbius group, the covering of $\mathcal{M}(p,q)$ by $\mathbf{Pin}(p+1,q+1)$ is four-fold. The elements mapped to the identity are $\pm I$ and $\pm P$ where I is the identity of $\mathbf{Pin}(p+1,q+1)$ and

$$P = \pm \begin{pmatrix} \mathbf{e}_1 \cdots \mathbf{e}_n & 0 \\ 0 & (-1)^{n+1}\mathbf{e}_1 \cdots \mathbf{e}_n \end{pmatrix}$$

is the pseudoscalar of $C\ell_{p+1,q+1}$, with pseudodeterminant equal to $(-1)^{n+q+1} = (-1)^{p+1}$. In classical notation $P = \mathbf{e}_1 \cdots \mathbf{e}_n \mathbf{e}_+ \mathbf{e}_-$. We then have different possibilities to obtain a double covering of $\mathcal{M}(p,q)$:

- If n is odd, then P is odd. Thus, for an arbitrary element A of $\mathbf{Pin}(p+1,q+1)$ either A or AP is in the $\mathbf{Spin}$ group. Hence $\mathbf{Spin}(p+1,q+1)$ gives a double covering of $\mathcal{M}(p,q)$. This was to be expected, since $SO(p+1,q+1)$ was a single covering of $\mathcal{M}(p,q)$ in this case.

- If p is even, then $\lambda(P)$ is -1 and for A arbitrary we have $\lambda(A) = -\lambda(AP)$. It makes sense to use the group $\mathbf{Pin}^+(p+1,q+1)$ of elements of $\mathbf{Pin}(p+1,q+1)$ with pseudodeterminant $+1$, and this gives a double covering of $\mathcal{M}(p,q)$. This case includes the important case of Euclidean space where $p = 0$.

If both n is even and p is odd (and hence q is odd) then the curve in $\mathbf{Pin}(p+1,q+1)$ given by

$$\theta \to e^{\theta \mathbf{e}_p \mathbf{e}_+} e^{\theta \mathbf{e}_q \mathbf{e}_-} \prod_{i=1}^{(p-1)/2} e^{\theta \mathbf{e}_{2i-1} \mathbf{e}_{2i}} \prod_{i=1}^{(q-1)/2} e^{\theta \mathbf{e}_{p+2i-1} \mathbf{e}_{p+2i}}$$

links I to P (θ going from 0 to $\pi/2$), while $e^{\theta \mathbf{e}_p \mathbf{e}_+}$ links I to $-I$ (θ going to π). In this case all four elements covering the identity of $\mathcal{M}(p,q)$ are connected, and a similar construction of a double covering of $\mathcal{M}(p,q)$ is not possible.

References

L. Ahlfors: 1986, 'Möbius transformations in $\mathbb{R}^n$ expressed through 2×2 matrices of Clifford numbers', *Complex Variables* **5**, pp. 215–224.

R. Delanghe, F. Sommen, V. Souček: 1992, *Clifford analysis and spinor valued functions*, Kluwer Acad. Publ., Dordrecht.

J. Elstrodt, F. Grunewald, J. Mennicke: 1987, 'Vahlen's group of Clifford matrices and spin groups', *Math. Z.*, **196**, pp. 369–390.

J. Fillmore and A. Springer: 1990, 'Möbius groups over general fields using Clifford algebras associated with spheres', *Int. J. Theo. Phys.*, **29**, pp. 225–246.

J. Haantjes: 1937, 'Conformal representations of an n-dimensional Euclidean space with a non-definite fundamental form on itself', *Proc. Kon. Nederl. Acad. Wetensch.*, **40**, pp. 700–705.

P. Lounesto: 1986, 'Clifford algebras and spinors', in J. S. R. Chisholm and A. K. Common (eds): *Clifford algebras and their applications in mathematical physics*, D. Reidel Publ. Co. Dordrecht, pp. 25–37.

P. Lounesto and E. Latvamaa: 1980, 'Conformal transformations and Clifford algebras', *Proc. Am. Math. Soc.*, **79**, pp. 533–538.

J. Maks: 1989, *Modulo $(1,1)$-periodicity of Clifford algebras and generalized (anti-)Möbius transformations*, Ph.D. Thesis, Delft University of Technology.

I. R. Porteous: 1981, *Topological geometry*, 2nd edition, Cambridge University Press.

3

Generalized Clifford Algebras and Number Systems, Projective Geometry and Crystallography

ON CLIFFORD ALGEBRAS OF A BILINEAR FORM WITH AN ANTISYMMETRIC PART

RAFAL ABLAMOWICZ
Department of Mathematics
Gannon University
Erie, PA 16541, U.S.A.
e-mail: ablamowicz@cluster.gannon.edu

and

PERTTI LOUNESTO
Institute of Mathematics
Helsinki University of Technology
FIN-02150 Espoo, Finland
e-mail: lounesto@dopey.hut.fi

Abstract. We explicitly demonstrate with a help of a computer that Clifford algebra $C\ell(B)$ of a bilinear form B with a non-trivial antisymmetric part A is isomorphic as an associative algebra to the Clifford algebra $C\ell(Q)$ of the quadratic form Q induced by the symmetric part of B [in characteristic $\neq 2$]. However, the multivector structure of $C\ell(B)$ depends on A and is therefore different than the one of $C\ell(Q)$. Operation of reversion is still an anti-automorphism of $C\ell(B)$. It preserves a new kind of gradation in $\bigwedge V$ determined by A but it does not preserve the gradation in $\bigwedge V$. The demonstration is given for Clifford algebras in real and complex vector spaces of dimension ≤ 9 with a help of a Maple package 'Clifford'. The package has been developed by one of the authors to facilitate computations in Clifford algebras of an arbitrary bilinear form B.

Key words: Clifford algebra – contraction – exterior algebra – multilinear structure – reversion

1. Introduction

Let Q be a quadratic form on a vector space V over a field K, and let B be a bilinear form on V such that $B(\mathbf{x}, \mathbf{x}) = Q(\mathbf{x})$. In this paper we follow approach due to Chevalley 1954 in which the Clifford algebra $C\ell(Q)$ is embedded into the endomorphism algebra $\text{End}(\bigwedge V)$ of the exterior algebra $\bigwedge V$ via the formula

$$\mathbf{x}u = \mathbf{x} \wedge u + \mathbf{x} \lrcorner u \tag{1}$$

where $\mathbf{x} \in V$, $u \in \bigwedge V$, and $\mathbf{x} \lrcorner u$ denotes the left contraction of u by $\mathbf{x}$. In general, the contraction is introduced on the exterior algebra $\bigwedge V$ as a mapping $\bigwedge V \times \bigwedge V \to \bigwedge V$, $(u, v) \to u \lrcorner v$ (Lounesto 1993, 1995). While it is well known that Chevalley introduced his construction to study Clifford algebras over fields of characteristic 2, here we study the case when the characteristic is not 2.

The main objective of this paper is to consider $C\ell(B)$ when B has a non-trivial antisymmetric part A in addition to its symmetric part g so that

$$B(\mathbf{x}, \mathbf{y}) = g(\mathbf{x}, \mathbf{y}) + A(\mathbf{x}, \mathbf{y}), \quad B(\mathbf{x}, \mathbf{x}) = g(\mathbf{x}, \mathbf{x}) = Q(\mathbf{x}) \tag{2}$$

for any $\mathbf{x}, \mathbf{y} \in V$. The symmetric part g gives rise to the natural decomposition

$$\mathbf{x}\mathbf{y} = \mathbf{x} \wedge \mathbf{y} + g(\mathbf{x}, \mathbf{y}) \tag{3}$$

where $g(\mathbf{x}, \mathbf{y}) = \frac{1}{2}(\mathbf{x}\mathbf{y} + \mathbf{y}\mathbf{x})$ is a scalar and $\mathbf{x} \wedge \mathbf{y} = \frac{1}{2}(\mathbf{x}\mathbf{y} - \mathbf{y}\mathbf{x})$ is a bivector. However, in the presence of the antisymmetric part A, the bivector part in (3) generalizes to a new kind of a bivector $\mathbf{x} \dot{\wedge} \mathbf{y} = \mathbf{x} \wedge \mathbf{y} + A(\mathbf{x}, \mathbf{y})$, which is clearly dependent on A, and we have

$$\mathbf{x}\mathbf{y} = \mathbf{x} \dot{\wedge} \mathbf{y} + g(\mathbf{x}, \mathbf{y}) = \mathbf{x} \wedge \mathbf{y} + A(\mathbf{x}, \mathbf{y}) + g(\mathbf{x}, \mathbf{y}) \tag{4}$$

where $\mathbf{x} \dot{\wedge} \mathbf{y} \in \dot{\bigwedge}^2 V$ and $\mathbf{x} \wedge \mathbf{y} \in \bigwedge^2 V$.

Oziewicz 1986 considered a Clifford algebra $C\ell(B)$ of an *arbitrary*, not necessarily symmetric, bilinear form $B(\mathbf{x}, \mathbf{y}) = g(\mathbf{x}, \mathbf{y}) + A(\mathbf{x}, \mathbf{y})$ in an attempt to generalize the Clifford algebra $C\ell(Q)$ of the quadratic from Q, the latter being uniquely determined in characteristic $\neq 2$ by g. It was pointed out earlier in Lounesto 1995 that Oziewicz's $C\ell(B)$ is determined up to an (associative algebra) isomorphism by the symmetric part of B and in that sense is no different from $C\ell(Q)$. However, as a consequence of (4), the multivector structure of $C\ell(B)$ is explicitly dependent on A and, as such, it varies with A. In particular, it is totally different from the one in $C\ell(Q)$ in the sense that the grades of $\bigwedge V$ are mixed under reversion [recall that reversion is the unique anti-automorphism of $C\ell(B)$ extending the identity mapping on V].

We intend to clarify the nature of the isomorphism $C\ell(Q) \simeq C\ell(B)$ by constructing an algebra homomorphism

$$\varphi : C\ell(Q) \to C\ell(B) \tag{5}$$

for an arbitrary A. We will prove that at least when $1 \leq \dim(V) \leq 9$, φ is an isomorphism in the category of associative algebras. Certainly, φ depends on A and it reduces to the identity when A vanishes. Furthermore, the definition of φ takes into account the new multivector structure of $C\ell(B)$ and is, in fact, related to it through A. We will investigate that new structure and how it is preserved by operation of reversion $\tilde{\ } : C\ell(B) \to C\ell(B)$ which preserves the grades in $\dot{\bigwedge} V$ [but not in $\bigwedge V$ unless B is symmetric.]

2. Reversion and contraction in $C\ell(B)$ depend on A

The properties of contraction $u \lrcorner v$ of $v \in \bigwedge V$ by $u \in \bigwedge V$ for an arbitrary Q were studied by Helmstetter 1982 and more recently by Lounesto 1995. Chevalley's formula (1) allows in fact introduction of the *Clifford product* of $\mathbf{x} \in V$ and $u \in \bigwedge V$, and can be extended by linearity and associativity to all of $\bigwedge V$ which then becomes, as an associative algebra, isomorphic to $C\ell(Q)$.

However, certain unexpected things happen when B has a nonzero antisymmetric part A. In that case both the reversion and the contraction depend on the choice of A.

Consider the reversion anti-automorphism of a Clifford algebra $C\ell(V, B)$ induced by a general, not necessarily symmetric bilinear form B. Let $e_1, e_2, \ldots, e_n$ be a basis of V, and B_{ij} be the matrix of B in that basis. If we decompose B into its symmetric and antisymmetric parts $B = g + A$ [in characteristic $\neq 2$], then

$$e_1 e_2 = e_1 \wedge e_2 + B_{12} = \underbrace{e_1 \wedge e_2 + A_{12}}_{e_1 \dot{\wedge} e_2} + g_{12} \tag{6}$$

and

$$e_2 e_1 = e_2 \wedge e_1 + B_{21} = \underbrace{-e_1 \wedge e_2 - A_{12}}_{-e_1 \dot{\wedge} e_2} + g_{12} \tag{7}$$

where $B_{21} = -A_{12} + g_{12}$. Thus, since $e_1 \dot{\wedge} e_2 = e_1 \wedge e_2 + A_{12} \in K + \bigwedge^2 V$, the new exterior product $\dot{\wedge}$ does not give the same multivector structure as $\wedge$ unless B is symmetric.

Take the reversion of $e_2 e_1$ to obtain $(e_2 e_1)^{\sim} = \tilde{e}_1 \tilde{e}_2 = e_1 e_2$. So reversion preserves scalars and sends the new bivector $e_1 \dot{\wedge} e_2$ to its opposite $-e_1 \dot{\wedge} e_2$ (it is this induced multivector structure of $C\ell(V, B)$ which is preserved by reversion). To put all this in a nutshell, we have

$$\begin{aligned} e_1 e_2 &= e_1 \dot{\wedge} e_2 + A_{12} + g_{12} = e_1 \dot{\wedge} e_2 + B_{12} \\ (e_1 e_2)^{\sim} &= -e_1 \dot{\wedge} e_2 - A_{12} + g_{12} = -e_1 \dot{\wedge} e_2 + B_{21} \\ &= -e_1 \dot{\wedge} e_2 + (B^{\mathsf{T}})_{12} \end{aligned} \tag{8}$$

where B^{T} is the transpose of B; or more generally, in a basis free form,

$$\begin{aligned} \mathbf{x}\mathbf{y} &= \mathbf{x} \dot{\wedge} \mathbf{y} + B(\mathbf{x}, \mathbf{y}), \\ (\mathbf{x}\mathbf{y})^{\sim} &= -\mathbf{x} \dot{\wedge} \mathbf{y} + B^{\mathsf{T}}(\mathbf{x}, \mathbf{y}) = \mathbf{y} \dot{\wedge} \mathbf{x} + B(\mathbf{y}, \mathbf{x}). \end{aligned} \tag{9}$$

This shows that reversion, unlike the grade involution, depends on the choice of the antisymmetric part of B, and cannot be defined, in the case of a non-symmetric B, in terms of the dimension grading as

$$\tilde{u} = \langle u \rangle_0 + \langle u \rangle_1 - \langle u \rangle_2 - \langle u \rangle_3 + + - - \ldots \tag{10}$$

for $u = \langle u \rangle_0 + \langle u \rangle_1 + \langle u \rangle_2 + \langle u \rangle_3 + \ldots$ with k-vector parts of u belonging to appropriate k-vector subspaces of $C\ell(Q) = K + V + \bigwedge^2 V + \ldots + \bigwedge^n V$.

Let us observe now how the contraction $\lrcorner$ and Clifford product in $C\ell(B)$ depend on B. In order to make that dependence explicit, from now on we write $\underset{B}{\lrcorner}$ instead of $\lrcorner$ and $\underset{B}{uv}$ instead of uv for any u and v in $C\ell(B)$. Notice first that from (1), which is of course true for any B,

$$\underset{B}{\mathbf{x}u} = \mathbf{x} \wedge u + \mathbf{x} \underset{B}{\lrcorner} u \tag{1}$$

(see also Lounesto 1995, page 138) one cannot conclude that

$$\underset{B}{\hat{u}\mathbf{x}} = \mathbf{x} \wedge u - \mathbf{x} \underset{B}{\lrcorner} u \tag{11}$$

in the case of a non-symmetric bilinear form B (see Equation (2) on page 139 of Lounesto 1995). Here $\hat{u}$ denotes grade involution [recall that grade involution is the unique automorphism of $C\ell(B)$ extending the mapping on V which changes signs of vectors.]

In fact, (11) follows from (1) only if B is symmetric. For example, let $\mathbf{x} = \mathbf{e}_1$ and $u = \mathbf{e}_2$. Then, from (1) we have

$$\mathbf{x}\underset{B}{u} = \mathbf{e}_1\mathbf{e}_2 = \mathbf{e}_1 \wedge \mathbf{e}_2 + \mathbf{e}_1 \underset{B}{\lrcorner} \mathbf{e}_2 = \mathbf{e}_1 \wedge \mathbf{e}_2 + B_{12} = \mathbf{e}_1 \wedge \mathbf{e}_2 + g_{12} + A_{12} \qquad (12)$$

and from the left hand side of (11) we have

$$\hat{u}\underset{B}{\mathbf{x}} = \hat{\mathbf{e}}_2\underset{B}{\mathbf{e}}_1 = -\mathbf{e}_2\underset{B}{\mathbf{e}}_1 = -\mathbf{e}_2 \wedge \mathbf{e}_1 - B_{21} = \mathbf{e}_1\mathbf{e}_2 \underset{B}{-} B_{21}$$
$$= \mathbf{e}_1 \wedge \mathbf{e}_2 - A_{21} - g_{21} = \mathbf{e}_1 \wedge \mathbf{e}_2 + A_{12} - g_{12} \qquad (13)$$

while from the right hand side of (11) we have

$$\mathbf{x} \wedge u - \mathbf{x}\underset{B}{\lrcorner}u = \mathbf{e}_1 \wedge \mathbf{e}_2 - \mathbf{e}_1 \underset{B}{\lrcorner} \mathbf{e}_2 = \mathbf{e}_1 \wedge \mathbf{e}_2 - B_{12}$$
$$= \mathbf{e}_1 \wedge \mathbf{e}_2 - A_{12} - g_{12} \neq \mathbf{e}_1 \wedge \mathbf{e}_2 + A_{12} - g_{12} \qquad (14)$$

unless $A_{12} = 0$. Thus, in general, (11) follows from (1) only when $A = 0$.

Lounesto's error in deducing (11) from (1) can be corrected either by restricting to a symmetric B or by making explicit that the contraction depends on B. In the latter case, denote by $B^{\mathrm{op}} = -B^{\mathsf{T}}$ or

$$B^{\mathrm{op}}(\mathbf{x}, \mathbf{y}) = -B(\mathbf{y}, \mathbf{x})$$

the bilinear form inducing the opposite metric, that is, $B^{\mathrm{op}}(\mathbf{x}, \mathbf{x}) = Q^{\mathrm{op}}(\mathbf{x}) = -Q(\mathbf{x})$, $Q(\mathbf{x}) = B(\mathbf{x}, \mathbf{x})$. Thus, $B^{\mathrm{op}} = -B^{\mathsf{T}} = -g + A$ since $B = g + A$, $A^{\mathsf{T}} = -A$, and $g^{\mathsf{T}} = g$. Then, we have

$$\mathbf{x}\underset{B}{u} = \mathbf{x} \wedge u + \mathbf{x}\underset{B}{\lrcorner}u \qquad (1')$$

and, since $\mathrm{op}(\mathbf{x}u) = \hat{u}\mathbf{x}$, instead of (11) we have

$$\hat{u}\underset{B}{\mathbf{x}} = \mathbf{x} \wedge u + \mathbf{x}\underset{B^{\mathrm{op}}}{\lrcorner}u \qquad (11')$$

which, in the case of a symmetric B, can be denoted as (1) and (11) [without showing B.] Since $B = g + A$, we can split the two contraction terms in $(1')$ and $(11')$ as

$$\mathbf{x}\underset{B}{\lrcorner}u = \mathbf{x}\underset{A}{\lrcorner}u + \mathbf{x}\underset{g}{\lrcorner}u, \qquad \mathbf{x}\underset{B^{\mathrm{op}}}{\lrcorner}u = \mathbf{x}\underset{A}{\lrcorner}u - \mathbf{x}\underset{g}{\lrcorner}u, \qquad (15)$$

where $\mathbf{x}\underset{A}{\lrcorner}u$ and $\mathbf{x}\underset{g}{\lrcorner}u$ denote parts of $\mathbf{x}\underset{B}{\lrcorner}u$ dependent on g and A respectively. Then, $(1')$ and $(11')$ can be written as:

$$\mathbf{x}\underset{B}{u} = \mathbf{x} \wedge u + \mathbf{x}\underset{B}{\lrcorner}u = \mathbf{x} \wedge u + \mathbf{x}\underset{g}{\lrcorner}u + \mathbf{x}\underset{A}{\lrcorner}u = \mathbf{x}\underset{g}{u} + \mathbf{x}\underset{A}{\lrcorner}u$$
$$= \mathbf{x} \wedge u + \mathbf{x}\underset{A}{\lrcorner}u + \mathbf{x}\underset{g}{\lrcorner}u = \mathbf{x} \dot{\wedge} u + \mathbf{x}\underset{g}{\lrcorner}u \qquad (1'')$$

and

$$\hat{u}\mathbf{x} = \mathbf{x} \wedge u + \mathbf{x} \underset{B^{\mathrm{op}}}{\lrcorner} u = \mathbf{x} \wedge u - \mathbf{x} \underset{g}{\lrcorner} u + \mathbf{x} \underset{A}{\lrcorner} u = \underset{-g}{\mathbf{x}u} + \mathbf{x} \underset{A}{\lrcorner} u. \tag{11''}$$

This implies in turn that

$$\frac{1}{2}(\underset{B}{\mathbf{x}u} + \underset{B}{\hat{u}\mathbf{x}}) = \mathbf{x} \wedge u + \mathbf{x} \underset{A}{\lrcorner} u = \mathbf{x} \dot{\wedge} u \tag{16}$$

$$\frac{1}{2}(\underset{B}{\mathbf{x}u} - \underset{B}{\hat{u}\mathbf{x}}) = \frac{1}{2}(\mathbf{x} \underset{B}{\lrcorner} u - \mathbf{x} \underset{B^{\mathrm{op}}}{\lrcorner} u) = \mathbf{x} \underset{g}{\lrcorner} u \tag{17}$$

where the right hand side of (16) may be viewed as a definition of the new gradation $\bigwedge V$ [compare (16) with formula (3) on page 139 in Lounesto 1995 where symbol $\wedge$ was used in the sense of $\dot{\wedge}$.]

Notice that the right hand side of (17) depends only on the symmetric part g of B. Consider these two examples.

Example 1 Let $\mathbf{x} = e_1$ and $u = e_2$. Then, the left hand side of (17) is

$$\tfrac{1}{2}(\underset{B}{\mathbf{x}u} - \underset{B}{\hat{u}\mathbf{x}}) = \tfrac{1}{2}(\underset{B}{e_1 e_2} - \underset{B}{\hat{e}_2 e_1}) = \tfrac{1}{2}(\underset{B}{e_1 e_2} + \underset{B}{e_2 e_1})$$

$$= \tfrac{1}{2}(e_1 \wedge e_2 + B_{12} + e_2 \wedge e_1 + B_{21})$$

$$= \tfrac{1}{2}(e_1 \wedge e_2 + g_{12} + A_{12} - e_1 \wedge e_2 + g_{21} + A_{21}) = g_{12}$$

which equals $\mathbf{x} \underset{g}{\lrcorner} u = e_1 \underset{g}{\lrcorner} e_2$.

Example 2 Let $\mathbf{x} = e_1$ and $u = e_2 \wedge e_3$. Then, the left hand side of (17) is

$$\tfrac{1}{2}(\underset{B}{e_1(e_2 \wedge e_3)} - \underset{B}{(e_2 \wedge e_3)\hat{\ } e_1}) = \tfrac{1}{2}(e_1 \wedge e_2 \wedge e_3 + B_{12}e_3 - B_{13}e_2 - \underset{B}{(e_2 \wedge e_3)e_1})$$

$$= \tfrac{1}{2}(e_1 \wedge e_2 \wedge e_3 + B_{12}e_3 - B_{13}e_2 - e_1 \wedge e_2 \wedge e_3 + B_{21}e_3 - B_{31}e_2)$$

$$= \tfrac{1}{2}(g_{12}e_3 + A_{12}e_3 - g_{13}e_2 - A_{13}e_2 + g_{21}e_3 + A_{21}e_3 - g_{31}e_2 - A_{31}e_2)$$

$$= g_{12}e_3 - g_{13}e_2$$

which again equals $\mathbf{x} \underset{g}{\lrcorner} u = e_1 \underset{g}{\lrcorner} (e_2 \wedge e_3)$.

Thus, in the sense that the right hand side of (17) depends only on the symmetric part g of B, it is not necessary to require in (17) that B be symmetric or even explicitly show dependence of the contraction in the left hand side of (17) on B. In short, (17) can be simplified to

$$\mathbf{x} \lrcorner u = \frac{1}{2}(\mathbf{x}u - \hat{u}\mathbf{x}). \tag{18}$$

which gives formula (4) on page 139 in Lounesto 1995.

3. Isomorphism $C\ell(Q) \simeq C\ell(B)$ in dimension 3

Formula (1) was used by Chevalley 1954, pp. 38–42, to introduce a linear operator $\gamma_{\mathbf{x}} \in \text{End}(\bigwedge V)$ such that (cf. Oziewicz 1986, page 252 line 3 of formula (23))

$$\gamma_{\mathbf{x}}(u) = \mathbf{x} \wedge u + \mathbf{x} \lrcorner u \qquad \text{for} \quad \mathbf{x} \in V, \ u \in \bigwedge V. \tag{19}$$

From the derivation rule

$$\mathbf{x} \lrcorner (\mathbf{x} \wedge u) = (\mathbf{x} \lrcorner \mathbf{x}) \wedge u - \mathbf{x} \wedge (\mathbf{x} \lrcorner u) \tag{20}$$

and the identities

$$\mathbf{x} \wedge \mathbf{x} \wedge u = 0, \quad \mathbf{x} \lrcorner (\mathbf{x} \lrcorner u) = 0 \tag{21}$$

(cf. Lounesto 1995) one may conclude that $(\gamma_{\mathbf{x}})^2 = Q(\mathbf{x})$. Therefore, Chevalley's map $V \to \text{End}(\bigwedge V)$, $\mathbf{x} \to \gamma_{\mathbf{x}}$, was a Clifford map and as such could be extended to an algebra homomorphism $\psi : C\ell(Q) \to \text{End}(\bigwedge V)$ whose image evaluated at $1 \in \bigwedge V$ yielded a map $\varphi : \text{End}(\bigwedge V) \to \bigwedge V$. The composite linear map $\theta = \varphi \circ \psi$ was the right inverse of the natural map $\bigwedge V \to C\ell(Q)$ and

$$\bigwedge V \to C\ell(Q) \xrightarrow{\psi} \text{End}(\bigwedge V) \xrightarrow{\varphi} \bigwedge V \tag{22}$$

was the identity mapping on $\bigwedge V$. The faithful representation ψ sent $C\ell(Q)$ onto an isomorphic subalgebra of $\text{End}(\bigwedge V)$. As it was remarked in Lounesto 1995, Chevalley's identification also works well with a contraction defined by an arbitrary, **not necessarily symmetric**, bilinear form B such that $B(\mathbf{x}, \mathbf{x}) = Q(\mathbf{x})$ and

$$\mathbf{x} \lrcorner \mathbf{y} = B(\mathbf{x}, \mathbf{y}) \quad \text{for} \quad \mathbf{x}, \mathbf{y} \in V \tag{23.a}$$

$$\mathbf{x} \lrcorner (u \wedge v) = (\mathbf{x} \lrcorner u) \wedge v + \hat{u} \wedge (\mathbf{x} \lrcorner v) \tag{23.b}$$

$$(u \wedge v) \lrcorner w = u \lrcorner (v \lrcorner w) \quad \text{for} \quad u, v, w \in \bigwedge V \tag{23.c}$$

(see Helmstetter 1982). As before, $\mathbf{x} \lrcorner \mathbf{a} \in \bigwedge^{k-1} V$ for $\mathbf{a} \in \bigwedge^k V$ and

$$\mathbf{x} \lrcorner (\mathbf{x}_1 \wedge \mathbf{x}_2 \wedge \ldots \wedge \mathbf{x}_k)$$
$$= \sum_{i=1}^{k} (-1)^{i-1} B(\mathbf{x}, \mathbf{x}_i) \mathbf{x}_1 \wedge \mathbf{x}_2 \wedge \ldots \wedge \mathbf{x}_{i-1} \wedge \mathbf{x}_{i+1} \wedge \ldots \wedge \mathbf{x}_k, \tag{24}$$

and the faithful representation ψ sends the Clifford algebra $C\ell(Q)$ onto an isomorphic subalgebra of $\text{End}(\bigwedge V)$ which, however, as a subspace depends on B.

As a concrete example we present detailed construction of the algebra isomorphism $C\ell(Q) \simeq C\ell(B)$ in the special case when $n = 3$ and $V = \mathbb{R}^3$ and the symmetric part of B is obtained by polarizing

$$x_1^2 + x_2^2 + x_3^2 = B(\mathbf{x}, \mathbf{x}) = g(\mathbf{x}, \mathbf{x}) = Q(\mathbf{x}).$$

In the following, we use the basis $\{1, \mathbf{e}_1, \mathbf{e}_2, \mathbf{e}_3, \mathbf{e}_1 \wedge \mathbf{e}_2, \mathbf{e}_1 \wedge \mathbf{e}_3, \mathbf{e}_2 \wedge \mathbf{e}_3, \mathbf{e}_1 \wedge \mathbf{e}_2 \wedge \mathbf{e}_3\}$ of $\bigwedge \mathbb{R}^3$. In order to shorten display, rather than consider an arbitrary symbolic form B we define the bilinear form B as

$$B(\mathbf{x}, \mathbf{y}) = x_1 y_1 + x_2 y_2 + x_3 y_3 + a(x_1 y_2 - x_2 y_1) + b(x_1 y_3 - x_3 y_1) + c(x_2 y_3 - x_3 y_2)$$

which has the following matrix ($B_{ij} = e_i \lrcorner e_j$)

$$B = \begin{pmatrix} 1 & a & b \\ -a & 1 & c \\ -b & -c & 1 \end{pmatrix}. \tag{25}$$

First, using (19), we determine the matrix γ_{e_1} of e_1 by the following computation:

$e_1 1 = e_1 \wedge 1 = e_1$ (first column   = 0100 0000)

$e_1 e_1 = e_1 \lrcorner e_1 = 1$ (second column   = 1000 0000)

$e_1 e_2 = e_1 \wedge e_2 + e_1 \lrcorner e_2 = e_1 \wedge e_2 + a$ (third column   = a000 1000)

$e_1 e_3 = e_1 \wedge e_3 + b$ (fourth column   = b000 0100)

$e_1 (e_1 \wedge e_2) = e_1 \lrcorner (e_1 \wedge e_2) = (e_1 \lrcorner e_1) \wedge e_2 - e_1 \wedge (e_1 \lrcorner e_2) = e_2 - a e_1$

$e_1 (e_1 \wedge e_3) = e_3 - b e_1$

$e_1 (e_2 \wedge e_3) = (e_1 \lrcorner e_2) e_3 - e_2 \wedge (e_1 \lrcorner e_3) + e_1 \wedge e_2 \wedge e_3 = a e_3 - b e_2 + e_1 \wedge e_2 \wedge e_3$

$e_1 (e_1 \wedge e_2 \wedge e_3) = (e_1 \lrcorner e_1) \wedge (e_2 \wedge e_3) - e_1 \wedge (e_1 \lrcorner (e_2 \wedge e_3))$

   $= e_2 \wedge e_3 - e_1 \wedge ((e_1 \lrcorner e_2) \wedge e_3 - e_2 \wedge (e_1 \lrcorner e_3))$

   $= e_2 \wedge e_3 - a e_1 \wedge e_3 + b e_1 \wedge e_2$

and find

$$e_1 \simeq \gamma_{e_1} = \begin{pmatrix} 0 & 1 & a & b & 0 & 0 & 0 & 0 \\ 1 & 0 & 0 & 0 & -a & -b & 0 & 0 \\ 0 & 0 & 0 & 0 & 1 & 0 & -b & 0 \\ 0 & 0 & 0 & 0 & 0 & 1 & a & 0 \\ 0 & 0 & 1 & 0 & 0 & 0 & 0 & b \\ 0 & 0 & 0 & 1 & 0 & 0 & 0 & -a \\ 0 & 0 & 0 & 0 & 0 & 0 & 0 & 1 \\ 0 & 0 & 0 & 0 & 0 & 0 & 1 & 0 \end{pmatrix}. \tag{26}$$

Similarly we obtain the matrices for other basis elements in the standard basis for $\bigwedge \mathbb{R}^3$ (only part of the computation is displayed):

$e_2 (e_1 \wedge e_3) = e_2 \lrcorner (e_1 \wedge e_3) + e_2 \wedge e_1 \wedge e_3 = (e_2 \lrcorner e_1) e_3 - e_1 (e_2 \lrcorner e_3) - e_1 \wedge e_2 \wedge e_3$

     $= -a e_3 - c e_1 - e_1 \wedge e_2 \wedge e_3$ (sixth column)

$$e_2 \simeq \gamma_{e_2} = \begin{pmatrix} 0 & -a & 1 & c & 0 & 0 & 0 & 0 \\ 0 & 0 & 0 & 0 & -1 & -c & 0 & 0 \\ 1 & 0 & 0 & 0 & -a & 0 & -c & 0 \\ 0 & 0 & 0 & 0 & 0 & -a & 1 & 0 \\ 0 & -1 & 0 & 0 & 0 & 0 & 0 & c \\ 0 & 0 & 0 & 0 & 0 & 0 & 0 & -1 \\ 0 & 0 & 0 & 1 & 0 & 0 & 0 & -a \\ 0 & 0 & 0 & 0 & 0 & -1 & 0 & 0 \end{pmatrix} \tag{27}$$

$e_3 (e_1 \wedge e_2 \wedge e_3) = (e_3 \lrcorner e_1) \wedge (e_2 \wedge e_3) - e_1 \wedge (e_3 \lrcorner (e_2 \wedge e_3))$

   $= -b e_2 \wedge e_3 - e_1 \wedge (-c e_3 - e_2) = -b e_2 \wedge e_3 + c e_1 \wedge e_3 + e_1 \wedge e_2$

$$e_3 \simeq \gamma_{e_3} = \begin{pmatrix} 0 & -b & -c & 1 & 0 & 0 & 0 & 0 \\ 0 & 0 & 0 & 0 & c & -1 & 0 & 0 \\ 0 & 0 & 0 & 0 & -b & 0 & -1 & 0 \\ 1 & 0 & 0 & 0 & 0 & -b & -c & 0 \\ 0 & 0 & 0 & 0 & 0 & 0 & 0 & 1 \\ 0 & -1 & 0 & 0 & 0 & 0 & 0 & c \\ 0 & 0 & -1 & 0 & 0 & 0 & 0 & -b \\ 0 & 0 & 0 & 0 & 1 & 0 & 0 & 0 \end{pmatrix} \tag{28}$$

$(e_1 \wedge e_2)(e_2 \wedge e_3) = (e_1 e_2 - e_1 \lrcorner e_2)(e_2 \wedge e_3) = e_1(e_2(e_2 \wedge e_3)) - a e_2 \wedge e_3$

$= e_1((e_2 \lrcorner e_2) \wedge e_3 - e_2 \wedge (e_2 \lrcorner e_3)) - a e_2 \wedge e_3 = e_1(e_3 - c e_2) - a e_2 \wedge e_3$

$= b + e_1 \wedge e_3 - ac - c e_1 \wedge e_2 - a e_2 \wedge e_3$

$$e_1 \wedge e_2 \simeq \gamma_{e_1 \wedge e_2} = \begin{pmatrix} 0 & 0 & 0 & 0 & -1-a^2 & -ab-c & b-ac & 0 \\ 0 & -a & 1 & c & 0 & 0 & 0 & b-ac \\ 0 & -1 & -a & -b & 0 & 0 & 0 & c+ab \\ 0 & 0 & 0 & 0 & 0 & 0 & 0 & -1-a^2 \\ 1 & 0 & 0 & 0 & -2a & -b & -c & 0 \\ 0 & 0 & 0 & 0 & 0 & -a & 1 & 0 \\ 0 & 0 & 0 & 0 & 0 & -1 & -a & 0 \\ 0 & 0 & 0 & 1 & 0 & 0 & 0 & -2a \end{pmatrix}$$

$(e_1 \wedge e_3)(e_1 \wedge e_2 \wedge e_3) = (e_1 e_3 - b)e_1 \wedge e_2 \wedge e_3 = e_1(e_3(e_1 \wedge e_2 \wedge e_3)) - b e_1 \wedge e_2 \wedge e_3$

$= e_1(e_1 \wedge e_2 + c e_1 \wedge e_3 - b e_2 \wedge e_3) - b e_1 \wedge e_2 \wedge e_3$

$= (e_2 - a e_1 + c e_3 - bc e_1 - ab e_3 + b^2 e_2 - b e_1 \wedge e_2 \wedge e_3) - b e_1 \wedge e_2 \wedge e_3$

$$e_1 \wedge e_3 \simeq \gamma_{e_1 \wedge e_3} = \begin{pmatrix} 0 & 0 & 0 & 0 & c-ab & -1-b^2 & -a-bc & 0 \\ 0 & -b & -c & 1 & 0 & 0 & 0 & -a-bc \\ 0 & 0 & 0 & 0 & 0 & 0 & 0 & 1+b^2 \\ 0 & -1 & -a & -b & 0 & 0 & 0 & c-ab \\ 0 & 0 & 0 & 0 & -b & 0 & -1 & 0 \\ 1 & 0 & 0 & 0 & -a & -2b & -c & 0 \\ 0 & 0 & 0 & 0 & 1 & 0 & -b & 0 \\ 0 & 0 & -1 & 0 & 0 & 0 & 0 & -2b \end{pmatrix}$$

$(e_2 \wedge e_3)(e_2 \wedge e_3) = (e_2 e_3 - c)(e_2 \wedge e_3) = e_2(e_3(e_2 \wedge e_3)) - c e_2 \wedge e_3$

$= e_2((e_3 \lrcorner e_2)e_3 - e_2(e_3 \lrcorner e_3)) - c e_2 \wedge e_3 = e_2(-c e_3 - e_2) - c e_2 \wedge e_3$

$= -c e_2 \wedge e_3 - c^2 - 1 - c e_2 \wedge e_3 = -1 - c^2 - 2c e_2 \wedge e_3$

$$e_2 \wedge e_3 \simeq \gamma_{e_2 \wedge e_3} = \begin{pmatrix} 0 & 0 & 0 & 0 & -b-ac & a-bc & -1-c^2 & 0 \\ 0 & 0 & 0 & 0 & 0 & 0 & 0 & -1-c^2 \\ 0 & -b & -c & 1 & 0 & 0 & 0 & -a+bc \\ 0 & a & -1 & -c & 0 & 0 & 0 & -b-ac \\ 0 & 0 & 0 & 0 & -c & 1 & 0 & 0 \\ 0 & 0 & 0 & 0 & -1 & -c & 0 & 0 \\ 1 & 0 & 0 & 0 & -a & -b & -2c & 0 \\ 0 & 1 & 0 & 0 & 0 & 0 & 0 & -2c \end{pmatrix}$$

$$(e_1 \wedge e_2 \wedge e_3)(e_1 \wedge e_2 \wedge e_3) = ((e_1 \wedge e_2) \wedge e_3)(e_1 \wedge e_2 \wedge e_3) = ((e_1 e_2 - a) \wedge e_3)(e_1 \wedge e_2 \wedge e_3)$$

$$= (-ae_3 + (e_1 e_2) \wedge e_3)(e_1 \wedge e_2 \wedge e_3) = (-ae_3 + ((e_1 e_2)e_3 - (e_1 e_2) \llcorner e_3))(e_1 \wedge e_2 \wedge e_3)$$

$$= (-ae_3 + e_1 e_2 e_3 - (e_1 \wedge e_2 + a) \llcorner e_3)(e_1 \wedge e_2 \wedge e_3)$$

$$= (-ae_3 + e_1 e_2 e_3 - (ce_1 - be_2 + 0))(e_1 \wedge e_2 \wedge e_3)$$

$$= (-ce_1 + be_2 - ae_3 + e_1 e_2 e_3)(e_1 \wedge e_2 \wedge e_3)$$

$$= -ce_1 \lrcorner (e_1 \wedge e_2 \wedge e_3) + be_2 \lrcorner (e_1 \wedge e_2 \wedge e_3) - ae_3 \lrcorner (e_1 \wedge e_2 \wedge e_3) + e_1(e_2(e_3(e_1 \wedge e_2 \wedge e_3)))$$

$$= -1 - a^2 - b^2 - c^2 - 2ae_1 \wedge e_2 - 2be_1 \wedge e_3 - 2ce_2 \wedge e_3 \qquad \text{(last column)}$$

$$e_1 \wedge e_2 \wedge e_3 \simeq \begin{pmatrix} 0 & 0 & 0 & 0 & 0 & 0 & 0 & -1-a^2-b^2-c^2 \\ 0 & 0 & 0 & 0 & -b-ac & a-bc & -1-c^2 & 0 \\ 0 & 0 & 0 & 0 & -c+ab & 1+b^2 & a+bc & 0 \\ 0 & 0 & 0 & 0 & -1-a^2 & -c-ab & b-ac & 0 \\ 0 & -b & -c & 1 & 0 & 0 & 0 & -2a \\ 0 & a & -1 & -c & 0 & 0 & 0 & -2b \\ 0 & 1 & a & b & 0 & 0 & 0 & -2c \\ 1 & 0 & 0 & 0 & -2a & -2b & -2c & 0 \end{pmatrix}$$

The multiplication table of the Clifford algebra $C\ell(B)$ is then seen to be (only some parts will be shown):

$C\ell(B)$	e_1	e_2
e_1	1	$e_1 \wedge e_2 + a$
e_2	$-e_1 \wedge e_2 - a$	1
e_3	$-e_1 \wedge e_3 - b$	$-e_2 \wedge e_3 - c$
$e_1 \wedge e_2$	$-ae_1 - e_2$	$e_1 - ae_2$
$e_1 \wedge e_3$	$-be_1 - e_3$	$-e_1 \wedge e_2 \wedge e_3 - ce_1 - ae_3$
$e_2 \wedge e_3$	$e_1 \wedge e_2 \wedge e_3 - be_2 + ae_3$	$-ce_2 - e_3$
$e_1 \wedge e_2 \wedge e_3$	$-be_1 \wedge e_2 + ae_1 \wedge e_3 + e_2 \wedge e_3$	$-ce_1 \wedge e_2 - e_1 \wedge e_3 + ae_2 \wedge e_3$

... the multiplication table continues ...

$C\ell(B)$	e_3	$e_1 \wedge e_2$	
e_1	...	$-ae_1 + e_2$	...
e_2	...	$-e_1 - ae_2$	...
e_3	...	$e_1 \wedge e_2 \wedge e_3 + ce_1 - be_2$	...
$e_1 \wedge e_2$	...	$-1 - a^2 - 2ae_1 \wedge e_2$	...
$e_1 \wedge e_3$	...	$e_2 \wedge e_3 - be_1 \wedge e_2 - ae_1 \wedge e_3 + c - ab$	...
$e_2 \wedge e_3$	...	$-e_1 \wedge e_3 - ce_1 \wedge e_2 - ae_2 \wedge e_3 - b - ac$	...
$e_1 \wedge e_2 \wedge e_3$	...	$(-b - ac)e_1 + (-c + ab)e_2 - (1 + a^2)e_3 - 2ae_1 \wedge e_2 \wedge e_3$	...

... and continues ...

For instance the square of $e_1 \wedge e_2 \wedge e_3$ in $C\ell(B)$ is

$$(e_1 \wedge e_2 \wedge e_3)^2 = -1 - a^2 - b^2 - c^2 - 2ae_1 \wedge e_2 - 2be_1 \wedge e_3 - 2ce_2 \wedge e_3.$$

Finally, we compute the matrix of $e_1 e_2 e_3$

$$e_1 e_2 e_3 \simeq \begin{pmatrix} 0 & c & -b & a & 0 & 0 & 0 & -1-a^2-b^2-c^2 \\ c & 0 & 0 & 0 & -ac & -bc & -1-c^2 & 0 \\ -b & 0 & 0 & 0 & ab & 1+b^2 & bc & 0 \\ a & 0 & 0 & 0 & -1-a^2 & -ab & -ac & 0 \\ 0 & 0 & 0 & 1 & 0 & 0 & 0 & -a \\ 0 & 0 & -1 & 0 & 0 & 0 & 0 & -b \\ 0 & 1 & 0 & 0 & 0 & 0 & 0 & -c \\ 1 & 0 & 0 & 0 & -a & -b & -c & 0 \end{pmatrix}$$

and verify that indeed $e_1 e_2 e_3 = e_1 \wedge e_2 \wedge e_3 + ce_1 - be_2 + ae_3$.

When the antisymmetric part of B vanishes, that is, B is the identity matrix $B = I$, we have the multiplication table of $C\ell(I)$

$C\ell(I)$	e_1	e_2	e_3	$e_1 \wedge e_2$	...	$e_1 \wedge e_2 \wedge e_3$
e_1	1	$e_1 \wedge e_2$	$e_1 \wedge e_3$	e_2	...	$e_2 \wedge e_3$
e_2	$-e_1 \wedge e_2$	1	$e_2 \wedge e_3$	$-e_1$	...	$-e_1 \wedge e_3$
e_3	$-e_1 \wedge e_3$	$-e_2 \wedge e_3$	1	$e_1 \wedge e_2 \wedge e_3$	...	$e_1 \wedge e_2$
$e_1 \wedge e_2$	$-e_2$	e_1	$e_1 \wedge e_2 \wedge e_3$	-1	...	$-e_3$
$e_1 \wedge e_3$	$-e_3$	$-e_1 \wedge e_2 \wedge e_3$	e_1	$e_2 \wedge e_3$	...	e_2
$e_2 \wedge e_3$	$e_1 \wedge e_2 \wedge e_3$	$-e_3$	e_2	$-e_1 \wedge e_3$	...	$-e_1$
$e_1 \wedge e_2 \wedge e_3$	$e_2 \wedge e_3$	$-e_1 \wedge e_3$	$e_1 \wedge e_2$	$-e_3$	...	-1

The following correspondences establish an isomorphism φ of the associative algebras $C\ell(I)$ and $C\ell(B)$:

$C\ell(I)$	$C\ell(B)$
1	1
e_1	e_1
e_2	e_2
e_3	e_3
$e_1 e_2 = e_1 \wedge e_2$	$e_1 e_2 = e_1 \dot\wedge e_2 = e_1 \wedge e_2 + a$
$e_1 e_3 = e_1 \wedge e_3$	$e_1 e_3 = e_1 \dot\wedge e_3 = e_1 \wedge e_3 + b$
$e_2 e_3 = e_2 \wedge e_3$	$e_2 e_3 = e_2 \dot\wedge e_3 = e_2 \wedge e_3 + c$
$e_1 e_2 e_3 = e_1 \wedge e_2 \wedge e_3$	$e_1 e_2 e_3 = e_1 \dot\wedge e_2 \dot\wedge e_3 = e_1 \wedge e_2 \wedge e_3 + ce_1 - be_2 + ae_3$

In the above table we have used the dotted wedge $\dot\wedge$ defined in (16) and we are using the same names for the basis monomials including the 1-vectors $\{e_1, e_2, e_3\}$ in both algebras $C\ell(I)$ and $C\ell(B)$. It should be clear however that the product in $C\ell(I)$ depends, via its contraction component, only on the diagonal part of B whereas the same product in $C\ell(B)$ depends on both the diagonal part and the antisymmetric part of B. We are justified in using the same notation for the basis monomials in both algebras since, following Chevalley's approach, both algebras are the same (up to an isomorphism) associative subalgebra in the ring of endomorphisms $\operatorname{End}(\bigwedge V)$ of the same exterior algebra $\bigwedge V$.

This shows that Clifford algebras $C\ell(I)$ and $C\ell(B)$ (at least when $V = \mathbb{R}^3$) are isomorphic as associative algebras for all B with the symmetric part I.

4. Mathematical structure built into 'Clifford'

Computations in Section (3) were carried out by hand and so were computations in dimensions 2 and 4 (not shown here). They proved the point that at least in dimensions 2, 3 and 4 associative algebras $C\ell(Q)$ and $C\ell(B)$, $Q(\mathbf{x}) = B(\mathbf{x}, \mathbf{x})$, were the same object in the category of associative algebras with unit, and also that they were the same object in the category of algebras of quadratic forms (even when B had a non-trivial antisymmetric part.) They also helped elucidate the fact that the isomorphism φ did not preserve grades of a multivector in $\bigwedge V$ (it did preserve grades in $\bigwedge V$) and that the multivector structures of these algebras were different.

Thus, a conjecture was formed that in fact this isomorphism also manifested itself in higher dimensions. In view of complexity of computations in $C\ell(B)$ it became clear that a computer program capable of symbolic computations in Clifford algebra $C\ell(B)$ of any arbitrary bilinear form B was needed.

CLICAL, written by Lounesto, Mikkola and Vierros in 1987, works with orthogonal Clifford algebras $C\ell(Q)$ and it has proven itself useful in finding counterexamples in such algebras (see article by Lounesto elsewhere in this collection). However, since it is not capable of handling computations with an arbitrary bilinear form, it couldn't be used to verify/falsify the conjecture. Thus, a package 'Clifford' for Maple V from Waterloo Maple Software 1994 has been written by one of the authors (Abłamowicz 1995, 1996).

4.1. MATHEMATICAL FOUNDATIONS OF 'Clifford'

The basic mathematical foundation for 'Clifford' is Chevalley's treatment of the Clifford algebra $C\ell(Q)$ as a subalgebra of the ring $\operatorname{End}(\bigwedge V)$ of endomorphisms of the exterior algebra $\bigwedge V$ (see Section 1). His approach can be applied to $C\ell(B)$ where B is an arbitrary, not necessarily symmetric bilinear form such that $B(\mathbf{x}, \mathbf{x}) = Q(\mathbf{x})$. Formula (1) giving the Clifford product of a vector $\mathbf{x} \in V$ and $u \in \bigwedge V$ is then extended recursively with respect to $\mathbf{x}$ by associativity of Clifford multiplication to any product of vectors and then, by linearity, to any element of $C\ell(B)$.

We now illustrate a few basic features of the package. First, we load 'Clifford' and define, as an example, a standard ordered basis for $C\ell(B)$ in dimension 3 ('Clifford' works in dimensions ≤ 9.) Any multivector in $C\ell(B)$ is then viewed as a multivariate *Clifford polynomial* in all basis monomials. For more information about 'Clifford' see Abłamowicz 1995, 1996.

```
> restart:with(Clifford); #list of all procedures in 'Clifford'
```

$[LC, LCQ, builtm, c\_conjug, cbasis, cexp, cinv, clicollect, clisort, cliterms, cmul,$
$\quad cmulQ, conjugation, extract, gradeinv, init, isVahlenmatrix, isproduct,$
$\quad ord, pseudodet, q\_conjug, qdisplay, qinv, qmul, qnorm, remove, reorder,$
$\quad reversion, rmulm, rot3d, scalarpart, specify\_constants, vectorpart,$
$\quad version, wedge, wexp]$

```
> clibasis:=cbasis(3);
```
$$clibasis := [\, Id, e1, e2, e3, e1we2, e1we3, e2we3, e1we2we3\,]$$

In the above, $e1we2$ denotes $e_1 \wedge e_2$ and Id denotes the unit of $\bigwedge V$ (identified with the unit in $C\ell(B)$). In $'Clifford'$, the Clifford product uv is given by $\mathbf{cmul(u, v)}$ and by its infix notation $u \mathbin{\&c} v$, the wedge product $u \wedge v$ is given by $\mathbf{wedge(u, v)}$ and by its infix notation $u \mathbin{\&w} v$, and the B-dependent left contraction $\mathbf{x} \lrcorner u$ is given by $LC(\mathbf{x}, \mathbf{u})$. Procedures $'clicollect'$ and $'clisort'$ may be used to collect and sort answers with respect to the standard basis.

```
> clicollect(cmul(2*Id+3*e2we3, e2-2*e1));
```
$$-6\,e1we2we3 - 4\,e1 + (-6\,B_{3,1} + 2 + 3\,B_{3,2})\,e2 + (6\,B_{2,1} - 3\,B_{2,2})\,e3$$

```
> clicollect(wedge(2*Id+3*e2we3, e2-2*e1));
```
$$-6\,e1we2we3 - 4\,e1 + 2\,e2$$

```
> LC(e1+2*e2,Id+2*e1we2);
```
$$2\,B_{1,1}\,e2 - 2\,B_{1,2}\,e1 + 4\,B_{2,1}\,e2 - 4\,B_{2,2}\,e1$$

```
> clicollect(cmul(eiwej+Id+2*ek,2*emwen+el));
```
$$2\,B_{j,m}\,eiwen + 2\,eiwejwemwen + 2\,B_{i,n}\,ejwem$$
$$+ (2\,B_{j,m}\,B_{i,n} + 2\,B_{k,l} - 2\,B_{j,n}\,B_{i,m})\,Id + B_{j,l}\,ei - B_{i,l}\,ej + 2\,emwen + el$$
$$+ eiwejwel + 4\,ekwemwen - 4\,B_{k,n}\,em + 4\,B_{k,m}\,en - 2\,B_{j,n}\,eiwem$$
$$- 2\,B_{i,m}\,ejwen + 2\,ekwel$$

Notice that since the bilinear form B has not been specified, the output above contains symbolic entries of B and that computations with purely symbolic indices are possible. It is not difficult in Maple to express B as a sum of its symmetric part g and its antisymmetric part A as it was done in (2).

As an example of some computations that can be done with $'Clifford'$, we will verify formulas (12) – (17). We will proceed as follows. First, we define two functions: one antisymmetrizing $'antisymmetric'$ and one symmetrizing $'symmetric1'$ which will allow us to create symbolic matrices g and A (these functions are not part of the package). The dimension of V will be set to 4 by assigning 4 to a global Maple name $'dim'$:

```
> dim:=4:  #dimension of V
> #defining antisymmetrizing function
> antisymmetric:=proc(a1:posint,a2:posint) local index1,index2;
> if args[1] = args[2] then RETURN(0) fi;
> if args[1] < args[2] then RETURN(cat(A,args[1],args[2])) else
>                           RETURN(-cat(A,args[2],args[1])) fi;end:
> #defining symmetrizing function
> symmetric1:=proc(a1:posint,a2:posint) local index1,index2;
> if args[1] <= args[2] then RETURN(cat(g,args[1],args[2])) else
```

```
>                              RETURN(cat(g,args[2],args[1])) fi;end:
> #defining the alternating part A of B
> A:=linalg[matrix](dim,dim,antisymmetric):
> g:=linalg[matrix](dim,dim,symmetric1):  #defining the symmetric part of B
> B:=matrix(maxdim,maxdim,[]):B:=evalm(g+A);  #defining B as g + A
```

$$B := \begin{bmatrix} g11 & g12 + A12 & g13 + A13 & g14 + A14 \\ g12 - A12 & g22 & g23 + A23 & g24 + A24 \\ g13 - A13 & g23 - A23 & g33 & g34 + A34 \\ g14 - A14 & g24 - A24 & g34 - A34 & g44 \end{bmatrix}$$

We have to let $'Clifford'$ know that the symbolic entries of **g** and **A** are to be treated as constants:

```
> gparameters:=op({op(convert(g,mlist))});
> Aparameters:={}:x:='x': for x in {op(convert(A,mlist))} minus {0} do
>                     Aparameters:=Aparameters union {x/sign(x)} od:
> Aparameters:=op(Aparameters);
> specify_constants(gparameters,Aparameters):
```

$$gparameters := g11, g12, g13, g14, g22, g23, g24, g33, g34, g44$$

$$Aparameters := A24, A34, A14, A23, A12, A13$$

Notice that even though the antisymmetric part of B might be nontrivial, the fundamental identity

$$\mathbf{e}_i \mathbf{e}_j + \mathbf{e}_j \mathbf{e}_i = 2g_{ij}, \quad i,j = 1, \ldots, \dim(V) \tag{29}$$

is still fulfilled by the basis 1-vectors $\{\mathbf{e}_1, \mathbf{e}_2, \ldots, \mathbf{e}_n\} \subset C\ell(B)$ as it can be verified by hand or with a computer. We show how to verify (29) in dimension 3 and leave similar verification in all other dimensions ≤ 9 to the reader.

```
> basis1:=cbasis(3,1);   #standard basis in dimension 3
```
$$basis1 := [\, e1, e2, e3 \,]$$

```
> printlevel:=2:
> for x in basis1 do for y in basis1 do
> 'cmul('.x.','.y.')+cmul('.y.','.x.')'=clisort(simplify(x &c y + y &c x))
> od od;
```

$$2\, cmul(e1, e1) = 2\, g11\, Id$$

$$cmul(e1, e2) + cmul(e2, e1) = 2\, g12\, Id$$

$$cmul(e1, e3) + cmul(e3, e1) = 2\, g13\, Id$$

$$cmul(e1, e2) + cmul(e2, e1) = 2\, g12\, Id$$

$$2\, cmul(e2, e2) = 2\, g22\, Id$$

$$cmul(e2, e3) + cmul(e3, e2) = 2\, g23\, Id$$

$$cmul(e1, e3) + cmul(e3, e1) = 2\, g13\, Id$$

$$cmul(e2, e3) + cmul(e3, e2) = 2\, g23\, Id$$

$$2\, cmul(e3, e3) = 2\, g33\, Id$$

When the basis vectors $\{e_1, e_2, \ldots, e_n\}$ form a pseudo-orthogonal set in V, then it is much faster, especially in higher dimensions, to use a simplified version of `'cmul'` called `'cmulQ'`. Procedure `'cmulQ'` gives Clifford product in $C\ell(Q)$ and it uses only diagonal elements of B (or g). Similarly, `'LCQ'` gives the left contraction in $C\ell(Q)$. Thus, the above verification of (29) will give in $C\ell(Q)$:

```
> for x in basis1 do for y in basis1 do
> 'cmulQ('.x.','.y.') + cmulQ('.y.','.x.')'=
>       clisort(simplify(x &cQ y + y &cQ x)) od od;
```

$$2\, cmulQ(e1,\, e1) = 2\, g11\ Id$$

$$cmulQ(e1,\, e2) + cmulQ(e2,\, e1) = 0$$

$$cmulQ(e1,\, e3) + cmulQ(e3,\, e1) = 0$$

$$cmulQ(e1,\, e2) + cmulQ(e2,\, e1) = 0$$

$$2\, cmulQ(e2,\, e2) = 2\, g22\ Id$$

$$cmulQ(e2,\, e3) + cmulQ(e3,\, e2) = 0$$

$$cmulQ(e1,\, e3) + cmulQ(e3,\, e1) = 0$$

$$cmulQ(e2,\, e3) + cmulQ(e3,\, e2) = 0$$

$$2\, cmulQ(e3,\, e3) = 2\, g33\ Id$$

We now turn our attention to reversion. It was demonstrated in Section 2 that when B has an antisymmetric part, reversion of a monomial e.g., $(e_1 \wedge e_2 \wedge \cdots \wedge e_n)^\sim$ cannot be calculated as $(-1)^{n(n-1)/2} e_1 \wedge e_2 \wedge \cdots \wedge e_n$ as it is often done in $C\ell(Q)$. One has to express the monomial in terms of the Clifford products of the basis vectors $\{e_1, e_2, \ldots, e_n\}$, reverse their order in each product, and then re-express these products in terms of the standard basis. Thus, one has to use the standard definition of reversion (see Crumeyrolle 1990). All these steps are accomplished with a procedure `'reversion'`. As we now show, reversion is always a linear anti-automorphism of degree 2. Here is an example in dimension 3 (which can be easily extended to all other dimensions):

```
> clibasis:=cbasis(3);
```

$$clibasis := [\, Id, e1, e2, e3, e1we2, e1we3, e2we3, e1we2we3\,]$$

```
> for x in clibasis do 'reversion('.x.')' = reversion(x) od;
```

$$reversion(Id) = Id$$

$$reversion(e1) = e1$$

$$reversion(e2) = e2$$

$$reversion(e3) = e3$$

$$reversion(e1we2) = -e1we2 - 2\, Id\, A12$$

$$reversion(e1we3) = -e1we3 - 2\, Id\, A13$$

$$reversion(e2we3) = -e2we3 - 2\, Id\, A23$$

$$reversion(e1we2we3) = -2\, e1\, A23 + 2\, e2\, A13 - e1we2we3 - 2\, e3\, A12$$

Notice that when B is symmetric (but not necessarily diagonal), reversion preserves the grading and may only change the sign of the monomial. However, when B has a non-zero antisymmetric part then reversion mixes grades.

Reversion is an automorphism of degree 2 :

```
> for x in clibasis do
>         'reversion(reversion('.x.'))' = reversion(reversion(x)) od;
```

$$reversion(reversion(Id)) = Id$$

$$reversion(reversion(e1)) = e1$$

$$reversion(reversion(e2)) = e2$$

$$reversion(reversion(e3)) = e3$$

$$reversion(reversion(e1we2)) = e1we2$$

$$reversion(reversion(e1we3)) = e1we3$$

$$reversion(reversion(e2we3)) = e2we3$$

$$reversion(reversion(e1we2we3)) = e1we2we3$$

Reversion is an anti-automorphism even when B is not symmetric. This means that $(uv)\tilde{} = \tilde{v}\tilde{u}$ for any $u, v \in C\ell(B)$. For example:

```
> u:=Id + e1 + e3we2;v:=e3we1+ 2*e4;
```

$$u := e1 + Id + e3we2$$

$$v := e3we1 + 2\,e4$$

```
> evalb(simplify(reversion(u &c v) =
>                     simplify(reversion(v) &c reversion(u))));
                          true
```

```
> reversion(reversion(u)); #reversion is an automorphism of degree 2
> reversion(reversion(v));
```

$$e1 + Id - e2we3$$

$$-e1we3 + 2\,e4$$

4.2. VERIFICATION OF SOME FORMULAS WITH $'$Clifford$'$

We will show now how to verify with $'$Clifford$'$ computations $(12)-(14)$ for $\mathbf{x} = \mathbf{e}_1$ and $u = \mathbf{e}_2$.

```
> x:=e1:u:=e2:
> 'x &c u' = x &c u; #formula (12)
> #left hand side of (11) / formula (13)
> 'gradeinv(u) &c x' = gradeinv(u) &c x;
> #right hand side of (11) / formula (14)
> 'x &w u - LC(x,u)' = x &w u - LC(x,u);
> #the difference is 0 only if A = 0
> 'gradeinv(u) &c x - (x &w u - LC(x,u))' =
>        simplify(gradeinv(u) &c x - (x &w u - LC(x,u)));
```

$$x\,\&c\,u = e1we2 + (\,g12 + A12\,)\,Id$$

$$gradeinv(\,u\,)\,\&c\,x = e1we2 - (\,g12 - A12\,)\,Id$$

$$(\,x\,\&w\,u\,) - LC(\,x,u\,) = e1we2 - (\,g12 + A12\,)\,Id$$

$$(\,gradeinv(\,u\,)\,\&c\,x\,) - (\,x\,\&w\,u\,) + LC(\,x,u\,) = 2\,Id\,A12$$

Of course, using 'Clifford' the above computations can be performed with ease for any $x \in V$ and $u \in C\ell(B)$:

```
> x:=2*e1+3*e3:u:=2*e2-e3we2:
> 'x &c u' = clicollect(x &c u);  #formula (12)
> #left hand side of (11) / formula (13)
> 'gradeinv(u) &c x' = clicollect(gradeinv(u) &c x);
> #right hand side of (11) / formula (14)
> 'x &w u - LC(x,u)' = clicollect(x &w u - LC(x,u));
> #the difference is 0 only if A = 0
> 'gradeinv(u) &c x - (x &w u - LC(x,u))' =
>         simplify(gradeinv(u) &c x -·(x &w u - LC(x,u)));
```

$$x \,\&c\, u = 2\,e1we2we3 + (-2\,g13 - 2\,A13 - 3\,g33)\,e2$$
$$+ (2\,g12 + 2\,A12 + 3\,g23 - 3\,A23)\,e3$$
$$+ (4\,g12 + 4\,A12 + 6\,g23 - 6\,A23)\,Id + 4\,e1we2 - 6\,e2we3$$

$$\text{gradeinv}(u)\,\&c\, x = 2\,e1we2we3 + (2\,g13 - 2\,A13 + 3\,g33)\,e2$$
$$+ (2\,A12 - 2\,g12 - 3\,g23 - 3\,A23)\,e3$$
$$+ (-4\,g12 + 4\,A12 - 6\,g23 - 6\,A23)\,Id + 4\,e1we2 - 6\,e2we3$$

$$(x \,\&w\, u) - \text{LC}(x,u) = 2\,e1we2we3 + (2\,A13 + 2\,g13 + 3\,g33)\,e2$$
$$+ (-2\,g12 + 3\,A23 - 2\,A12 - 3\,g23)\,e3$$
$$+ (-6\,g23 + 6\,A23 - 4\,g12 - 4\,A12)\,Id + 4\,e1we2 - 6\,e2we3$$

$$(\text{gradeinv}(u)\,\&c\, x) - (x \,\&w\, u) + \text{LC}(x,u) =$$
$$-12\,Id\,A23 + 4\,e3\,A12 - 6\,e3\,A23 - 4\,e2\,A13 + 8\,Id\,A12$$

In order to verify formulas (15) – (17) and follow notation used there, we define 'LCB' to be the contraction in $C\ell(B)$, 'LCBop' to be the contraction in $C\ell(B^{\text{op}})$, 'LCg' to be the symmetric part and 'LCA' to be the antisymmetric part of 'LCB'. Even though procedures 'LCBop', 'LCg', and 'LCA' are not part of the package, it is very easy to create them and to experiment with them.

```
> LCB:=proc() RETURN(simplify(LC(args[1],args[2]))) end:
>
> LCBop:=proc() local negset, x; global gparameters;
> negset:={}:for x in gparameters do negset:={op(negset),x = -x} od:
>       RETURN(simplify(subs(negset,LC(args[1],args[2]))));
> end:
>
> LCg:=proc() local zerosetA, x; global Aparameters;
> zerosetA:={}:for x in Aparameters do zerosetA:={op(zerosetA),x=0} od:
>       RETURN(simplify(subs(zerosetA,LC(args[1],args[2]))));
> end:
>
> LCA:=proc() local zerosetg, x; global gparameters;
> zerosetg:={}:
> for x in gparameters do zerosetg:={op(zerosetg),x=0} od:zerosetg;x:='x':
>       RETURN(simplify(subs(zerosetg,LC(args[1],args[2]))));
> end:
```

Now we proceed with formulas (15) for $x = 2e_1 + 3e_3$ and $u = e_2 \wedge e_3 + 2e_2$:

```
> x := 2*e1 + 3*e3;u:=e3we2 + 2*e2;
```

$$x := 2\,e1 + 3\,e3$$

$$u := e3we2 + 2\,e2$$

```
> clicollect(LCB(x,u));  #the left hand side of the first formula in (15)
```

$$(2\,A13 + 2\,g13 + 3\,g33)\,e2 + (-2\,g12 + 3\,A23 - 2\,A12 - 3\,g23)\,e3$$
$$+ (4\,g12 + 4\,A12 + 6\,g23 - 6\,A23)\,Id$$

```
> #the right hand side of the first formula in (15)
> clicollect(LCA(x,u) + LCg(x,u));
```

$$(2\,A13 + 2\,g13 + 3\,g33)\,e2 + (-2\,g12 + 3\,A23 - 2\,A12 - 3\,g23)\,e3$$
$$+ (4\,g12 + 4\,A12 + 6\,g23 - 6\,A23)\,Id$$

```
> clicollect(LCBop(x,u));  #the left hand side of the second formula in (15)
```

$$(-2\,g13 + 2\,A13 - 3\,g33)\,e2 + (2\,g12 - 2\,A12 + 3\,g23 + 3\,A23)\,e3$$
$$+ (-4\,g12 + 4\,A12 - 6\,g23 - 6\,A23)\,Id$$

```
> #the right hand side of the second formula in (15)
> clicollect(LCA(x,u) - LCg(x,u));
```

$$(-2\,g13 + 2\,A13 - 3\,g33)\,e2 + (2\,g12 - 2\,A12 + 3\,g23 + 3\,A23)\,e3$$
$$+ (-4\,g12 + 4\,A12 - 6\,g23 - 6\,A23)\,Id$$

Thus, both parts of (15) have been verified. Finally, we verify formulas (16) and (17) for the same values of x and u as above:

```
> 'x &dw u ' = x &w u + LCA(x,u); #the right hand side of (16)
```

$$x\,\&dw\,u = 4\,e1we2 - 6\,e2we3 - 2\,e1we2we3 + 4\,Id\,A12 - 6\,Id\,A23 + 2\,e2\,A13$$
$$- 2\,e3\,A12 + 3\,e3\,A23$$

where by $\&dw$ we have denoted the dotted wedge $\dot\wedge$. Then we have:

```
> #the left hand side of (16)
> 'x &dw u ' = simplify((1/2)*(x &c u + gradeinv(u) &c x));
```

$$x\,\&dw\,u = 4\,e1we2 - 6\,e2we3 - 2\,e1we2we3 + 4\,Id\,A12 - 6\,Id\,A23 + 2\,e2\,A13$$
$$- 2\,e3\,A12 + 3\,e3\,A23$$

```
> #the middle part of (17)
> 'LCg(x,u)' = (1/2)*(LCB(x,u) - LCBop(x,u));
```
$$LCg(\,x,u\,) = 4\,Id\,g12 + 6\,Id\,g23 + 2\,e2\,g13 - 2\,e3\,g12 + 3\,g33\,e2 - 3\,e3\,g23$$

```
> #the left hand side of (17)
> 'LCg(x,u)' = simplify((1/2)*(x &c u - gradeinv(u) &c x));
```
$$LCg(\,x,u\,) = 4\,Id\,g12 + 6\,Id\,g23 + 2\,e2\,g13 - 2\,e3\,g12 + 3\,g33\,e2 - 3\,e3\,g23$$

The above computations can be performed effortlessly with 'Clifford' for any $x = x_1e_1 + x_2e_2 + \ldots + x_ne_n \in V$, $x_i \in \mathbb{R}$ or $x_i \in \mathbb{C}$, $1 \le i \le n$, $1 \le n \le 9$, and any arbitrary multivector/Clifford polynomial $u \in C\ell(B)$. Verification is left to the reader.

The ability of 'Clifford' to perform computations with any symbolic form B is based on observation that multiplication of multivectors in any Clifford algebra in any dimension is essentially done the same way and that it only depends on the entries of B. By this we mean, for example, that irrespective of the dimension of V the product of vectors e_1 and e_2 is completely determined either by B_{12} or by B_{21} [depending on the order in which e_1 and e_2 are multiplied]. However, B

may easily be assigned numerical as well as symbolic values. For example, we may specify the bilinear form (25) as follows:

```
> B:=linalg[matrix](3,3,[1,a,b,-a,1,c,-b,-c,1]);
```

$$B := \begin{bmatrix} 1 & a & b \\ -a & 1 & c \\ -b & -c & 1 \end{bmatrix}$$

Then, the computations leading to finding $\gamma_{e_1 \wedge e_2 \wedge e_3}$ in Section 3 can be easily done as follows:

```
> clicollect(simplify(e1we2we3 &c e1we2we3));   #last column
```

$$(-1 - a^2 - c^2 - b^2)\, Id - 2\,a\, e1we2 - 2\,b\, e1we3 - 2\,c\, e2we3$$

```
> clicollect(simplify(e1we2we3 &c e1we2));   #the 5th column
```

$$-2\,a\, e1we2we3 + (-b - c\,a)\, e1 + (a\,b - c)\, e2 + (-a^2 - 1)\, e3$$

In fact, matrix $\gamma_{e_1 \wedge e_2 \wedge e_3}$ of $e_1 \wedge e_2 \wedge e_3$ in the left regular representation of $C\ell(B)$ on itself with respect to the ordered basis 'clibasis' can be computed with one single procedure 'builtm' (for examples of how 'builtm' can be used to find spinor representations see Abłamowicz 1995):

```
> clibasis:=cbasis(3);
```

$$clibasis := [\,Id, e1, e2, e3, e1we2, e1we3, e2we3, e1we2we3\,]$$

```
> builtm(e1we2we3,clibasis);
```

$$\begin{bmatrix} 0 & 0 & 0 & 0 & 0 & 0 & 0 & -1 - a^2 - c^2 - b^2 \\ 0 & 0 & 0 & 0 & -b - c\,a & -b\,c + a & -c^2 - 1 & 0 \\ 0 & 0 & 0 & 0 & a\,b - c & 1 + b^2 & b\,c + a & 0 \\ 0 & 0 & 0 & 0 & -a^2 - 1 & -c - a\,b & -c\,a + b & 0 \\ 0 & -b & -c & 1 & 0 & 0 & 0 & -2\,a \\ 0 & a & -1 & -c & 0 & 0 & 0 & -2\,b \\ 0 & 1 & a & b & 0 & 0 & 0 & -2\,c \\ 1 & 0 & 0 & 0 & -2\,a & -2\,b & -2\,c & 0 \end{bmatrix}$$

Finally, we verify that $e_1 e_2 e_3$ equals $e_1 \wedge e_2 \wedge e_3 + c e_1 - b e_2 + a e_3$:

```
> e1 &c e2 &c e3;
```

$$c\, e1 + e1we2we3 + a\, e3 - b\, e2$$

4.3. DEFINING AND VERIFYING THE ISOMORPHISM $\varphi : C\ell(I) \to C\ell(B)$ WHEN $\dim V = 3$

In Section 3 the isomorphism φ was constructed between $C\ell(I)$ and $C\ell(B)$ where B was given in (25). In preparation for a proof in the next section when $1 \leq \dim(V) \leq 9$, we now explain the construction of φ done with 'Clifford' in dimension 3.

We assume that Maple definitions of B and A from Section 4.1 remain all in force and we assume that the symmetric part g of B has been diagonalized (if g were not diagonal then the definition of φ would be different). We define φ as a Maple procedure 'phi' as:

```
> phi:=proc(a1:algebraic) local j,ind,expr; global dim;
> option remember;
> if type(args[1],{'*','+'}) then RETURN(map(phi,args[1])) fi;      #line 3
> if type(args[1],{mathfunc,numeric,rational,constant,             #line 4
>               complex,indexed,function}) or                      #line 5
>    member(args[1],{Id,e1,e2,e3,e4,e5,e6,e7,e8,e9}) or            #line 6
>    Clifford[cliterms](args[1])={} then RETURN(args[1]) fi;       #line 7
> ind:=Clifford[extract](args[1],'integers');                      #line 8
> expr:=simplify(eval(&c(seq(cat(e,ind[j]),j=1..nops(ind)))));     #line 9
> RETURN(expr);
> end:
```

Notice that lines 3–5 and 7 in 'phi' assure that 'phi' is linear and that it maps scalars to scalars; line 6 assures that $\varphi(Id) = Id$, and $\varphi(e_i) = e_i$, $1 \leq i \leq 9$; in line 8 subscripts of a monomial are extracted, e.g., **extract**$(e1we2we3) = [1, 2, 3]$; and in line 9 the Clifford product of 1-vectors with the extracted indices is calculated, e.g., &c$(e1, e2, e3)$ [see (30) below].

The definition of 'phi' is based on observation that the correspondences between the basis monomials of $C\ell(I)$ and $C\ell(B)$ found in Section 4.1 define φ as:

$$\varphi : C\ell(I) \to C\ell(B)$$
$$\varphi(Id) = Id, \ \varphi(e_i) = e_i, \ \varphi(e_i e_j) = \varphi(e_i \wedge e_j) = \varphi(e_i)\varphi(e_j), \ i \neq j$$
$$\varphi(e_i e_j e_k) = \varphi(e_i \wedge e_j \wedge e_k) = \varphi(e_i)\varphi(e_j)\varphi(e_k), \ i \neq j \neq k, \tag{30}$$
$$\cdots \cdots \cdots$$

and so on, for any orthogonal basis $e_1, e_2, \ldots, e_n \in V$, and with (30) extended by linearity to the whole algebra. Thus, by evaluating 'phi' on the monomials in the standard basis for $C\ell(I)$ we find the following in agreement with Section 3:

```
> for x in clibasis do 'phi('.x.')' = phi(x) od;
```

$$phi(Id) = Id$$
$$phi(e1) = e1$$
$$phi(e2) = e2$$
$$phi(e3) = e3$$
$$phi(e1we2) = e1we2 + a\,Id$$
$$phi(e1we3) = e1we3 + b\,Id$$
$$phi(e2we3) = e2we3 + c\,Id$$
$$phi(e1we2we3) = e1we2we3 + a\,e3 - b\,e2 + c\,e1$$

Using 'Clifford' we can verify that 'phi' is an algebra homomorphism. This can be accomplished by first checking the homomorphic property

$$\mathbf{phi}(x \ \&cQ \ y) = \mathbf{phi}(x) \ \&c \ \mathbf{phi}(y) \tag{31}$$

for any two basis monomials x and y found in 'clibasis' [by using the orthogonal Clifford multiplication in the left hand side of (31) we assume that the symmetric part of B has been diagonalized]. Then, because dimensions of $\varphi(C\ell(I))$ and $C\ell(B)$

are equal, it is clear that 'phi' is an isomorphism of the finite algebras. Using the package we will explicitly demonstrate the homomorphic property (31) for any two elements x and y of $C\ell(I)$ and show that $\ker(\varphi) = \{Id\}$.

The following computation results in verification of (31) on basis monomials (the loop below results in a sequence of 64 zeros which are not displayed):

```
> printlevel:=2:
> for x in clibasis do for y in clibasis do
>        simplify(phi(x &cQ y) - (phi(x) &c phi(y))) od od:
```

Thus, by linearity of 'phi' we may conclude that 'phi' is a homomorphism of algebras and, in fact, it is an isomorphism. In order to prove (31) for any two elements x and y in $C\ell(I)$, we first expand x and y with respect to the standard basis as follows (coefficients x_i, y_i, $i = 0, \ldots, 7$, can be real or complex):

```
> x:=sum(x.(i-1) * clibasis[i],'i'=1..nops(clibasis));
> y:=sum(y.(i-1) * clibasis[i],'i'=1..nops(clibasis));
```

$$x := x0\ Id + x1\ e1 + x2\ e2 + x3\ e3 + x4\ e1we2 + x5\ e1we3 + x6\ e2we3$$
$$+ x7\ e1we2we3$$
$$y := y0\ Id + y1\ e1 + y2\ e2 + y3\ e3 + y4\ e1we2 + y5\ e1we3 + y6\ e2we3$$
$$+ y7\ e1we2we3$$

Now we calculate the Clifford product of x and y in $C\ell(I)$, assign it to xy, apply 'phi' to it and assign $phi(xy)$ to 'left' (the left hand side of (31)). Similarly we calculate the right hand side of (31) and assign it to 'right'. Since these outputs are very long ('right' has 112 terms) we won't display these intermediate steps. Finally we verify that 'left' = 'right'

```
> xy:=clicollect(simplify(cmulQ(x,y))):
> left:=phi(xy):  #we apply 'phi' to xy
> right:=simplify(phi(x) &c phi(y)): #we won't display 112 terms in 'right'
> simplify(left - right);
                            0
```

Thus we have verified through the direct computation that $\varphi(x \& cQy) = \varphi(x) \& c\varphi(y)$ [when the symmetric part of B has been diagonalized]. Likewise we can check that if $\varphi(x) = Id$ then $x = Id$ and if $\varphi(y) = 0$ then $y = 0$. In Maple, we just apply 'phi' to x and y, make their images equal to Id and 0 respectively, and then solve two resulting systems of 8 equations in 8 unknowns each. We set $\mathbf{exprx} = \mathbf{phi(x)} - \mathbf{Id}$, find the equations and solve the system (and similarly for y):

```
> exprx:=clicollect(simplify(phi(x)-Id));
> expry:=clicollect(simplify(phi(y)));
```

$$exprx := x5\ e1we3 + x4\ e1we2 + x6\ e2we3 + (\,x5\,b + x4\,a + x6\,c + x0 - 1\,)\,Id$$
$$+ (\,x1 + x7\,c\,)\,e1 + (\,x2 - x7\,b\,)\,e2 + (\,x3 + x7\,a\,)\,e3 + x7\ e1we2we3$$
$$expry := y5\ e1we3 + y4\ e1we2 + y6\ e2we3 + (\,y5\,b + y6\,c + y0 + y4\,a\,)\,Id$$
$$+ (\,y1 + y7\,c\,)\,e1 + (\,y2 - y7\,b\,)\,e2 + (\,y3 + y7\,a\,)\,e3 + y7\ e1we2we3$$

```
> #system of equations in variables 'varsx' is called 'sysx'
> for i from 1 to 8 do eqx.i:=coeff(exprx,clibasis[i]) od:
> sysx:={eqx.(1..8)};varsx:={x.(0..7)};
```

$$sysx := \{\,x5\,b + x4\,a + x6\,c + x0 - 1, x3 + x7\,a, x1 + x7\,c, x2 - x7\,b, x4, x5, x6, x7\,\}$$
$$varsx := \{\,x0, x4, x5, x6, x7, x1, x2, x3\,\}$$

```
> #system of equations in variables 'varsy' is called 'sysy'
> for i from 1 to 8 do eqy.i:=coeff(expry,clibasis[i]) od:
> sysy:={eqy.(1..8)};varsy:={y.(0..7)};
```

$$sysy := \{\, y5, y7, y6, y3 + y7\,a, y2 - y7\,b, y1 + y7\,c, y5\,b + y6\,c + y0 + y4\,a, y4 \,\}$$
$$varsy := \{\, y5, y7, y6, y0, y1, y2, y3, y4 \,\}$$

```
> solx:=solve(sysx,varsx);subs(solx,x);
```

$$solx := \{\, x6 = 0, x7 = 0, x5 = 0, x4 = 0, x0 = 1, x2 = 0, x3 = 0, x1 = 0 \,\}$$
$$Id$$

```
> soly:=solve(sysy,varsy);subs(soly,y);
```

$$soly := \{\, y3 = 0, y2 = 0, y6 = 0, y4 = 0, y0 = 0, y1 = 0, y5 = 0, y7 = 0 \,\}$$
$$0$$

Thus, since $\varphi(x) = Id$ implies $x = Id$ and $\varphi(y) = 0$ implies $y = 0$, we have established explicitly that φ is an algebra isomorphism.

5. Isomorphism $\varphi : C\ell(Q) \to C\ell(B)$

In this final section we state the main result of this paper and describe its verification with $'$Clifford$'$.

Theorem 1.   *Suppose that B is a non-degenerate bilinear form in V, $\dim(V) \leq 9$, with a non-trivial antisymmetric part and let Q be the quadratic form associated with the symmetric part g of B, i.e., $Q(\mathbf{x}) = g(\mathbf{x}, \mathbf{x}) = B(\mathbf{x}, \mathbf{x})$. Then, in the category of associative algebras, there exists an isomorphism φ from the Clifford algebra $C\ell(Q)$ to the Clifford algebra $C\ell(B)$ of the bilinear form B. However, algebras $C\ell(Q)$ and $C\ell(B)$ are not isomorphic as Clifford algebras due to different multivector structure.*

Proof: The verification of this theorem is essentially an extension to higher dimensions of the verification described in Section 4.3. Due to the linearity of φ, it is enough to establish the homomorphic property (31) for 2^k monomials in the standard basis β_k in each dimension k, $1 \leq k \leq 9$. First, we define φ as in (30) and then evaluate $'$phi$'$ on all 512 basis monomials in β_9, the standard basis of the Clifford algebra $C\ell(B)$ in dimension 9 :

$$\begin{aligned}
\beta_9 = \{Id, \mathbf{e}_1, \ldots, \mathbf{e}_1 \wedge \mathbf{e}_2, \ldots, \mathbf{e}_1 \wedge \mathbf{e}_2 \wedge \mathbf{e}_3, \ldots, \\
\mathbf{e}_1 \wedge \mathbf{e}_2 \wedge \mathbf{e}_3 \wedge \mathbf{e}_4, \ldots, \mathbf{e}_1 \wedge \mathbf{e}_2 \wedge \mathbf{e}_3 \wedge \mathbf{e}_4 \wedge \mathbf{e}_5, \ldots, \\
\mathbf{e}_1 \wedge \mathbf{e}_2 \wedge \mathbf{e}_3 \wedge \mathbf{e}_4 \wedge \mathbf{e}_5 \wedge \mathbf{e}_6, \ldots, \mathbf{e}_1 \wedge \mathbf{e}_2 \wedge \mathbf{e}_3 \wedge \mathbf{e}_4 \wedge \mathbf{e}_5 \wedge \mathbf{e}_6 \wedge \mathbf{e}_7, \ldots, \\
\mathbf{e}_1 \wedge \mathbf{e}_2 \wedge \mathbf{e}_3 \wedge \mathbf{e}_4 \wedge \mathbf{e}_5 \wedge \mathbf{e}_6 \wedge \mathbf{e}_7 \wedge \mathbf{e}_8, \ldots, \\
\mathbf{e}_1 \wedge \mathbf{e}_2 \wedge \mathbf{e}_3 \wedge \mathbf{e}_4 \wedge \mathbf{e}_5 \wedge \mathbf{e}_6 \wedge \mathbf{e}_7 \wedge \mathbf{e}_8 \wedge \mathbf{e}_9 \}.
\end{aligned} \qquad (32)$$

Since the verification of (31) for all elements in β_9 would require computation of a prohibitive number of $374,544$ products in $C\ell(Q)$, another as many in $C\ell(B)$ and some having over 500 symbolic terms, a simplification of the method was desirable.

Let's observe that since the standard bases form an ascending chain

$$\beta_1 = \{Id, \mathbf{e}_1\} \subset \beta_2 = \{Id, \mathbf{e}_1, \mathbf{e}_2, \mathbf{e}_1 \wedge \mathbf{e}_2\} \subset \beta_3 \subset \cdots \subset \beta_9,$$

one obvious way to reduce the number of computations is to not repeat computations with those elements in β_k which also belong to β_{k-1}. Furthermore, in each β_k, it is enough to limit verification to one or two elements of the given grade: once (31) has been verified for two monomials, say one belonging to $\bigwedge^m V$ and another belonging to $\bigwedge^n V$ in $C\ell(Q)$, validity of (31) for all monomials in $\bigwedge^m V \cup \bigwedge^n V$ may be concluded by simple permutation of the monomial indices. This way the verification of (31) has been done in all dimensions $1 \leq \dim(V) \leq 9$ for all basis monomials which, since $\ker(\varphi) = \{Id\}$, also proves that φ is the anticipated algebra isomorphism from $C\ell(Q)$ to $C\ell(B)$, $Q(\mathbf{x}) = B(\mathbf{x}, \mathbf{x})$. $\square$

Acknowledgements

One of the authors (R. A.) expresses gratitude to Yvon Siret, Université Joseph Fourier, Grenoble, France, for his generous help and computer time to run many Maple worksheets under 'Clifford'. These tests were needed to verify results of this paper as well as to debug, expand, and improve the code therein.

References

R. Abłamowicz: 'Clifford algebra computations with Maple', Proceedings of CAP Summer School in Theoretical Physics, "Geometric (Clifford) Algebras in Physics," Banff, 1995 (to appear). *Technical Report* No. 1995-1, Department of Mathematics, Gannon University, October 1995.

R. Abłamowicz: 1996, 'Clifford' - *Maple V package for Clifford algebra computations*, ver. 2, available at: http://www.gannon.edu/service/dept/mathdept.

C. Chevalley: 1954, 'The Algebraic Theory of Spinors', Columbia University Press, New York.

A. Crumeyrolle: 1990, 'Orthogonal and Symplectic Clifford Algebras: Spinor Structures', Kluwer, Dordrecht.

J. Helmstetter: 1982, 'Algèbres de Clifford et algèbres de Weyl', Cahiers Math. **25**, Montpellier.

P. Lounesto, R. Mikkola, and V. Vierros: 1987, 'CLICAL User Manual', Helsinki University of Technology, Institute of Mathematics, Research Reports **A248**, Helsinki.

P. Lounesto: 1993, 'What is a bivector?', in 'Spinors, Twistors, Clifford Algebras and Quantum Deformations', Proceedings of the Second Max Born Seminar Series, Wrocław, Poland, 1992; eds. Z. Oziewicz, B. Jancewicz, and A. Borowiec Kluwer, Dordrecht, pp. 153–158.

P. Lounesto: 1995, 'Crumeyrolle's bivectors and spinors', pp. 137-166 in R. Abłamowicz, P. Lounesto (eds.): *Clifford Algebras and Spinor Structures, A Special Volume Dedicated to the Memory of Albert Crumeyrolle (1919-1992)*. Kluwer, Dordrecht, 1995.

'Maple V Release 3 for DOS and Windows,' 1994, Waterloo Maple Software, Waterloo, Ontario.

Z. Oziewicz: 1986, 'From Grassmann to Clifford', pp. 245-255 in J.S.R. Chisholm, A.K. Common (eds.): *Clifford Algebras and their Applications in Mathematical Physics (Canterbury, 1985)*. Reidel, Dordrecht, 1986.

A UNIPODAL ALGEBRA PACKAGE FOR MATHEMATICA

GARRET SOBCZYK *
Universidad de las Americas
Departamento de Fisico-Matematicas
Apartado Postal #100, Santa Catarina Mártir
72820 Puebla, Pue., México
e-mail: sobczyk@udlapvms.pue.udlap.mx

Abstract. This article discusses unipodal numbers, a kind of hyperbolic version of the complex plane, where the imaginary unit is replaced by a unipotent $u^2 = 1$, $u \neq \pm 1$. A function theory for unipodal numbers is sketched, the values of functions are evaluated using minimal polynomials, and a related computer package for MATHEMATICA is described. Unipodal numbers offer a nice first step for teachers of Clifford algebra.

Key words: Unipodal numbers – cubic equation – minimal polynomial

1. Introduction

The MATHEMATICA Package UALGEBRA.M is designed to work in the unipodal number system. This number system $\mathbb{U} = \mathbb{C}[u]$ is the complex number system $\mathbb{C}$ extended to include the *unipotent* u, where $u^2 = 1$ but $u \neq \pm 1$, and is one of the few examples of a *commutative* Clifford algebra. The package extends entire functions of a complex variable to entire functions of a unipodal variable. This package also contains formulas for the solutions of the classical cubic equation (Sobczyk, 1995). A separate package of tools VDET.M is included which is useful in solving the *structure equation* of an arbitrary Clifford number once its minimal polynomial is known.

2. The unipodal number system

In the *standard basis* $\{1, u\}$ a *unipodal number* $w \in \mathbb{U}$ has the form $w = w_0 + uw_1$ where $u^2 = 1$ but $u \neq \pm 1$, and $w_0, w_1 \in \mathbb{C}$. The *idempotent basis* $\{u_+, u_-\}$ is defined by

$$u_+ = \frac{1}{2}(1 + u) \quad \text{and} \quad u_- = \frac{1}{2}(1 - u),$$

and satisfies $u_+ + u_- = 1$ and $u_+ - u_- = u$. We say that u_+ and u_- are *idempotents* because $u_+^2 = u_+$ and $u_-^2 = u_-$, and they are *mutually annihilating* because $u_+ u_- = 0$. Using the projective properties of the idempotent basis we can write

$$w = w(u_+ + u_-) = w_+ u_+ + w_- u_-, \tag{1}$$

---

* I gratefully acknowledge the support given by INIP of the Universidad de las Americas, and CONACYT (The Mexican National Council for Science and Technology) grant 3803-E.

where $w_+ = w_0 + w_1$ and $w_- = w_0 - w_1$. Conversely, given $w = w_+u_+ + w_-u_-$ for $w_+, w_- \in \mathbb{C}$, we can recover the coordinates with respect to the standard basis by

$$w_0 = \frac{1}{2}(w_+ + w_-) \quad \text{and} \quad w_1 = \frac{1}{2}(w_+ - w_-). \tag{2}$$

The special properties of the idempotent basis make calculations particularly simple. For example, the binomial theorem takes the simple form

$$(w_+u_+ + w_-u_-)^k = (w_+)^k u_+^k + (w_-)^k u_-^k = (w_+)^k u_+ + (w_-)^k u_- \tag{3}$$

This formula is valid for *all* real numbers $k \in \mathbb{R}$, and not just the positive integers. For example, for $k = -1$ we find that

$$1/w = w^{-1} = (1/w_+)u_+ + (1/w_-)u_-,$$

a valid formula for the inverse of $w \in \mathbb{U}$ provided that $w_+w_- \neq 0$.

Indeed, the validity of (3) allows us to extend the definitions of *all* of the elementary functions in the complex plane to the corresponding elementary functions in the unipodal plane. If $f(w)$ is such a function for $w = w_+u_+ + w_-u_-$, we define

$$f(w) \equiv f(w_+)u_+ + f(w_-)u_- \tag{4}$$

provided that $f(w_+)$ and $f(w_-)$ are defined. For example, using the MATHEMATICA Package UALGEBRA.M, we can find $\log(2 + 3u)$ by entering the line

$$\log[2 + 3u]$$

getting the answer $w = 1/2 \log 5 + \pi i/2 + u(1/2 \log 5 - \pi i/2)$. As a check, we enter the line

$$\exp[w]$$

to get back the original $2 + 3u$. We can equally well find the symbolic expression for $\log(x0 + x1u)$.

The unipodal numbers have been studied in (Hestenes al., 1991). They are used in the next section to find the solutions of the cubic equation.

3. Solutions to cubic equation

Historically, the complex numbers were grudgingly accepted because of their utility in solving the cubic equation. We demonstrate here the usefulness of unipodal numbers by finding a formula for the solutions of the venerated reduced cubic equation

$$x^3 + 3ax + b = 0. \tag{5}$$

(The coefficient of the x^2 term of the more general cubic $x^3 + p_2x^2 + p_1x + p_0 = 0$, can be eliminated by making the linear substitution $x \rightarrow x - \frac{1}{3}p_2$.)

The basic unipodal equation $w^n = r$ can easily be solved using the idempotent basis, with the help of equation (3). Writing $w = w_+u_+ + w_-u_-$, and $r = r_+u_+ + r_-u_-$ we get

$$w^n = w_+^n u_+ + w_-^n u_- = r_+u_+ + r_-u_-, \tag{6}$$

so $w_+^n = r_+$ and $w_-^n = r_-$. It follows that $w_+ = r_+^{\frac{1}{n}}\alpha^j$ and $w_- = r_-^{\frac{1}{n}}\alpha^k$ for some integers $0 \le j,k \le n-1$, where α is a primitive n^{th} root of unity and $r_+^{\frac{1}{n}}$ and $r_-^{\frac{1}{n}}$ are arbitrary but fixed n^{th}-roots of r_+ and r_-, respectively. This proves the following theorem.

Theorem 1 *For any positive integer n, the unipodal equation $w^n = r$ has n^2 solutions $w = \alpha^j r_+^{\frac{1}{n}} u_+ + \alpha^k r_-^{\frac{1}{n}} u_-$ for $j,k = 0,1,\ldots,n-1$, where $\alpha \equiv \exp(2\pi i/n)$.*

The number of roots to the equation $w^n = r$ can be reduced by adding constraints. The following corollary follows immediately from the theorem, by noting that $w_+ w_- = \rho \ne 0$ is equivalent to $w_- = \rho/w_+$.

Corollary. The unipodal equation $w^n = r$, subject to the constraint $w_+ w_- = \rho$, for a nonzero complex number ρ, has the n solutions

$$w = \alpha^j r_+^{\frac{1}{n}} u_+ + \frac{\rho}{\alpha^j r_+^{\frac{1}{n}}} u_-,$$

for $j = 0,1,\ldots,n-1$, where $\alpha \equiv \exp(2\pi i/n)$, and $r_+^{\frac{1}{n}}$ denotes any n^{th} root of the complex number r_+.

We are now prepared to solve the reduced cubic equation (5), (Sobczyk, 1995).

Theorem 2 *The reduced cubic equation $x^3 + 3ax + b = 0$, has the solutions, for $j = 0,1,2$,*

$$x = \frac{1}{2}\left(\alpha^j \sqrt[3]{s+t} + \frac{\rho}{\alpha^j \sqrt[3]{s+t}}\right), \tag{7}$$

where $\alpha = \exp(2\pi i/3) = -\frac{1}{2} + \frac{1}{2}i\sqrt{3}$ is a primitive cube root of unity, and $\rho = -4a$, $s = -4b$, and $t = \sqrt{s^2 - \rho^3} = 4\sqrt{b^2 + 4a^3}$.

Proof. The unipodal equation $w^3 = r$, where $r = s + ut$, is equivalent in the standard basis to $(x + yu)^3 = s + tu$, or $(x^3 + 3xy^2) + u(y^3 + 3x^2 y) = s + ut$. Equating the complex scalar parts gives

$$x^3 + 3xy^2 - s = 0. \tag{8}$$

Making the additional constraint that $w_+ w_- = x^2 - y^2 = \rho$, we can eliminate y^2 from (8), getting the equivalent equation

$$x^3 - \frac{3}{4}\rho x - \frac{1}{4}s = 0. \tag{9}$$

The constraint $w_+ w_- = \rho$ further implies that

$$\rho^3 = (w_+ w_-)^3 = w_+^3 w_-^3 = r_+ r_- = s^2 - t^2,$$

which gives $t = \sqrt{s^2 - \rho^3}$. By letting $\rho = -4a$, and $s = -4b$, so $t = \sqrt{s^2 - \rho^3} = 4\sqrt{b^2 + 4a^3}$, the equation (9) becomes the reduced cubic equation $x^3 + 3ax + b = 0$.

Since $r_+ = s + t$, the desired solution (7) is then obtained by taking the complex scalar part of the solution given in the corollary.   Q.E.D.

The MATHEMATICA Package UALGEBRA.M has a built in routine to solve the cubic equation $x^3 + p_2 x^2 + p_1 x + p_0 = 0$ based on Theorem 2. For example, to solve the cubic equation $x^3 + x^2 + 2x + 3 = 0$, open the package "ualgebra" by typing $<<$ ualgebra.m   and then enter the line

$$p2 = 1; p1 = 2; p0 = 3;$$

The roots are obtained by entering

$$roots$$

These roots may be checked by entering

$$f[roots]$$

4. Solutions of the structure equation

Let $\mathbb{C}_n$ denote the *complex Clifford algebra* of an n-dimensional complex vector space $\mathbb{C}^n$. Let $a \in \mathbb{C}_n$ denote an arbitrary element of this Clifford algebra. By the *minimal polynomial* of a we mean the unique monic polynomial $\psi(\alpha)$ of *least* degree,

$$\psi(\alpha) \equiv \prod_{i=1}^{r} (\alpha - \alpha_i)^{m_i}$$

for the distinct complex numbers $\alpha_1, \alpha_2, \ldots, \alpha_r \in \mathbb{C}$, for which $\psi(a) = 0$. For example, in the complex Clifford algebra $\mathbb{C}_2$ generated by the orthonormal unit vectors e_1, e_2, a general element has the form $a = \alpha_0 + \alpha_1 e_1 + \alpha_2 e_2 + \alpha_3 e_1 e_2$. Noting that

$$(a - \alpha_0)^2 = (\alpha_1 e_1 + \alpha_2 e_2 + \alpha_3 e_1 e_2)^2 = \alpha_1^2 + \alpha_2^2 - \alpha_3^2 = z^2,$$

we can conclude that a will always satisfy the polynomial equation

$$(\alpha - \alpha_0 - z)(\alpha - \alpha_0 + z) = 0.$$

It follows that the minimal polynomial of an arbitrary element $a \in \mathbb{C}_2$ has degree ≤ 2.

The *structure equation* of $a \in \mathbb{C}_n$ with the minimal polynomial $\psi(\alpha)$ given above is defined by

$$a = \sum_{i=1}^{r} (\alpha_i + q_i) p_i, \tag{10}$$

where p_i and q_j satisfy the rules

$$\{p_1 + \ldots + p_r = 1, \ p_i p_j = \delta_{ij} p_i, \ q_k^{m_k-1} \neq 0 \text{ but } q_k^{m_k} = 0, \ p_k q_k = q_k\}. \tag{11}$$

The p_i's are *mutually annihilating idempotents* which partition unity and the q_j's are *nilpotents* with the respective *indexes* m_j's. The algebra $\mathbb{C}\{m_1, m_2, \ldots, m_r\}$ generated by the set $\{p_i, q_i\}$ is commutative and it is isomorphic to the factor ring $\mathbb{C}[\alpha]/ < \psi(\alpha) >$ of the principal ideal $< \psi(\alpha) >$, (Sobczyk, 1993).

The $s \times s$ dimensional *Vandermonde determinant* V, for $s = \sum_{j=1}^{r} m_j$, can be efficiently defined in terms of the column function

$$f_x \equiv f(x) = \{1, x, x^2, \ldots, x^{s-1}\}^T.$$

We have

$$V \equiv \det\{f(\alpha_1), f(\alpha_2), \ldots, f(\alpha_s)\} = \prod_{1 \leq i < j \leq s} (\alpha_j - \alpha_i), \tag{12}$$

where the *normalized* derivatives $f^{(k)}(\alpha)$ are given by

$$f_\alpha^{(k)} \equiv f^{(k)}(\alpha) = \frac{1}{k!} D_\alpha f(\alpha).$$

The *confluent* or *generalized Vandermonde matrix* W is defined by

$$W \equiv \{ \underbrace{f_1 f_1^{(1)} \ldots f_1^{(m_1-1)}}_{m_1 \ columns} \ \underbrace{f_2 f_2^{(1)} \ldots f_1^{(m_2-1)}}_{m_2 \ columns} \ \cdots \ \underbrace{f_r f_r^{(1)} \ldots f_r^{(m_r-1)}}_{m_r \ columns} \}, \tag{13}$$

and has the determinant

$$\det W = \prod_{1 \leq i < j \leq r} (\alpha_j - \alpha_i)^{m_i m_j}. \tag{14}$$

We can now state the *structure theorem* for the element a, (Sobczyk, 1993).

Theorem 3 *If $\psi(\alpha)$ is the minimal polynomial of the Clifford number $a \in \mathbb{C}_n$, then a can be expressed in the eigenprojector form*

$$a = \sum_{j=1}^{r} (\alpha_j + q_j) p_j,$$

where the idempotents and nilpotents p_i and q_i, satisfying (11), are the unique polynomials in a of degree $< s = m_1 + \ldots m_r$, specified by

$$p_i = \frac{\det\{ \overbrace{f_1 \ldots f_1^{(m_1-1)}}^{m_1} \cdots \overbrace{f_a f_i^{(1)} f_i^{(2)} \ldots f_i^{(m_1-1)}}^{m_i} \cdots \overbrace{f_r \ldots f_r^{(m_r-1)}}^{m_r}\}}{\det W},$$

and

$$q_i = \frac{\det\{\overbrace{f_1 \ldots f_1^{(m_1-1)}}^{m_1} \ldots \overbrace{f_i f_a f_i^{(2)} \ldots f_i^{(m_i-1)}}^{m_i} \ldots \overbrace{f_r \ldots f_r^{(m_r-1)}}^{m_r}\}}{\det W},$$

where the $\det W$ *is given in (14).*

As an application of this theorem, consider any Clifford number $a \in \mathbb{C}_n$ with the minimal polynomial $\psi(\alpha) = (\alpha - \alpha_1)(\alpha - \alpha_2)^2$. The structure equation of this element is

$$a = \alpha_1 p_1 + (\alpha_2 + q_2)p_2, \tag{15}$$

with the unique solution, for $f(\alpha) = (1, \alpha, \alpha^2)^T$, given by

$$p_1 = \frac{\det\{f_a, f_2, f_2'\}}{\det W} = \frac{D_{\alpha_3} V|_{\alpha_1 \to a, \alpha_3 \to \alpha_2}}{\det W} = \frac{(\alpha_2 - a)^2}{(\alpha_2 - \alpha_1)^2}, \quad q_1 \equiv 0,$$

$$p_2 = \frac{\det\{f_1, f_a, f_2'\}}{\det W} = \frac{D_{\alpha_3} V|_{\alpha_2 \to a, \alpha_3 \to \alpha_2}}{\det W} = \frac{(\alpha_1 - a)(a + \alpha_1 - 2\alpha_2)}{(\alpha_2 - \alpha_1)^2}$$

$$q_2 = \frac{\det\{f_1, f_2, f_a\}}{\det W} = \frac{V|_{\alpha_3 \to a}}{\det W} = \frac{(\alpha_1 - a)(\alpha_2 - a)}{(\alpha_2 - \alpha_1)},$$

where the Vandermonde determinant of this system is $V = (\alpha_3 - \alpha_2)(\alpha_3 - \alpha_1)(\alpha_2 - \alpha_1)$, and

$$\det W = D_{\alpha_3} V|_{\alpha_3 \to \alpha_2} = (\alpha_2 - \alpha_1)^2.$$

The calculations for p_1, p_2, and q_2 can be carried out with the help of the MATH-EMATICA Package VDET.M, as is included as an example at the end of this package.

An important application of eigenprojector form is evaluating any function of a Clifford number. For our example, the *inverse* of a is

$$a^{-1} = \frac{1}{\alpha_1} p_1 + \frac{1}{\alpha_2 + q_2} p_2 = \frac{1}{\alpha_1} p_1 + \frac{\alpha_2 - q_2}{\alpha_2^2} p_2,$$

and just as easily we find

$$\log(a) = \log(\alpha_1) p_1 + \log(\alpha_2 + q_2) p_2 = \log(\alpha_1) p_1 + [\log(\alpha_2) + \frac{q_2}{\alpha_2}] p_2.$$

More generally, let $g(z)$ be any function which has a convergent Taylor series around each of the eigenvalues $z = \alpha_j \in \mathbb{C}$ of the Clifford number a given in (10). Then we can define $g(a)$ by

where

$$g(\alpha_i + q_i) = g(\alpha_i) + g^{(1)}(\alpha_i)q_i + \ldots + \frac{1}{(m_i - 1)!}g^{(m_i - 1)}(\alpha_i)q_i^{m_i - 1}$$

is the Taylor series expansion of $g(z)$ around $z = \alpha_i$. We say that (10) determines the *analytic structure* of a, because $g(a)$ is well defined for any function g which is analytic on the spectrum of a.

The relationship between Clifford algebra and algebras of endomorphisms is discussed in (Sobczyk,).

Appendix

A.

```
(* SOLVING THE STRUCTURE EQUATION *)

Off[General::spell1]
Off[General::spell]

(* Gives Vandermonde Determinants up until 10x10 *)

vdet[1] := 1
vdet[2] := x2 - x1
vdet[3] := (x3 - x2)*(x3 - x1)*(x2 - x1)
vdet[4] := (x4 - x3)*(x4 - x2)*(x4 - x1)*vdet[3]
vdet[5] := (x5 - x4)*(x5 - x3)*(x5 - x2)*(x5 - x1)*vdet[4]
vdet[6] := (x6 - x5)*(x6 - x4)*(x6 - x3)*(x6 - x2)*(x6 - x1)*vdet[5]
vdet[7] := (x7 - x6)*(x7 - x5)*(x7 - x4)*(x7 - x3)*(x7 - x2)*(x7 - x1)*
vdet[6]
vdet[8] := (x8 - x7)*(x8 - x6)*(x8 - x5)*(x8 - x4)*(x8 - x3)*(x8 - x2)*
(x8 - x1)*vdet[7]
vdet[9] := (x9 - x8)*(x9 - x7)*(x9 - x6)*(x9 - x5)*(x9 - x4)*(x9 - x3)*
(x9 - x2)*(x9 - x1)*vdet[8]
vdet[10] := (x10 - x9)*(x10 - x8)*(x10 - x7)*(x10 - x6)*(x10 - x5)*
(x10 - x4)*(x10 - x3)*(x10 - x2)*(x10 - x1)*vdet[9]

(* Confluent or Generalized Vandermonde Determinants with up
to 5 blocks of sizes (m1,m2,m3,m4,m5). *)

gvdet[1] := vdet[1]
gvdet[2] := 1
gvdet[3] := 1
gvdet[1, 1] := vdet[2]
gvdet[1, 2] := (x2 - x1)^2*vdet[1]
gvdet[1, 3] := (x2 - x1)^3
gvdet[1, 4] := (x2 - x1)^4
gvdet[2,2] :=vdet[2]^4
```

```
gvdet[2,3]:=vdet[2]^6
gvdet[1, 1, 1]:= vdet[3]
gvdet[1, 1, 2]:= (x3 - x2)^2*(x3 - x1)^2*vdet[2]
gvdet[1, 1, 3]:= (x3 - x2)^3*(x3 - x1)^3*vdet[2]
gvdet[1, 1, 4]:= (x3 - x2)^4*(x3 - x1)^4*vdet[2]
gvdet[1, 2, 3]:= (x3 - x2)^6*(x3 - x1)^3*(x2 - x1)^2
gvdet[2,2,2]:=vdet[3]^4
gvdet[1, 1, 1, 1]:= vdet[4]
gvdet[1, 1, 1, 2]:= (x4 - x3)^2*(x4 - x2)^2*(x4 - x1)^2*vdet[3]
gvdet[1, 1, 1, 3]:= (x4 - x3)^3*(x4 - x2)^3*(x4 - x1)^3*vdet[3]
gvdet[1, 1, 1, 4]:= (x4 - x3)^4*(x4 - x2)^4*(x4 - x1)^4*vdet[3]
gvdet[1, 1, 2, 3]:=
(x4 - x3)^6*(x4 - x2)^3*(x4 - x1)^3*(x3 - x2)^2*(x3 - x1)^2*vdet[2]
gvdet[1, 1, 1, 1, 1]:= vdet[5]
gvdet[1, 1, 1, 1, 2]:=
(x5 - x4)^2*(x5 - x3)^2*(x5 - x2)^2*(x5 - x1)^2*vdet[4]
gvdet[1, 1, 1, 1, 3]:=
(x5 - x4)^3*(x5 - x3)^3*(x5 - x2)^3*(x5 - x1)^3*vdet[4]
gvdet[1, 1, 1, 1, 4]:=
(x5 - x4)^4*(x5 - x3)^4*(x5 - x2)^4*(x5 - x1)^4*vdet[4]


(* ROW VECTOR FUNCTIONS & DERIVATIVES - USEFUL
FOR WRITING DOWN CONFLUENT VANDERMONDE MATRICES *)

f[n_, x_]:= Table[x^k, {k, 0, n - 1}]
d[0, f_, x_]:= f
d[1, f_, x_]:= D[f, x]
d[2, f_, x_]:= (1*D[D[f, x], x])/2
d[3, f_, x_]:= (1*D[D[D[f, x], x], x])/6
d[4, f_, x_]:= (1*D[D[D[D[f, x], x], x], x])/24
d[5, f_, x_]:= (1*D[D[D[D[D[f, x], x], x], x], x])/120


(* EXAMPLE for MINIMAL POLYNOMIAL PSI = (x-x1)(x-x2)^2  *)


vdet3:=vdet[3]
gvdet12:=gvdet[1,2]
p1:= (d[1,vdet3,x3] /. {x1->a,x3->x2})/gvdet12
q1:=0
p2:= (d[1,vdet3,x3] /. {x2->a,x3->x2})/gvdet12
q2:= (vdet3 /. {x3->a})/gvdet12


B.


(* UNIT.M  July 4, 1995 *)
(* Geometric Algebra G[1,C] with basis {1,u} over C *)
```

```
Off[General::spell1]
Off[General::spell]
Unprotect[Power,Times]

u*u=1
u^(n_):=If[Mod[n,2]==0,1,u,HoldForm[u^n]]
Protect[Power,Times]

(* Special operations *)

dtb[a_]:=Distribute[a]
cll[a_]:=Collect[a,u]
epd[a_]:=Expand[a]
sim[a_]:=Simplify[a]
rat[a_]:=Rationalize[a]

(* Sample function and geometric numbers *)

a=a0+a1 u
b=b0+b1 u
c=c0+c1 u
d=d0+d1 u
x=x0+x1 u

(* Geometric variables *)

w = w0 + w1 u
w$ = w0 - w1 u ;

(* Vector derivatives *)

dd[w_] := cll[epd[D[w, w0] ]]
dir[w_,a_]:= cll[epd[csp[a] D[w, w0] + csp[a u] D[w,w1]]]

(* Elementary Geometric Functions *)

cvp[a_]:= a - csp[a]
csp[a_]:= epd[a] /. u -> 0
rvs[a_]:=2 csp[a]-a

con[a_ + b_]:= con[a]+con[b]
con[a_ b_]:=con[a] con[b]
con[a_]:=Conjugate[a] /; NumberQ[a]
con[a_]:=a /; !NumberQ[a]

re[a_]:=cll[1/2(a+con[a])]
```

```
im[a_]:=cll[I/2(con[a]-a)]

vp[a_]:=1/2(cll[cvp[a]+con[cvp[a]]])
bp[a_]:=1/2(cll[cvp[a]-con[cvp[a]]])
sp[a_]:=1/2(cll[csp[a]+con[csp[a]]])
tp[a_]:=1/2(cll[csp[a]-con[csp[a]]]);

sti[a_,b_]:= csp[a rvs[b]]
sti[a_]:=sti[a,a]
csmag[a_]:= Sqrt[sti[a,a]]
csmag[a_,b_]:=Sqrt[sti[a,b]]
mag[a_]:=Sqrt[csp[a con[a]]]

up=1/2 (1+u)
un=1/2 (1-u)
upp[x_]:=csp[x]+D[x,u]
unn[x_]:=csp[x]-D[x,u]

exp[a_]:=cll[Exp[upp[a]] up+Exp[unn[a]] un]
sinh[a_]:=cll[Sinh[upp[a]] up+Sinh[unn[a]] un]
cosh[a_]:=cll[Cosh[upp[a]] up+Cosh[unn[a]] un]
log[a_]:=cll[Log[upp[a]] up+Log[unn[a]] un]

root[a_,n_,i_,j_]:=cll[epd[exp[log[a]/n] exp[2 Pi I (i+j u)/n]]]
root[a_,n_,i_]:=root[a,n,i,0]
root[a_,n_]:=root[a,n,0,0]
root[a_]:=root[a,2,0,0]

power[w_,a_]:=cll[epd[exp[cll[a log[w]]]]];

(* SOLUTIONS OF CUBIC EQUATION   June 1995 *)
(* Solution of Cubic Equation: x0^3 + p2 x0^2 + p1 x0 + p0 = 0 *)
(* Equivalent to (x-a0)^3=b where x=x0+b1 u, b=b0+b1 u   *)

a0:= -p2/3
b0:= (-108*p0 + 36*p1*p2 - 8*p2^3)/27
b1:=4*( p0^2+(4*p1^3)/27-(2*p0*p1*p2)/3-(p1^2*p2^2)/27+
(4*p0*p2^3)/27 )^(1/2)
r:= 1/9 (4 p2^2-12 p1)
f[x_] := x^3 + p2*x^2 + p1*x + p0
x0[k_]:=1/2 N[2 a0+(root[b0+b1,3,k]+r/root[b0+b1,3,k])]
rootf[j_,k_] := csp[a0 + sim[rat[N[root[b0 + b1*u, 3, j, k]]]]]
f[k1_,k2_]:=Chop[f[x0[k1,k2]]]
roots:={x0[0],x0[1],x0[2]}
```

References

D. Hestenes, P. Reany, G. Sobczyk: 1991, 'Unipodal algebra and roots of polynomials', *Advances in Applied Clifford Algebras*, Vol. 1, No. 1, pp. 31–51.

G. Sobczyk: 1995, 'Hyperbolic number plane', *The College Mathematics Journal*, September.

G. E. Sobczyk: 1993, 'Jordan form in Clifford algebras', *Clifford Algebras and their Applications in Mathematica Physics*, Proceedings of the Third International Clifford Algebras Workshop, Edited by Fred Brackx, Richard Delanghe, and Herman Serras, Kluwer, Dordrecht.

G. E. Sobczyk: 'Structure of factor algebras and Clifford algebra', *Journal of Linear Algebra and Applications*, to appear.

OCTONION X-PRODUCT ORBITS

GEOFFREY DIXON
Department of Mathematics or Physics
Brandeis University
Waltham, MA 02254
e-mail: dixon@binah.cc.brandeis.edu

and

Department of Mathematics
University of Massachusetts
Boston, MA 02125

Abstract. The octonionic X-product gives to the octonions a flexibility not found in the other real division algebras (reals, complexes, quaternions). The pattern of this flexibility is investigated here.

Key words: Division algebras, octonions.

1. Introduction

The inspiration for this article arose from three sources (Cederwall and Preitschopf, 1993), (Schray and Manogue, 1994), and (Dixon, 1994). In (Cederwall and Preitschopf, 1993) the octonionic X-product was introduced, and it was pointed out that although the 7-sphere (S^7) is the unique parallelizable manifold not also a group manifold, with the aid of the X-product S^7 gains an almost group structure.

In (Schray and Manogue, 1994) the X-product was applied to the 480 renumberings of the octonionic basis, and it was pointed out that this set of renumberings actually splits into two sets of 240 renumberings via an X-product equivalence. These two sets were dubbed opposites. In both (Cederwall and Preitschopf, 1993) and (Schray and Manogue, 1994) the ultimate goal was an application of the octonions to string theory, in which context the octonions play a natural role.

In (Dixon, 1994) I presented my own view of division algebra theory and how it connects to physics. The presentation of octonion theory in that monograph is pragmatic, a kind of get-down-and-get-dirty mathematics that I find comprehensible and useful. This article is a result of the application of my methods to the octonionic X-product.

To facilitate this work I used a computer spread sheet (I imagine just about any would work). Two columns of eight numbers can be easily multiplied into a third column once one has an octonion multiplication rule on which to build. This idea can then be elaborated into products of arbitrary many elements (for example, it is not difficult to compute the associator of three columns that represent distinct octonions).

2. Four octonionic basis numberings

In all that follows the symbols e_a, $a = 1, ..., 7$, will represent an orthogonal basis for the 7-dimensional imaginary (pure) octonions, and $e_0 = 1$ will be the identity. There are 7! permutations of the indices of the pure octonions, and each gives rise to a modified copy of $\mathbb{O}$ (the real octonion division algebra) with an altered, but still octonionic, multiplication table. As it turns out, however, these index rearrangements are not all unique. In the end we will find that there are only 480 distinct multiplication tables for which

$$e_a e_b = \pm e_c$$

for all $a, b \in \{0, ..., 7\}$ and some a, b-dependent c.

Of all these 480 distinct copies of $\mathbb{O}$, there are 4 that are singled out for their elegance and symmetry. These four arise from the following 8×8 array of binary numbers (see (Dixon, 1994)):

$$O = \begin{bmatrix} 0 & 0 & 0 & 0 & 0 & 0 & 0 & 0 \\ 0 & 1 & 0 & 0 & 1 & 0 & 1 & 1 \\ 0 & 1 & 1 & 0 & 0 & 1 & 0 & 1 \\ 0 & 1 & 1 & 1 & 0 & 0 & 1 & 0 \\ 0 & 0 & 1 & 1 & 1 & 0 & 0 & 1 \\ 0 & 1 & 0 & 1 & 1 & 1 & 0 & 0 \\ 0 & 0 & 1 & 0 & 1 & 1 & 1 & 0 \\ 0 & 0 & 0 & 1 & 0 & 1 & 1 & 1 \end{bmatrix}. \tag{1}$$

Let OR_a be the ath row and OC_a be the ath column of O. The set of rows and the set of columns are individually closed under binary vector addition (denoted $\oplus_2$). For example,

$$OR_1 \oplus_2 OR_2 = OR_6;$$

$$OC_1 \oplus_2 OC_2 = OC_4.$$

Taking advantage of this closure, I now let the set of rows, or the set of columns, be bases for 8-dimensional real algebras, and define the following 4 products therefrom ($a \in \{0, 1, ..., 7\}$):

$$\begin{aligned} OR_a \circ OR_b &= (-1)^{O_{ab}}(OR_a \oplus_2 OR_b); \\ OR_a * OR_b &= (-1)^{O_{ba}}(OR_a \oplus_2 OR_b); \\ OC_a \diamond OC_b &= (-1)^{O_{ab}}(OC_a \oplus_2 OC_b); \\ OC_a \star OC_b &= (-1)^{O_{ba}}(OC_a \oplus_2 OC_b). \end{aligned} \tag{2}$$

The power of (-1) out front determines the sign of the result.

Each of the four products in (2) defines an 8-dimensional real algebra from the array O and each is isomorphic to the octonion algebra $\mathbb{O}$. Let e_a be a basis for $\mathbb{O}$, $a \in \{0, 1, ..., 7\}$. Given any of the products in (2), the e_a automatically satisfy the following useful properties for distinct indices $a, b, c \in \{1, ..., 7\}$:

$$\boxed{\text{if } e_a e_b = \pm e_c, \text{ then } e_{a+1} e_{b+1} = \pm e_{c+1};} \tag{3}$$

and

$$\boxed{\text{if } e_a e_b = \pm e_c, \text{ then } e_{2a} e_{2b} = \pm e_{2c},}$$ (4)

where the indices in (3) and (4) are understood to cycle from 1 to 7 modulo 7. The multiplication laws (3) (index cycling) and (4) (index doubling) will only both be valid for octonion multiplication rules derived from (2).

Some more general laws, valid for all the 480 renumberings of the e_a we will consider here, are

$$\boxed{e_a e_b = \pm e_c \Longrightarrow e_c e_a = \pm e_b}$$ (5)

(that is, $\{e_a, e_b, e_c\}$ are a quaternionic triple in this case), and

$$\boxed{e_a^2 = -1,}$$ (6)

where again we are assuming $a, b, c \in \{1, ..., 7\}$ in (5) and (6).

The multiplication laws (3 - 6) are enough to completely determine the octonion multiplication tables resulting from the following four product rules (which arise from the four respective products rules defined in (2)):

$$\boxed{\begin{array}{ll} \mathbb{O}^{+5} : & e_a e_{a+1} = e_{a+5}; \\[2mm] \mathbb{O}^{-5} : & e_a e_{a+1} = -e_{a+5}; \\[2mm] \mathbb{O}^{+3} : & e_a e_{a+1} = e_{a+3}; \\[2mm] \mathbb{O}^{-3} : & e_a e_{a+1} = -e_{a+3}. \end{array}}$$ (7)

The four copies of $\mathbb{O}$ that result from the rules (7) and the laws (3 - 6) are the cornerstones upon which I shall built the tower of 480 renumberings, and the cement holding it all together will be the X-product (Cederwall and Preitschopf, 1993).

3. The 480 renumberings

Given any of the 480 distinct copies of $\mathbb{O}$, a complete multiplication table is determined once one has listed 7 quaternionic index triples. For example, for $\mathbb{O}^{+5}$, $e_a e_b = e_c$ if (abc) is one of the following triples, or any cyclic permutation thereof:

$$\mathbb{O}^{+5} \text{ triples} : \{(126), (237), (341), (452), (563), (674), (715)\}.$$

There are simpler, more schematic ways of indicating the same information. A common method uses septagons, but I find the idea of making such a figure with

LATEXdaunting, so I will use the following more concise diagrams:

$$\begin{array}{ll}
\mathbb{O}^{+5}: & \boxed{1}\,2\,\boxed{3}\,\boxed{4}\,5\,6\,7; \\[4pt]
\mathbb{O}^{-5}: & \boxed{7}\,\boxed{6}\,5\,\boxed{4}\,3\,2\,1; \\[4pt]
\mathbb{O}^{+3}: & \boxed{1}\,\boxed{2}\,3\,\boxed{4}\,5\,6\,7; \\[4pt]
\mathbb{O}^{-3}: & \boxed{7}\,6\,\boxed{5}\,\boxed{4}\,3\,2\,1.
\end{array} \tag{8}$$

In each case, the quaternionic index triples of the four respective octonionic algebras are obtained via a cyclic shifting of the pattern of boxes given for the algebra. So for $\mathbb{O}^{+5}$, (134) is followed by (245), (356), (467), (571), (612), (723), and back to (134).

So, we have 7 index triples, and each triple has 3 cyclic permutations, so there are $3 \times 7 = 21$ pairs of distinct indices $a, b \in \{1, ..., 7\}$ such that

$$e_a e_b = +e_c$$

for some $c \in \{1, ..., 7\}$. For each such pair a and b there is a boxed sequence beginning with that pair from which the algebra multiplication table is derivable. For example, in $\mathbb{O}^{+5}$, we have $e_5 e_7 = e_1$, and the sequence

$$\mathbb{O}^{+5}: \quad \boxed{5}\,7\,\boxed{2}\,\boxed{4}\,6\,1\,3;$$

results in the same set of index triples for $\mathbb{O}^{+5}$ as does that in (8).

Therefore, via this type of rearrangement it is always possible to begin the boxed sequences of any of the 480 rearrangements of the indices of $\mathbb{O}$ with either the pair (1,2,...) or (2,1,...). There are

$$\frac{7!}{21} = 240 = 2(5!)$$

such reorderings. And there are *two inequivalent* ways of boxing the triples (modulo cyclic shifts), those shown it (8). So there are

$$2(240) = 480$$

different multiplication tables resulting from rearrangements of the indices of the e_a. As was pointed out in (Schray and Manogue, 1994), these fall into two groups of 240 (different groups than those above related by box pattern) related by the octonionic X-product (Cederwall and Preitschopf, 1993).

I will consider two other boxed sequences before finishing this section. In (Schray and Manogue, 1994) the following septuplet of index triples is introduced:

$$\mathbb{O}^{[2]}: \{(123), (145), (167), (264), (257), (347), (356)\}. \tag{9}$$

Once you get the hang of it, determining the boxed sequence from this is easy:

$$\mathbb{O}^{[2]}: \quad \boxed{1}\,\boxed{2}\,6\,\boxed{3}\,4\,5\,7. \tag{10}$$

In (Porteous, 1981) (which was my introduction to the octonions), a frequently encountered octonion multiplication is used with the following septuplet of triples:

$$\mathbb{O}^{[4]} : \{(123), (174), (275), (376), (165), (246), (354)\}. \tag{11}$$

Its boxed sequence is

$$\mathbb{O}^{[4]} : \boxed{1}\, 2\, \boxed{7}\, \boxed{4}\, 6\, 3\, 5. \tag{12}$$

4. The X-product

In what follows I shall use the $\mathbb{O}^{+5}$ product as a starting point for all calculations and X-product variations unless explicitly stated otherwise.

Let $A, B, X \in \mathbb{O}$, with X a unit octonion ($XX^\dagger = 1$) (see (Cederwall and Preitschopf, 1993) or (Dixon, 1994)). Define

$$\boxed{A \circ_X B = (AX)(X^\dagger B) = (A(BX))X^\dagger = X((X^\dagger A)B),} \tag{13}$$

the X-product of A and B. Because of the non associativity of $\mathbb{O}$, $A \circ_X B \neq AB$ in general. But remarkably, for fixed X, the algebra $\mathbb{O}_X$ ($\mathbb{O}$ endowed with the X-product) is isomorphic to $\mathbb{O}$ itself. Modulo sign change each X gives rise to a distinct copy of $\mathbb{O}$, so the orbit of copies of $\mathbb{O}$ arising from any given starting copy is

$$S^7/Z_2 = RP^7, \tag{14}$$

the manifold obtained from the 7-sphere by identifying opposite points. Moreover, composition of X-products is yet another X-product. That is, if

$$X, Y \in S^7 \subset \mathbb{O},$$

then

$$
\begin{aligned}
AB \xrightarrow{X} A \circ_X B &= (AX)(X^\dagger B) \xrightarrow{Y} \\
(A \circ_X Y) \circ_X (Y^\dagger \circ_X B) &= [((AX)(X^\dagger Y))X][X^\dagger((Y^\dagger X)(X^\dagger B))] \\
&= [((A(YX))X^\dagger)X][X^\dagger(X((X^\dagger Y^\dagger)B))] \\
&= [A(YX)][(X^\dagger Y^\dagger)B] \\
&= A \circ_{(YX)} B,
\end{aligned}
\tag{15}
$$

using the fact that for $X \in S^7$,

$$(UX^\dagger)X = U = X^\dagger(XU)$$

for all $U \in \mathbb{O}$. (This would seem to endow RP^7 with a Lie group structure, but in composing with yet a third element of S^7 one runs into the non associativity of $\mathbb{O}$, which spoils the game.)

Clearly in general the result of the X-product $e_a \circ_X e_b$, $0 \neq a \neq b \neq 0$, will be a linear combination of e_c, $c \in \{1, ..., 7\}$ (it is not difficult to prove that such a product can not have any terms linear in the identity). There are some X, however, such that for all $a, b \in \{1, ..., 7\}$, there will be a particular $c \in \{1, ..., 7\}$ satisfying

$$e_a \circ_X e_b = e_c.$$

Starting from $\mathbb{O}^{+5}$, any such X resides in one of the following sets:

$$
\begin{aligned}
\Xi_0^{+5} &= \{\pm e_a\}, \\
\Xi_1^{+5} &= \{(\pm e_a \pm e_b)/\sqrt{2} : a, b \text{ distinct}\}, \\
\Xi_2^{+5} &= \{(\pm e_a \pm e_b \pm e_c \pm e_d)/2 : a, b, c, d \text{ distinct}, \ e_a(e_b(e_c e_d)) = \pm 1\}, \\
\Xi_3^{+5} &= \{(\textstyle\sum_{a=0}^{7} \pm e_a)/\sqrt{8} : \text{ odd number of } +\text{'s}\}, \\
& \quad a, b, c, d \in \{0, ..., 7\}, e_a, e_b, e_c, e_d \in O^{+5}.
\end{aligned}
\tag{16}
$$

— NOTE: These sets are not general. They will work for $\mathbb{O}^{\pm 5}$, and certain of
   their X-product variants. See section 5. (Quite frankly, I don't know how many
   such sets there are for the total of 480 $\mathbb{O}$'s, but not 480.)

In (Dixon, 1994) I explicitly calculated the effect of the X-product (13). For
example, for a general

$$
X = X^0 + X^1 e_1 + X^2 e_2 + X^3 e_3 + X^4 e_4 + X^5 e_5 + X^6 e_6 + X^7 e_7 \in S^7,
$$

$$
\begin{aligned}
e_1 \circ_X e_2 = \\
((X^0)^2 + (X^1)^2 + (X^2)^2 + (X^6)^2 - (X^3)^2 - (X^4)^2 - (X^5)^2 - (X^7)^2)e_6 \\
+2(X^0 X^5 + X^1 X^7 - X^2 X^4 + X^3 X^6)e_3 \\
+2(-X^0 X^7 + X^1 X^5 + X^2 X^3 + X^4 X^6)e_4 \\
+2(-X^0 X^3 - X^1 X^4 - X^2 X^7 + X^5 X^6)e_5 \\
+2(X^0 X^4 - X^1 X^3 + X^2 X^5 + X^7 X^6)e_7.
\end{aligned}
\tag{17}
$$

Because (17) arises from $\mathbb{O}^{+5}$, all the other possible $e_a \circ_X e_b$ are derivable from
(17) via index cycling and doubling (in both cases, the index 0 is left out, and only
the indices $a = 1, ..., 7$ are subject to the cycling and doubling operations). Some
other products that will prove useful later are shown in TABLE I.

Clearly this is not a complete set, but it will suffice to construct a few examples.

— EXAMPLE 1: $X = (e_0 - e_1 - e_2 - e_3 - e_4 - e_5 - e_6 - e_7)/\sqrt{8}$.
   Note that because this X is invariant under both index cycling and index doubling, the product $e_a \circ_X e_b$ will have the index cycling and doubling properties
   shared by the products (7). Using the tables given above we can see that the
   index triples associated with this modification of the $\mathbb{O}^{+5}$ product are those of
   $\mathbb{O}^{+3}$. That is,

$$
\text{if } X = (e_0 - e_1 - e_2 - e_3 - e_4 - e_5 - e_6 - e_7)/\sqrt{8}, \text{ then } \mathbb{O}_X^{+5} = \mathbb{O}^{+3}.
\tag{18}
$$

TABLE I: X-Products for $\mathbb{O}^{+5}$.

$$e_3 \circ_X e_4 = ((X^0)^2 + (X^3)^2 + (X^4)^2 + (X^1)^2 - (X^2)^2 - (X^5)^2 - (X^6)^2 - (X^7)^2)e_1$$
$$+2(X^0 X^7 + X^3 X^2 - X^4 X^6 + X^5 X^1)e_5$$
$$+2(-X^0 X^2 + X^3 X^7 + X^4 X^5 + X^6 X^1)e_6$$
$$+2(-X^0 X^5 - X^3 X^6 - X^4 X^2 + X^7 X^1)e_7$$
$$+2(X^0 X^6 - X^3 X^5 + X^4 X^7 + X^2 X^1)e_2.$$

$$e_4 \circ_X e_5 = ((X^0)^2 + (X^4)^2 + (X^5)^2 + (X^2)^2 - (X^1)^2 - (X^3)^2 - (X^6)^2 - (X^7)^2)e_2$$
$$+2(X^0 X^1 + X^4 X^3 - X^5 X^7 + X^6 X^2)e_6$$
$$+2(-X^0 X^3 + X^4 X^1 + X^5 X^6 + X^7 X^2)e_7$$
$$+2(-X^0 X^6 - X^4 X^7 - X^5 X^3 + X^1 X^2)e_1$$
$$+2(X^0 X^7 - X^4 X^6 + X^5 X^1 + X^3 X^2)e_3.$$

$$e_5 \circ_X e_6 = ((X^0)^2 + (X^5)^2 + (X^6)^2 + (X^3)^2 - (X^2)^2 - (X^4)^2 - (X^7)^2 - (X^1)^2)e_3$$
$$+2(X^0 X^2 + X^5 X^4 - X^6 X^1 + X^7 X^3)e_7$$
$$+2(-X^0 X^4 + X^5 X^2 + X^6 X^7 + X^1 X^3)e_1$$
$$+2(-X^0 X^7 - X^5 X^1 - X^6 X^4 + X^2 X^3)e_2$$
$$+2(X^0 X^1 - X^5 X^7 + X^6 X^2 + X^4 X^3)e_4.$$

$$e_6 \circ_X e_7 = ((X^0)^2 + (X^6)^2 + (X^7)^2 + (X^4)^2 - (X^3)^2 - (X^5)^2 - (X^1)^2 - (X^2)^2)e_4$$
$$+2(X^0 X^3 + X^6 X^5 - X^7 X^2 + X^1 X^4)e_1$$
$$+2(-X^0 X^5 + X^6 X^3 + X^7 X^1 + X^2 X^4)e_2$$
$$+2(-X^0 X^1 - X^6 X^2 - X^7 X^5 + X^3 X^4)e_3$$
$$+2(X^0 X^2 - X^6 X^1 + X^7 X^3 + X^5 X^4)e_5.$$

Since $XX^\dagger = 1$,

$$\boxed{\text{if } X = (e_0 - e_1 - e_2 - e_3 - e_4 - e_5 - e_6 - e_7)/\sqrt{8}, \text{ then } \mathbb{O}^{+3}_{X^\dagger} = \mathbb{O}^{+5}.} \quad (19)$$

Therefore $\mathbb{O}^{+5}$ and $\mathbb{O}^{+3}$ are part of the same orbit of octonion X-product variants.

— EXAMPLE 2: $X = (e_1 - e_2 - e_4 - e_7)/2$.
In this case,

$$X^0 = X^3 = X^5 = X^6 = 0; \quad X^1 = -X^2 = -X^4 = -X^7 = \frac{1}{2}.$$

TABLE I: $\mathbb{O}^{+5}$ X-Products (continued).

$$
\begin{aligned}
e_1 \circ_X e_3 = \ & ((X^0)^2 + (X^1)^2 + (X^3)^2 + (X^4)^2 - (X^2)^2 - (X^5)^2 - (X^6)^2 - (X^7)^2)e_4 \\
& +2(X^0 X^2 + X^1 X^6 - X^3 X^7 + X^5 X^4)e_5 \\
& +2(-X^0 X^6 + X^1 X^2 + X^3 X^5 + X^7 X^4)e_7 \\
& +2(-X^0 X^5 - X^1 X^7 - X^3 X^6 + X^2 X^4)e_2 \\
& +2(X^0 X^7 - X^1 X^5 + X^3 X^2 + X^6 X^4)e_6.
\end{aligned}
$$

$$
\begin{aligned}
e_2 \circ_X e_4 = \ & ((X^0)^2 + (X^2)^2 + (X^4)^2 + (X^5)^2 - (X^1)^2 - (X^3)^2 - (X^6)^2 - (X^7)^2)e_5 \\
& +2(X^0 X^3 + X^2 X^7 - X^4 X^1 + X^6 X^5)e_6 \\
& +2(-X^0 X^7 + X^2 X^3 + X^4 X^6 + X^1 X^5)e_1 \\
& +2(-X^0 X^6 - X^2 X^1 - X^4 X^7 + X^3 X^5)e_3 \\
& +2(X^0 X^1 - X^2 X^6 + X^4 X^3 + X^7 X^5)e_7.
\end{aligned}
$$

$$
\begin{aligned}
e_3 \circ_X e_5 = \ & ((X^0)^2 + (X^3)^2 + (X^5)^2 + (X^6)^2 - (X^1)^2 - (X^2)^2 - (X^4)^2 - (X^7)^2)e_6 \\
& +2(X^0 X^4 + X^3 X^1 - X^5 X^2 + X^7 X^6)e_7 \\
& +2(-X^0 X^1 + X^3 X^4 + X^5 X^7 + X^2 X^6)e_2 \\
& +2(-X^0 X^7 - X^3 X^2 - X^5 X^1 + X^4 X^6)e_4 \\
& +2(X^0 X^2 - X^3 X^7 + X^5 X^4 + X^1 X^6)e_1.
\end{aligned}
$$

$$
\begin{aligned}
e_4 \circ_X e_6 = \ & ((X^0)^2 + (X^4)^2 + (X^6)^2 + (X^7)^2 - (X^1)^2 - (X^2)^2 - (X^3)^2 - (X^5)^2)e_7 \\
& +2(X^0 X^5 + X^4 X^2 - X^6 X^3 + X^1 X^7)e_1 \\
& +2(-X^0 X^2 + X^4 X^5 + X^6 X^1 + X^3 X^7)e_3 \\
& +2(-X^0 X^1 - X^4 X^3 - X^6 X^2 + X^5 X^7)e_5 \\
& +2(X^0 X^3 - X^4 X^1 + X^6 X^5 + X^2 X^7)e_2.
\end{aligned}
$$

Plug these values into the general form for $e_1 \circ_X e_2$ in (17) and get

$$
e_1 \circ_X e_2 = -e_3.
$$

Therefore (132) is a quaternionic index triple for this $\mathbb{O}_X$. Likewise, from the general $e_4 \circ_X e_5$ table,

$$
e_4 \circ_X e_5 = -e_1,
$$

so (154) is another such triple. Carrying on in this way leads to a complete set, with corresponding boxed sequence:

$$(321), (541), (761), (462), (752), (743), (653).$$

$$\boxed{2}\,1\,\boxed{7}\,\boxed{5}\,4\,3\,6.$$

(20)

Compare this set of triples to the set (9) for $\mathbb{O}^{[2]}$. Each of the triples in (20) is reversed with respect to a corresponding triple in (9). Following (Schray and Manogue, 1994) I shall call the copy of $\mathbb{O}$ generated from (20) the opposite of that generated from (9) ($\mathbb{O}^{[2]}$). Let's denote it $\underline{\mathbb{O}^{[2]}}$, the underline signifying opposite.

So

$$\underline{\mathbb{O}^{[2]}} = \mathbb{O}_X^{+5}.$$

That is, they are a part of the same orbit. What about $\mathbb{O}^{[2]}$ itself? It turns out that $\mathbb{O}^{+5}$ and $\mathbb{O}^{[2]}$ are not in the same orbit. Consider $\mathbb{O}^{+5}$ and $\mathbb{O}^{-5}$ as an example. The quaternionic triples of $\mathbb{O}^{-5}$ are the reverse of those for $\mathbb{O}^{+5}$. (So

$$\underline{\mathbb{O}^{+5}} = \mathbb{O}^{-5}.)$$

Therefore,

$$e_1 e_2 = e_6 \text{ in } \mathbb{O}^{+5} \text{ implies } e_1 e_2 = -e_6 \text{ in } \mathbb{O}^{-5}.$$

For this to happen via an X-product, we can see from (17) that

$$(X^0)^2 + (X^1)^2 + (X^2)^2 + (X^6)^2 - (X^3)^2 - (X^4)^2 - (X^5)^2 - (X^7)^2 = -1,$$

$$X^0 X^5 + X^1 X^7 - X^2 X^4 + X^3 X^6 = 0,$$

$$-X^0 X^7 + X^1 X^5 + X^2 X^3 + X^4 X^6 = 0,$$

$$-X^0 X^3 - X^1 X^4 - X^2 X^7 + X^5 X^6 = 0,$$

$$X^0 X^4 - X^1 X^3 + X^2 X^5 + X^7 X^6 = 0.$$

These equations can be satisfied in several ways. For example, set $X^7 = 1$, and $X^a = 0$, $a = 0, ..., 6$. But whatever value of X we choose must reverse not only this product, but every other $\mathbb{O}^{+5}$ product. In particular, this implies that each of the following eight equations must be satisfied:

$$
\begin{bmatrix}
1 & 1 & 1 & 1 & 1 & 1 & 1 & 1 \\
1 & -1 & -1 & -1 & 1 & -1 & 1 & 1 \\
1 & 1 & -1 & -1 & -1 & 1 & -1 & 1 \\
1 & 1 & 1 & -1 & -1 & -1 & 1 & -1 \\
1 & -1 & 1 & 1 & -1 & -1 & -1 & 1 \\
1 & 1 & -1 & 1 & 1 & -1 & -1 & -1 \\
1 & -1 & 1 & -1 & 1 & 1 & -1 & -1 \\
1 & -1 & -1 & 1 & -1 & 1 & 1 & -1
\end{bmatrix}
\begin{bmatrix}
(X^0)^2 \\
(X^1)^2 \\
(X^2)^2 \\
(X^6)^2 \\
(X^3)^2 \\
(X^4)^2 \\
(X^5)^2 \\
(X^7)^2
\end{bmatrix}
=
\begin{bmatrix}
1 \\
-1 \\
-1 \\
-1 \\
-1 \\
-1 \\
-1 \\
-1
\end{bmatrix}.
$$

The inverse of that square matrix is $\frac{1}{8}$ times its transpose. Therefore a solution is easily obtained, and in particular it implies that

$$(X_0)^2 = -\frac{3}{4},$$

which of course is not possible for the real algebra $\mathbb{O}$. (As it turns out, even the complexification of $\mathbb{O}$ wouldn't help in the end.)

—   EXAMPLE 3: $X = (e_0 - e_1 - e_2 - e_3 - e_4 - e_5 + e_6 + e_7)/\sqrt{8}$.
    In this case, using (17) and the other $\mathbb{O}^{+5}$ X-product tables, we find:

$$e_1 \circ_X e_2 = -e_3, \; e_3 \circ_X e_4 = e_5, \; e_5 \circ_X e_6 = e_1, \; e_6 \circ_X e_7 = e_3, \; e_2 \circ_X e_4 = -e_6.$$

By cyclically shifting the $e_4 \circ_X e_6$ table up by 1, and the $e_1 \circ_X e_2$ table down by 1, we get

$$e_5 \circ_X e_7 = e_2, \;\; e_7 \circ_X e_1 = e_4.$$

Therefore, a complete set of quaternionic triples for this case, with corresponding boxed sequence, is:

$$(321), (471), (572), (673), (561), (642), (453),$$

$$\boxed{2}\,\boxed{1}\,5\,\boxed{3}\,6\,4\,7.$$

(21)

These are all opposite those for $\mathbb{O}^{[4]}$ listed in (11). Therefore, in this case,

$$\mathbb{O}_X^{+5} = \underline{\mathbb{O}}^{[4]}.$$

—   NOTE: In general, if $\mathbb{O}$ and $\mathbb{O}'$ are in the same X-product orbit, then $\underline{\mathbb{O}}$ and $\underline{\mathbb{O}}'$ are not. There are two orbits all together, one arising from $\mathbb{O}^{+5}$ and containing $\mathbb{O}^{+3}$ ($\mathcal{O}rbit^+$), and one arising from $\mathbb{O}^{-5}$ and containing $\mathbb{O}^{-3}$ ($\mathcal{O}rbit^-$).

5. Two X-product orbits

Look again at the sets $\Xi_a^{+5}, a = 0, 1, 2, 3$, elements of which cause index rearrangements of $\mathbb{O}^{\pm 5}$ via X-product variation. Modulo sign change,

$$\text{order } \Xi_0^{+5} = 8,$$
$$\text{order } \Xi_1^{+5} = 56,$$
$$\text{order } \Xi_2^{+5} = 112,$$
$$\text{order } \Xi_3^{+5} = 64.$$

(22)

So there are 240 rearrangements arising from these elements all together. Hence each of the two X-product orbits, $\mathcal{O}rbit^\pm$, contains 240 of the 480 octonion index rearrangements. In each orbit there are 120 rearrangements using the $\mathbb{O}^{+5}$ (and $\mathbb{O}^{-3}$) pattern of boxes (denote these $\mathcal{O}rbit_0^{\pm}$, a discrete subset of $\mathcal{O}rbit^{\pm}$),

$$\boxed{}\!\!-\!\!\boxed{}\text{---}\,, \qquad (23)$$

and 120 using the $\mathbb{O}^{-5}$ (and $\mathbb{O}^{+3}$) pattern of boxes (denote these $\mathcal{O}rbit_1^{\pm}$),

$$\boxed{}\!\!-\!\!\boxed{}\text{---}\cdot \qquad (24)$$

Starting from $\mathbb{O}^{+5}$, the 120 distinct X-products arising from elements of $\Xi_0^{+5} \cup \Xi_2^{+5}$ will result in variants $\mathbb{O}_X^{+5}$ sharing $\mathbb{O}^{+5}$'s box pattern (23). For example, the boxed sequence of $\mathbb{O}_X^{+5}$, $X = (e_1 - e_2 - e_4 - e_7)/2 \in \Xi_2^{+5}$, given in (2), has the same box pattern as $\mathbb{O}^{+5}$ itself. (Remember that for any given reordering of the indices only one of the box patterns is possible.)

Starting from $\mathbb{O}^{+5}$, the 120 distinct X-products arising from elements of $\Xi_1^{+5} \cup \Xi_3^{+5}$ will result in variants $\mathbb{O}_X^{+5}$ having the box pattern (24). For example, the boxed sequence of $\mathbb{O}_X^{+5} = \mathbb{O}^{+3}$, $X = (1 - e_1 - e_2 - e_3 - e_4 - e_5 - e_6 - e_7)/\sqrt{8} \in \Xi_3^{+5}$, has pattern (24).

Schematically,

$$
\begin{array}{ccccc}
\mathcal{O}rbit_1^+ & \longleftarrow \Xi_1^{+5} \cup \Xi_3^{+5}\!\!- & \mathbb{O}^{+5} & -\Xi_0^{+5} \cup \Xi_2^{+5} \longrightarrow & \mathcal{O}rbit_0^+ \\[4pt]
\uparrow & & \uparrow & & \uparrow \\
\text{opposite} & & \text{opposite} & & \text{opposite} \qquad (25) \\
\downarrow & & \downarrow & & \downarrow \\[4pt]
\mathcal{O}rbit_0^- & \longleftarrow \Xi_1^{+5} \cup \Xi_3^{+5}\!\!- & \mathbb{O}^{-5} & -\Xi_0^{+5} \cup \Xi_2^{+5} \longrightarrow & \mathcal{O}rbit_1^-
\end{array}
$$

— NOTE: If $X \notin \Xi_0^{+5} \cup \Xi_1^{+5} \cup \Xi_2^{+5} \cup \Xi_3^{+5}$, then $\mathbb{O}_X^{+5}$ is not any of the simple index rearrangements of $\mathbb{O}^{+5}$, which are after all just discrete points in the full RP^7 orbit of $\mathbb{O}^{+5}$.

The set $\Xi_0^{+5} \cup \Xi_1^{+5} \cup \Xi_2^{+5} \cup \Xi_3^{+5}$ is $\mathbb{O}^{\pm 5}$ specific. Starting from a general $\mathbb{O}_X^{+5}$, for example, we know that

$$(\mathbb{O}_X^{+5})_{X\dagger} = \mathbb{O}_{(X\dagger X)}^{+5} = \mathbb{O}^{+5},$$

and

$$((\mathbb{O}_X^{+5})_{X\dagger})_Y = \mathbb{O}_Y^{+5} = (\mathbb{O}_X^{+5})_{(Y X\dagger)}$$

(see (15)). Therefore, for $(\mathbb{O}_X^{+5})_Z$ to be an index rearrangement of $\mathbb{O}_X^{+5}$, Z must satisfy

$$\boxed{Z \in (\Xi_0^{+5} \cup \Xi_1^{+5} \cup \Xi_2^{+5} \cup \Xi_3^{+5})X\dagger.} \qquad (26)$$

Therefore, for example, had we started with $\mathbb{O}^{+3} = \mathbb{O}_X^{+5}$, where $X = (1 - e_1 - \ldots - e_7)/\sqrt{8}$, so that

$$X^\dagger = (1 + e_1 + \ldots + e_7)/\sqrt{8},$$

then the set $(\Xi_0^{+5} \cup \Xi_1^{+5} \cup \Xi_2^{+5} \cup \Xi_3^{+5})X^\dagger$ can be broken up into subsets

$$
\begin{aligned}
\Xi_0^{+3} &= \{\pm e_a\}, \\[2mm]
\Xi_1^{+3} &= \{(\pm e_a \pm e_b)/\sqrt{2} : a, b \text{ distinct}\}, \\[2mm]
\Xi_2^{+3} &= \{(\pm e_a \pm e_b \pm e_c \pm e_d)/2 : a, b, c, d \text{ distinct}, \; e_a(e_b(e_c e_d)) = \pm 1\}, \\[2mm]
\Xi_3^{+3} &= \{(\textstyle\sum_{a=0}^7 \pm e_a)/\sqrt{8} : \text{ even number of } +\text{'s}\}, \\[2mm]
& a, b, c, d \in \{0, ..., 7\}, e_a, e_b, e_c, e_d \in \mathbb{O}^{+3}.
\end{aligned}
\tag{27}
$$

Everything that was said for $\mathbb{O}^{+5}$ and the Ξ_m^{+5} holds true in an analogous way for $\mathbb{O}^{+3}$ and the Ξ_m^{+3}.

6. Conclusions

In the case of the quaternions $\mathbb{H}$ there are also "opposites", but because $\mathbb{H}$ is associative there are no X-product variations. The quaternion basis with multiplication table determined by

$$q_1 q_2 = q_3$$

has an opposite representation with a multiplication table determined by

$$q_2 q_1 = -q_1 q_2 = q_3.$$

So we drop from 480 variations down to 2.

There is obviously no way to vary the complex numbers $\mathbb{C}$, the smallest of the three hypercomplex real division algebras.

Finally, in (Cederwall and Preitschopf, 1993), (Schray and Manogue, 1994), and (Dixon, 1994) interest in the division algebras arose from their evidently intimate connection to our physical reality. In a future article I will investigate the potential and consequences of gauging the X-product.

References

M. Cederwall, C. R. Preitschopf: 'S^7 and $\hat{S^7}$', hep-th-9309030.

G. M. Dixon, *Division Algebras: Octonions, Quaternions, Complex Numbers, and the Algebraic Design of Physics*, Kluwer, 1994.

I. R. Porteous, *Topological Geometry*, Cambridge UP, 2nd ed., 1981.

J. Schray and C. Manogue: 'Octonionic representations of Clifford algebras and triality', hep-th-9407179.

A COMMUTATIVE HYPERCOMPLEX ALGEBRA WITH ASSOCIATED FUNCTION THEORY

CLYDE M. DAVENPORT
4124 Guinn Road
Knoxville, TN 37931
e-mail: cmdaven@use.usit.net

Abstract. The group ring $\mathbb{C}^2$ of pairs (ξ, η) of classical complex numbers is used as an algebraic basis for a function theory for one four-dimensional variable of the form $\mathbf{Z} = \mathbf{1}x + \mathbf{i}y + \mathbf{j}z + \mathbf{k}ct$. All of the axioms and properties of the complex analysis are carried forward into four dimensions, including functions, derivatives, integrals, and even the notation. The new analysis properly subsumes and extends the classical complex analysis. The single unexpected divergence from traditional properties is that the sets $(\xi, 0)$ and $(0, \eta)$ of non invertible elements are interpreted as the zero element (zero set). The zero set can be constrained to lie in the relativistic limit of special relativity; consequently, the new analysis scheme may be applied in any real-world, physical problem. The four-dimensional Cauchy-Riemann equations provide a way to apply continuity conditions to the field equations of physics, while at the same time converting PDEs to easier-to-solve ODEs. Significant simplifications are achieved for many physics field equations. The technique is illustrated on the Emden's equation. Symbolic manipulations in four dimensions are easily handled by any computer algebra program that can perform classical complex algebraic operations.

Key words: Hypercomplex, commutative, four dimensions, function theory, Cauchy-Riemann conditions, partial differential equations, Emden's equation.

1. Introduction

We aim to create a straightforward extension of the classical complex analysis to treat a four-dimensional variable of the form $\mathbf{Z} = \mathbf{1}x + \mathbf{i}y + \mathbf{j}z + \mathbf{k}ct$ (not quaternions, as we shall explain), where $\mathbf{1}, \mathbf{i}, \mathbf{j}, \mathbf{k}$ are unit direction vectors and x, y, z, ct are coordinates in the physical four-space, with c being the speed of light in a vacuum and t being time. We shall construct our scheme of analysis out of elementary building blocks, such as group theory, ring theory, and matrix analysis. The result will not constitute new mathematics, but rather a new interpretation and application of some very old, very familiar concepts.

In the 1830s, Sir William Rowan Hamilton set out to create a new analysis of one four-dimensional variable, seeking to apply it to physics problems in mechanics and optics. Complex algebra and matrix theory were known only in rudimentary forms. Vector analysis and group theory had not been discovered. Eventually, Hamilton (1844) created the quaternion analysis, from which evolved the vector analysis of J. W. Gibbs and O. W. Heaviside, as discussed by Crowe (1967).

Because of the rudimentary state of multidimensional analysis in Hamilton's time, he could not have realized that he had chosen but one of five possible group rings of group order eight upon which to base his analysis. Group theory had not been discovered, so Hamilton examined the various multiplicative schemes by trial and error methods. In the author's opinion, Hamilton settled upon the quaternion case because of a belief that the algebra should contain a single zero element, just as in the real and complex algebras. Possibly, Gibbs and Heaviside later stayed with a quaternion algebra for the same reason, although they had access to group and ring theory, but not yet to Einstein's relativity. The other four group rings, once they were available, apparently were not seriously evaluated for use in scientific and engineering applications.

The quaternion multiplication is non commutative; consequently, it is not possible to construct upon quaternions a function theory having all the properties of the classical complex functions, as shown by Scheffers (1893). However, we shall show that Einstein's theory of special relativity allows us to not be limited to a single zero element, and that one of the other eighth-order group rings can be exceptionally useful.

The author established the basic ideas to be discussed, here, in a thesis [Davenport (1978)] and a monograph [Davenport (1991)]. Simultaneously with the release of the author's monograph, Price (1991) announced a text/reference work developing the same algebra, function theory, and analysis. The latter assumed a different perspective and used different techniques and notation, but produced the same analysis. No physics applications were developed.

2. An algebraic basis

2.1. PAIRS OF COMPLEX NUMBERS

Consider, first, the set S of all pairs of complex numbers (ξ, η). Under the pairwise operations

$$(\xi_1, \eta_1) + (\xi_2, \eta_2) \;=\; (\xi_1 + \xi_2, \, \eta_1 + \eta_2), \tag{1}$$

$$(\xi_1, \eta_1) \times (\xi_2, \eta_2) \;=\; (\xi_1 \times \xi_2, \, \eta_1 \times \eta_2), \tag{2}$$

the set S forms a commutative ring with unity, usually designated as $\mathbb{C}^2$ or $\mathbb{C} \times \mathbb{C}$. We define all other operations, such as function or derivative, in the same pairwise fashion,

$$oper(\xi, \eta) = (oper(\xi), oper(\eta)), \tag{3}$$

for our subsequent discussion; then we have, merely, two copies of the classical complex variables and associated analysis. Each pair of complex numbers, as an aggregate, has four components that could be assigned as coordinates in the real four-space, (x, y, z, ct); consequently, we have a well-defined analysis of one four-dimensional variable.

This analysis does not have all the properties of the classical complex analysis because this approach does not allow a division algebra. As a consequence of the pairwise definition, division is not defined when either element of the pair is zero; i.e., for all elements of the forms $(\xi, 0)$ or $(0, \eta)$. There are two entire hyperplanes of non invertible elements. However, there is a way around the difficulty. If we assign (define)

$$\xi \;=\; (x - ct) + i\,(y + z), \tag{4}$$

$$\eta = (x + ct) + i(y - z), \tag{5}$$

with x, y, z, ct real, i the classical complex imaginary, c the speed of light in a vacuum, and t the time; then the only way that either complex component can be zero is for the condition $x = \pm ct$ to hold. However, according to Einstein's special relativity, this is the relativistic limit, which no real, physical body can ever reach. Consequently, we can freely use the new analysis in this form for any scientific or engineering calculation. Moreover, not only do we not have to worry about violating relativistic principles, but can even let the mathematics point out where the relativistic limit is a problem.

Actually, c can just as well be interpreted as the upper limit for the speed of transmission of effects through the physical medium in which we are working; for example, the speed of sound in aerodynamics. Notice that (4),(5) represents an orthogonal transformation of the x, y, z, ct coordinates, when considered component by component.

We could proceed directly to applications, but there is considerable insight to be gained from two alternative representations of the algebra.

2.2. A FIELD ANALOGY

As stated earlier, the $\mathbb{C}^2$ algebra is a commutative ring with unity. It fails to be a field because it has manifold non invertible elements of the form $(\xi, 0)$ and $(0, \eta)$. Nevertheless, we will later define functions and analysis operations that have all the classical complex properties. We propose that $\mathbb{C}^2$ is analogous to a field if the definition of the "zero element" is broadened.

Consider the following properties of the classical complex zero element: If $\mathbb{C}$ is the classical complex field, θ is the zero element $(0,0)$ of $\mathbb{C}$, and $O = \{\theta\}$ (set notation), then

$$\text{(i)} \quad \text{Elements of } O \text{ have no multiplicative inverses in } \mathbb{C}. \tag{6}$$

$$\text{(ii)} \quad \text{If } z_1 \in O \text{ and } z_2 \in \mathbb{C}, \text{ then } z_1 z_2 \in O. \tag{7}$$

$$\text{(iii)} \quad \text{If } z_1, z_2 \in \mathbb{C}, \text{ then } z_1 z_2 \in O \text{ implies either} \tag{8}$$
$$z_1 \in O \text{ or } z_2 \in O, \text{ or both.}$$

The four-dimensional analogs of the above set relations are as follows: Let O be either of the sets $\{(0, \eta)\}, \{(\xi, 0)\}$, or their union throughout the following; then

$$\text{(i)} \quad \text{Elements of } O \text{ have no multiplicative inverses in } \mathbb{C}^2. \tag{9}$$

$$\text{(ii)} \quad \text{If } \mathbf{Z}_1 \in O \text{ and } \mathbf{Z}_2 \in \mathbb{C}^2, \text{ then } \mathbf{Z}_1 \mathbf{Z}_2 \in O. \tag{10}$$

$$\text{(iii)} \quad \text{If } \mathbf{Z}_1, \mathbf{Z}_2 \in \mathbb{C}^2, \text{ then } \mathbf{Z}_1 \mathbf{Z}_2 \in O \text{ implies either} \tag{11}$$
$$\mathbf{Z}_1 \in O \text{ or } \mathbf{Z}_2 \in O, \text{ or both.}$$

Moreover, if $\mathbf{Z} \in \mathbb{C}^2$ and O is defined as above, then in set notation one may write

$$O\mathbf{Z} = \mathbf{Z}O = O. \tag{12}$$

This appears to be a very natural extension of the concept of "zero element" to "zero set." If we accept this notion, then *the $\mathbb{C}^2$ algebra is analogous to a field.*

We know that the two sets $\{(\xi, 0)\}$ and $\{(0, \eta)\}$ for all ξ, η make up two maximal ideals in the $\mathbb{C}^2$ algebra. For physics applications, it is more instructive to state that they are *two separate subspaces*, distinguished from the rest of the four-space by the conditions $x = \pm ct$ (the relativistic limit).

2.3. 1, i, j, k Unit Basis Vectors

Consider the multiplication rules

$$\mathbf{ij} \;=\; \mathbf{k}, \qquad \mathbf{jk} = -\mathbf{i}, \qquad \mathbf{ki} = -\mathbf{j}, \tag{13}$$

$$\mathbf{ji} \;=\; \mathbf{k}, \qquad \mathbf{kj} = -\mathbf{i}, \qquad \mathbf{ik} = -\mathbf{j}, \tag{14}$$

$$\mathbf{ii} \;=\; \mathbf{jj} = -\mathbf{kk} = -1, \qquad \mathbf{ijk} = 1. \tag{15}$$

This has the superficial appearance of quaternion multiplication, but is expandable into an 8×8 multiplication table that is symmetrical about the principal diagonal. Multiplication is commutative. This is a representation for one of the five possible basis groups of order eight, this particular one being $\mathbb{C}_2 \times \mathbb{C}_4$, where $\mathbb{C}_n$ is the cyclic group of order n. We shall investigate it because it is the only one of the five that is both Abelian and has an element of cyclic order four, which is needed to reproduce the behavior of the classical complex imaginary, i.

If we construct a group ring upon this basis, then the typical element $\mathbf{Z}$ is of the form

$$\mathbf{Z} = \mathbf{1}x + \mathbf{i}y + \mathbf{j}z + \mathbf{k}ct, \tag{16}$$

with x, y, z, ct real. This form is already suggestive as an extension of the classical complex variable, because the first two terms have the properties of a complex variable. Addition is component-by-component, as for vectors; multiplication is term-by-term, with reduction of the cross terms by use of the multiplication table. We defer a discussion of division until later.

Having a ring, we are justified in rearranging the typical element into the form

$$\mathbf{Z} = \{(x - ct) + \mathbf{i}\,(y + z)\} \left[\frac{1 - \mathbf{k}}{2}\right] + \{(x + ct) + \mathbf{i}\,(y - z)\} \left[\frac{1 + \mathbf{k}}{2}\right]. \tag{17}$$

We introduce the following notation:

$$\xi \;=\; (x - ct) + \mathbf{i}\,(y + z), \tag{18}$$

$$\eta \;=\; (x + ct) + \mathbf{i}\,(y - z), \tag{19}$$

$$e_1 \;=\; \frac{(1 - \mathbf{k})}{2}, \tag{20}$$

$$e_2 \;=\; \frac{(1 + \mathbf{k})}{2}, \tag{21}$$

$$\mathbf{Z} \;=\; \xi\,e_1 + \eta\,e_2; \tag{22}$$

then the reader may notice the first two definitions, (18) and (19), are similar to (4) and (5). We are, in fact, dealing with equivalent objects, because the hypercomplex i and the classical complex imaginary i have the same properties. We shall call the above representation of $\mathbf{Z}$ the *canonical form*. Simple algebraic manipulation may be used to verify the following properties:

$$
\begin{aligned}
\xi, \eta &= \text{classical complex variables;} &(23)\\
\mathbf{e}_1 \bullet \mathbf{e}_2 &= 0, \quad \text{(Vector dot product)} &(24)\\
\mathbf{e}_1^n &= \mathbf{e}_1, \quad n = 1, 2, 3, \ldots &(25)\\
\mathbf{e}_2^n &= \mathbf{e}_2, \quad n = 1, 2, 3, \ldots &(26)\\
\mathbf{e}_1\,\mathbf{e}_2 &= (0, 0, 0, 0). &(27)
\end{aligned}
$$

In this notation, and consistent with (3), the algebraic operations take very simple forms:

$$
\begin{aligned}
\mathbf{Z}_1 + \mathbf{Z}_2 &= (\xi_1 + \xi_2)\,\mathbf{e}_1 + (\eta_1 + \eta_2)\,\mathbf{e}_2, &(28)\\
\mathbf{Z}_1\,\mathbf{Z}_2 &= \xi_1\xi_2\,\mathbf{e}_1 + \eta_1\eta_2\,\mathbf{e}_2, &(29)\\
\mathbf{Z}^{-1} &= \xi^{-1}\mathbf{e}_1 + \eta^{-1}\mathbf{e}_2. &(30)
\end{aligned}
$$

Of course, the inverse of $\mathbf{Z}$ exists only if (ξ, η) both are nonzero in the classical complex variable sense. Comparing (1), (2) with (28),(29), we see that a one-to-one correspondence of elements and operations can be made; consequently, *the* $\mathbf{1, i, j, k}$ *group ring is isomorphic to the ring* $\mathbb{C}^2$.

2.3.1. Axis Plane Interpretation

One concludes from (22) that the $\mathbb{C}^2$-space is coordinatized by two complex planes, which we shall call *axis planes*. Their orientation in the four-space may be deduced as follows: if a point (ξ, η) is to be in an axis plane, then either $\xi = 0$ or $\eta = 0$. If $\xi = 0$, then

$$
x = ct, \quad y = -z. \tag{31}
$$

In the x, y, z-space, this is a line in the y, z-plane, oriented at -45 degrees to the $+z$-axis and moving at speed c along the $+x$-direction.

Similarly, if $\eta = 0$, then $x = -ct$, $y = z$. This is a line in the y, z-plane, oriented at $+45$ degrees to the $+z$-axis and moving at speed c along the $-x$-direction.

A vector in the η-plane is characterized by (31); that is,

$$
\mathbf{Z}_\eta = \mathbf{1}\,x_1 + \mathbf{i}\,y_1 - \mathbf{j}\,y_1 + \mathbf{k}\,x_1; \tag{32}
$$

similarly, a vector in the ξ-plane is of the form

$$
\mathbf{Z}_\xi = \mathbf{1}\,x_2 + \mathbf{i}\,y_2 + \mathbf{j}\,y_2 - \mathbf{k}\,x_2. \tag{33}
$$

If we take the vector inner product (dot product), we obtain

$$
\mathbf{Z}_\xi \bullet \mathbf{Z}_\eta = x_1 x_2 + y_1 y_2 - y_1 y_2 - x_1 x_2 \equiv 0; \tag{34}
$$

consequently, *the axis planes are orthogonal.*

2.4. A Matrix Representation

Just as for the classical complex variables, the present algebra has a matrix representation, either 2×2 complex-valued or 4×4 real-valued. The motivation to investigate it is for the insights that it can provide about multiplicative inverses and permissible function definitions. Consider the following representations of a classical complex variable, followed by a simple extension to four dimensions:

$$x + i\,y \;\rightarrow\; \begin{bmatrix} x & -y \\ y & x \end{bmatrix} \tag{35}$$

$$\mathbf{1}x + \mathbf{i}y + \mathbf{j}z + \mathbf{k}ct = (x + i\,y) + \mathbf{i}(z + \mathbf{i}\,ct) \tag{36}$$

$$\rightarrow \begin{bmatrix} (x + i\,y) & -(z + i\,ct) \\ (z + i\,ct) & (x + i\,y) \end{bmatrix} \rightarrow \left[\begin{array}{cc|cc} x & -y & -z & ct \\ y & x & -ct & -z \\ \hline z & -ct & x & -y \\ ct & z & y & x \end{array} \right] \tag{37}$$

Note that the 4×4 matrix partitions into 2×2 complex-representational matrices. Now, we state without proof (the proof is by simple matrix manipulation) that (ξ, η) are *eigenvalues of the* 4×4 *real matrix representation* (37), and the matrix representation is consistent with a pairwise definition of operations, including the inverse as in (30), because all operations are performed on matrix eigenvalues. These are not typical real matrices, because the pattern of their elements profoundly influences their properties. Although much of our work will have a "vectorial" look, all the operations will obey matrix rules. In the matrix notation, the basis elements are:

$$\mathbf{1} \;\rightarrow\; \begin{bmatrix} 1 & 0 & 0 & 0 \\ 0 & 1 & 0 & 0 \\ 0 & 0 & 1 & 0 \\ 0 & 0 & 0 & 1 \end{bmatrix}, \quad \mathbf{i} \;\rightarrow\; \begin{bmatrix} 0 & -1 & 0 & 0 \\ 1 & 0 & 0 & 0 \\ 0 & 0 & 0 & -1 \\ 0 & 0 & 1 & 0 \end{bmatrix}, \tag{38}$$

$$\mathbf{j} \;\rightarrow\; \begin{bmatrix} 0 & 0 & -1 & 0 \\ 0 & 0 & 0 & -1 \\ 1 & 0 & 0 & 0 \\ 0 & 1 & 0 & 0 \end{bmatrix}, \quad \mathbf{k} \;\rightarrow\; \begin{bmatrix} 0 & 0 & 0 & 1 \\ 0 & 0 & -1 & 0 \\ 0 & -1 & 0 & 0 \\ 1 & 0 & 0 & 0 \end{bmatrix}. \tag{39}$$

These matrices are orthogonal; and they, together with their negatives, make up a group of order eight, that being $\mathbb{C}_2 \times \mathbb{C}_4$. This representation is verified by repeating the multiplication table (13)-(15) using the matrix forms.

2.5. A Clifford-Algebra Representation

The author has not undertaken any study of the $\mathbf{1}, \mathbf{i}, \mathbf{j}, \mathbf{k}$-algebra in relation to Clifford Algebras. The reason is that the $\mathbb{C}^2$ algebra is intrinsically commutative, necessary for classical complex function theories, while the Clifford Algebras incur anticommutative operations. However, Sobczyk (1988) pointed out to the author that the $\mathbb{C}^2$ algebra embeds in the 16-dimensional Clifford Algebra $\mathcal{G}(4,0)$ as follows. Let e_1, e_2, e_3, e_4 (distinct from the e_1, e_2 introduced earlier) be the orthonormal basis vectors of $\mathcal{G}(4,0)$, and make the following correspondences:

$$\mathbf{i} \;=\; \mathbf{e}_1\mathbf{e}_2, \tag{40}$$

$$\mathbf{j} \;=\; \mathbf{e}_3\mathbf{e}_4, \tag{41}$$

$$\mathbf{k} \;=\; \mathbf{e}_1\mathbf{e}_2\mathbf{e}_3\mathbf{e}_4; \tag{42}$$

then $\mathbb{C}^2$ and $\mathcal{G}(4,0)$ are equivalent. Sobczyk further notes that this identification requires that rotations in the $\mathbf{i}$ and $\mathbf{j}$ planes be handled differently than those for the $\mathbf{k}$ plane.

3. Analytic function theory

We now have a secure basis upon which to define an *analytic function* of one four-dimensional variable. Consistent with our pairwise definition of operation, (3), and consistent with past definitions of matrix functions [MacDuffee (1946)] as being operations upon matrix eigenvalues, we define

$$f(\mathbf{Z}) \;=\; f(\xi, \eta) \;=\; (f(\xi), f(\eta)), \tag{43}$$

where (ξ, η) are defined in the classical complex variable sense. If $f(\mathbf{Z})$ is to be analytic, then $f(\xi), f(\eta)$ must both be analytic in the classical sense. An example of an analytic function is the *exponential*:

$$\exp(\mathbf{Z}) = \exp(\xi)\mathbf{e}_1 + \exp(\eta)\mathbf{e}_2. \tag{44}$$

Any analytic function of the form

$$f(\mathbf{Z}) = f(\xi)\mathbf{e}_1 + f(\eta)\mathbf{e}_2 \tag{45}$$

can be expanded into four-component form as follows:

$$\begin{aligned}
f(\mathbf{Z}) = \;& \frac{1}{2} \left\{ \; \mathrm{Re}f(\xi) + \mathrm{Re}f(\eta) \right\} \mathbf{1} \\
& + \frac{1}{2} \left\{ \; \mathrm{Im}f(\xi) + \mathrm{Im}f(\eta) \right\} \mathbf{i} \\
& + \frac{1}{2} \left\{ \; \mathrm{Im}f(\xi) - \mathrm{Im}f(\eta) \right\} \mathbf{j} \\
& + \frac{1}{2} \left\{ \; -\mathrm{Re}f(\xi) + \mathrm{Re}f(\eta) \right\} \mathbf{k},
\end{aligned} \tag{46}$$

where Re,Im are the real and imaginary parts of the indicated functions by the classical complex rules.

For example, the exponential function expands to

$$\begin{aligned}
\exp(\mathbf{Z}) = \;& \frac{1}{2} \left\{ \; \exp(x-ct)\cos(y+z) + \exp(x+ct)\cos(y-z) \right\} \mathbf{1} \\
& + \frac{1}{2} \left\{ \; \exp(x-ct)\sin(y+z) + \exp(x+ct)\sin(y-z) \right\} \mathbf{i} \\
& + \frac{1}{2} \left\{ \; \exp(x-ct)\sin(y+z) - \exp(x+ct)\sin(y-z) \right\} \mathbf{j} \\
& + \frac{1}{2} \left\{ \; -\exp(x-ct)\cos(y+z) + \exp(x+ct)\cos(y-z) \right\} \mathbf{k}.
\end{aligned} \tag{47}$$

At this point, we could go on to explicitly develop the hypercomplex analysis (derivatives, integrals, etc.). However, it is not necessary. The whole story is told in the fundamental definition (3). Everything is done by pairwise classical complex operations on ordinary complex variables as defined in (18),(19). For example,

$$\frac{df}{d\mathbf{Z}} = \frac{df}{d\xi}\mathbf{e}_1 + \frac{df}{d\eta}\mathbf{e}_2, \tag{48}$$

$$\int f(\mathbf{Z})\,d\mathbf{Z} = \mathbf{e}_1 \int f(\xi)\,d\xi + \mathbf{e}_2 \int f(\eta)\,d\eta. \tag{49}$$

No new questions arise concerning existence, completeness, internal consistency, or similar requirements. No new mathematical theorems are necessary, and all classical complex theorems come over, unchanged. The entire scheme of classical complex analysis is subsumed.

We shall now develop the Cauchy-Riemann conditions in four dimensions and illustrate one of their most remarkable consequences, that being their simplification of certain field equations of physics. Given an analytic function such as (43), then we have two separate sets of Cauchy-Riemann equations in the independent complex variables ξ, η. Recall that the latter are both functions of x, y, z, ct. By straightforward, but tedious, algebraic manipulations, we may express the C-R conditions in terms of x, y, z, ct. Consider an analytic function expanded into the form

$$f(\mathbf{Z}) = U\,\mathbf{1} + V\,\mathbf{i} + W\,\mathbf{j} + S\,\mathbf{k}, \tag{50}$$

with U, V, W, S each functions of x, y, z, ct; then, in these terms,

$$U_x = V_y = W_z = S_{ct} \tag{51}$$
$$U_y = -V_x = W_{ct} = -S_z \tag{52}$$
$$U_z = V_{ct} = -W_x = -S_y \tag{53}$$
$$U_{ct} = -V_z = -W_y = S_x \tag{54}$$

are the C-R conditions in four dimensions. Note that the two upper-left conditions are the classical, two-dimensional expressions. These have many interesting consequences that are extensions of the classical complex case, and we present some of them here for the insights that they offer into the behavior of analytic functions in the x, y, z, ct four-space. If we use the following notation:

$$f(\mathbf{Z}) = U\,\mathbf{1} + V\,\mathbf{i} + W\,\mathbf{j} + S\,\mathbf{k}, \tag{55}$$

$$\square = \mathbf{1}\frac{\partial}{\partial x} + \mathbf{i}\frac{\partial}{\partial y} + \mathbf{j}\frac{\partial}{\partial z} + \mathbf{k}\frac{\partial}{\partial ct}, \tag{56}$$

$$\square^2 = \square \bullet \square, \qquad (scalar\,product) \tag{57}$$

then the following relations can be verified by simple substitution of the C-R conditions:

$$\frac{df}{d\mathbf{Z}} = \mathbf{1}\frac{\partial f}{\partial x} = -\mathbf{i}\frac{\partial f}{\partial y} = -\mathbf{j}\frac{\partial f}{\partial z} = \mathbf{k}\frac{\partial f}{\partial ct}, \tag{58}$$

$$\mathbf{1}\,\square\,U \;=\; -\mathbf{i}\,\square\,V = -\mathbf{j}\,\square\,W = \mathbf{k}\,\square\,S, \tag{59}$$

$$|\square\,U|^2 \;=\; |\square\,V|^2 = |\square\,W|^2 = |\square\,S|^2, \tag{60}$$

$$\square^2 U \;=\; \square^2 V \;=\; \square^2 W \;=\; \square^2 S \equiv 0. \tag{61}$$

These relationships are extensions of corresponding relationships for the classical analytic functions of one complex variable. For example, the leftmost two relations of (58) are the same as for the classical complex case. They indicate that *the derivative of an analytic function is the same in every direction from a given point.* We shall use this set of relations in our differential equations example, below.

For Eqns. (59), recall that $\mathbf{1}, \mathbf{i}, \mathbf{j}, \mathbf{k}$ have the orthogonal matrix representations (38),(39); then (59) and (60) indicate that *the four-gradients of the components U, V, W, S of an analytic function are of the same vector magnitude, but respectively rotated in the four-space.*

The last relations, (61), indicate that *the components U, V, W, S of an analytic function are harmonic.* This is tantamount to saying that they each obey a three-dimensional wave equation because the unitary transformation

$$x' = x, \quad y' = y, \quad z' = z, \quad ct' = ict, \tag{62}$$

where i is the classical imaginary, transforms each into a wave equation.

3.1. INVARIANCE PROPERTIES OF THE FUNCTION SPACE

Theoretical physics research that discovers some sort of invariance principle or invariant quantity or form is of exceptional importance, because nature hides many of her most profound secrets in the form of invariance. For example, the physics of relativity sprang from the insight that the form of the electromagnetic equations should be invariant under relative motion of the observer, and that the speed of light is constant and the same for all observers regardless of their relative motion.

The $\mathbb{C}^2$ algebra and function theory, because of the existence of an all-matrix representation, has many interesting rotational invariants, some of which may have important physical interpretations. To investigate them, we first need to construct orthogonal transformation matrices that are elements of the $\mathbb{C}^2$ algebra. Let $\mathbf{U} = (u, v, w, s)$ be an element of $\mathbb{C}^2$; then the standard orthogonality condition (in matrix form)

$$\mathbf{U}\,\mathbf{U}^T = \mathbf{I}, \tag{63}$$

where T indicates the transpose and $\mathbf{I}$ is the unit matrix, yields the following conditions on the matrix elements u, v, w, s for orthogonality:

$$u^2 + v^2 + w^2 + s^2 \;=\; 1, \tag{64}$$

$$2(us - vw) \;=\; 0. \tag{65}$$

The elements $\mathbf{U}$ of $\mathbb{C}^2$ which satisfy the orthogonality conditions make up a subring of $\mathbb{C}^2$.

Now, apply the orthogonal transformation $\mathbf{U}$ to each element in Eqns. (59):

$$\mathbf{1}\,\mathbf{U}\,\square\,U = -\mathbf{i}\,\mathbf{U}\,\square\,V = -\mathbf{j}\,\mathbf{U}\,\square\,W = \mathbf{k}\,\mathbf{U}\,\square\,S. \tag{66}$$

Every element in this set of relations has a 4 × 4 real matrix representation. $\mathbf{U}$ and $\mathbf{1}, \mathbf{i}, \mathbf{j}, \mathbf{k}$ are orthogonal. If we take the determinant of all four equivalent matrix products, then we see that *the respective eigenvalues* ξ, η, ξ^*, η^* *and their product* $\xi \eta \xi^* \eta^*$ *must remain unchanged by the transformation.* The products of the eigenvalues are, respectively (after much rearrangement)

$$| \square U |^2 - 2(U_x U_{ct} - U_y U_z) = | \square V |^2 - 2(V_x V_{ct} - V_y V_z) \tag{67}$$

$$= | \square W |^2 - 2(W_x W_{ct} - W_y W_z) \tag{68}$$

$$= | \square S |^2 - 2(S_x S_{ct} - S_y S_z). \tag{69}$$

The vector magnitude quantities are equal by Eqns. (60); consequently, the following quantities for a function $f(\mathbf{Z}) = \mathbf{1}U + \mathbf{i}V + \mathbf{j}W + \mathbf{k}S$ are not only equivalent, but are invariant under rotations:

$$U_x U_{ct} - U_y U_z = V_x V_{ct} - V_y V_z \tag{70}$$

$$= W_x W_{ct} - W_y W_z \tag{71}$$

$$= S_x S_{ct} - S_y S_z. \tag{72}$$

3.1.1. *Euclidean Norm of a Vector*
If we interpret an element $\mathbf{Z}$ as a vector, then its Euclidean norm (variously: magnitude, length, absolute value)

$$| \mathbf{Z} | = (x^2 + y^2 + z^2 + c^2 t^2)^{\frac{1}{2}} \tag{73}$$

is invariant under an orthogonal transformation of coordinates.

3.1.2. *Eigenvalues*
In the matrix form, an element $\mathbf{Z}$ has eigenvalues

$$\xi = (x - ct) + i(y + z), \tag{74}$$

$$\xi^* = (x - ct) - i(y + z), \tag{75}$$

$$\eta = (x + ct) + i(y - z), \tag{76}$$

$$\eta^* = (x + ct) - i(y - z). \tag{77}$$

Considering $\mathbf{Z}$ as a transformation operator in a matrix equation of the form $\mathbf{Y} = \mathbf{Z}\mathbf{X}$, where $\mathbf{X}$ and $\mathbf{Y}$ are vectors, then an orthogonal transformation of coordinates produces a similarity transformation, $\mathbf{Z}' = S^{-1}\mathbf{Z}S$, which, it is known from matrix theory, leaves the eigenvalues unchanged.

3.1.3. *Eigenvectors*

The eigenvectors e_1, e_2 of an element $\mathbf{Z}$ are fixed by the observer's arbitrary choice of coordinates:

$$e_1 = \frac{(1-\mathbf{k})}{2}, \tag{78}$$

$$e_2 = \frac{(1+\mathbf{k})}{2}, \tag{79}$$

hence will be moved to new directions in space if the coordinates are given an arbitrary rotation. However, if the orthogonal transformation matrix is an element of $\mathbb{C}^2$, then the rotation operation acts only upon the eigenvalues and leaves the original eigenvectors unchanged.

3.1.4. *Matrix Trace*

The trace of a square, non singular matrix is the sum of the principal diagonal elements. For an element $\mathbf{Z}$, this reads off directly as $4x$. (As an aside, the trace is also the sum of the four eigenvalues, which yields the same value). An orthogonal transformation leaves this invariant.

3.1.5. *The X Coordinate*

The trace being invariant and equal to $4x$, we conclude that the x coordinate is left invariant under any pure rotation in four dimensions (within $\mathbb{C}^2$. Note that the x coordinate is the "real" coordinate (coefficient of the $\mathbf{1}$ unit vector).

3.1.6. *Determinant*

The determinant of the matrix form of $\mathbf{Z}$ is the product of the four (invariant) eigenvalues, hence is invariant:

$$\det(\mathbf{Z}) = \{(x-ct)+i\,(y+z)\}\{(x-ct)-i\,(y+z)\} \tag{80}$$

$$\times\{(x+ct)+i\,(y-z)\}\{(x+ct)-i\,(y-z)\} \tag{81}$$

$$= \{(x-ct)^2+(y+z)^2\}\{(x+ct)^2+(y-z)^2\} \tag{82}$$

$$= (x^2+y^2+z^2+c^2t^2)^2 - 4\,(xct-yz)^2. \tag{83}$$

The determinant is invariant, as is the Euclidean norm, hence the quantity $(xct - yz)$ must also be invariant. Note that this result is a consequence of the matrix formulation of the algebra.

3.1.7. *Inner Product*

Matrix theory asserts that the inner product

$$\mathbf{Z}_1 \bullet \mathbf{Z}_2 = x_1 x_2 + y_1 y_2 + z_1 z_2 + ct_1 ct_2 \tag{84}$$

of two vectors is invariant under a pure rotation.

3.1.8. *Product With Transpose*

In the matrix form, taking the transpose $\mathbf{Z}^T$ of $\mathbf{Z}$ merely changes the sign of the y and z components. In the vector form, this is $\mathbf{Z}^T = x\mathbf{1} - y\mathbf{i} - z\mathbf{j} + ct\mathbf{k}$. Taking the product of $\mathbf{Z}$ with its transpose yields

$$\mathbf{Z}\,\mathbf{Z}^T \;=\; \mathbf{1}\,(x^2 + y^2 + z^2 + c^2t^2\,) - 2\,\mathbf{k}\,(xct - yz), \tag{85}$$

$$\;=\; \mathbf{1}\,|\,\mathbf{Z}\,|^2 - 2\,\mathbf{k}\,(xct - yz), \tag{86}$$

which, by the results for the determinant (above), must be invariant under any pure rotation.

3.1.9. *Gradient Inner Products*

The four-gradients of the components X, Y, Z, T of an analytic function $f(\mathbf{Z}) = X\mathbf{1} + Y\mathbf{i} + Z\mathbf{j} + T\mathbf{k}$ are of the form

$$\square U = \mathbf{1}U_x + \mathbf{i}\,U_y + \mathbf{j}\,U_z + \mathbf{k}\,U_{ct}, \tag{87}$$

where U can be any of the components X, Y, Z, T. The inner products of the form $\square U \bullet \square V$, where U, V are any mutually-exclusive components X, Y, Z, T of an analytic function, are invariant under pure rotation. Most of these combinations result in a zero-valued inner product, meaning that they are orthogonal in the vector interpretation. Those that do not produce explicitly zero results can be reduced to the form $2\,(U_x U_{ct} - U_y U_z)$, where U is any of X, Y, Z, T. By (70)-(72) these partial derivative forms must be invariant under pure rotation.

3.1.10. *Axis Planes Under Function Operations*

For the typical element $\mathbf{Z}$, an hypercomplex function operates only upon the eigenvalues, which themselves define separate complex planes. The function operation merely maps each complex plane upon itself, and does not affect the orientation of the planes with respect to the coordinate system. Consequently, the axis planes are invariant under the function operation.

4. Applications to partial differential equations

Analyticity is implicitly assumed in the derivations of the classical field equations of physics. It could not be otherwise, for there would be no way to proceed. That is not to say, however, that isolated discontinuities and singularities are not allowed; for example, shock discontinuities in aerodynamics. Although assumed, analyticity is not usually explicitly incorporated into the equations, themselves, as happens for classical complex functions. It so happens that, in (58), we have the means both to apply analyticity conditions to PDEs and to transform PDEs into greatly-simplified ODEs.

We illustrate the technique by finding a characteristic function that satisfies the three-dimensional Emden's equation:

$$\nabla^2 U + A^2 U^n = 0 \,, \quad n = 2, 3, 4, \dots \tag{88}$$

where ∇ is the scalar del operator of classical vector analysis, A is a scalar constant, and n is a positive integer. By use of (58), the scalar del operator converts as follows:

$$\nabla^2 U = U_{xx} + U_{yy} + U_{zz} \tag{89}$$

$$= \frac{d^2 U}{d\mathbf{Z}^2} - \frac{d^2 U}{d\mathbf{Z}^2} - \frac{d^2 U}{d\mathbf{Z}^2} \tag{90}$$

$$= -\frac{d^2 U}{d\mathbf{Z}^2}\,; \tag{91}$$

then (88) becomes

$$-\frac{d^2 U}{d\mathbf{Z}^2} + A^2 U^n = 0\,, \quad n = 2, 3, 4, \dots \tag{92}$$

We can treat this by traditional ODE methods. Why? Because of the way functions and operators are defined, $f(\mathbf{Z}) = f(\xi, \eta) = (f(\xi), f(\eta))$, it is equivalent to two classically-complex problems of the same form. A first integral is readily obtained:

$$\left(\frac{dU}{d\mathbf{Z}}\right)^2 = \left(\frac{2A^2}{n+1}\right) U^{n+1} + \mathbf{C}, \tag{93}$$

where $\mathbf{C}$ is a four-dimensional constant. The latter makes it difficult to find the general solution, but we are at liberty to set it to zero to simplify matters and find a particular solution, which is

$$U = \left\{ \frac{n-1}{2} \left[\pm \left(\frac{2A^2}{n+1} \right)^{\frac{1}{2}} \mathbf{Z} + \mathbf{D} \right] \right\}^{-\frac{2}{n-1}}, \quad n = 2, 3, 4, \dots \tag{94}$$

The independent variable is $\mathbf{Z}$, which has three space dimensions and one of time, hence $U(\mathbf{Z})$ shows the behavior in space and time. A form similar to this is easily found for the one-dimensional case of Emden's equation, but it does not give any indication of how the function behaves in three space dimensions.

4.1. HYPERCOMPLEX MANIPULATIONS WITH SYMBOLIC ALGEBRA PACKAGES

The definition $oper(\mathbf{Z}) = (oper(\xi), oper(\eta))$, where $oper()$ is any function or operation defined in the classical complex analysis, allows any expression involving a four-dimensional variable to be separated into two operations of the same type, except for a classical complex variable. That being so, any symbolic algebra computer program that can handle complex variables can perform four-dimensional calculations.

Certainly, the multiplication rules (13)-(15) and the canonical algebra operations (28)-(30) are easily handled by most symbolic algebra packages. If one wished, one could even test the full matrix representation of the algebra. There is no barrier to complete computer exploration of the algebra, but there is no reason to do so; it is just the venerable $\mathbb{C}^2$ of mathematical folklore.

A more useful and interesting application is for physics field analysis problems. The general approach for the typical physics analysis problem is as follows. Formulate the problem, by whatever means, into one or more integro-differential equations.

Use the C-R conditions or Eqns. (58)-(61) to convert partial derivatives to ordinary derivatives of a four-dimensional variable, and simplify where possible, as was done for the Emden's equation, above. If the result is not already easily solvable by hand, take the four-dimensional forms and solve them with a symbolic manipulation package as if the independent variable were classically complex. Denote the result(s) by the form(s) $f(z)$, with z complex. Now, z is unimportant; by the fundamental definition (3), we replace it with the four-dimensional $\mathbf{Z}$ to obtain the four-dimensional solution. If there are constants associated with $f(z)$, we must also allow that they may be four-dimensional along with $\mathbf{Z}$. Their exact value, as usual, is determined by boundary or initial conditions.

After the four-dimensional solution is found, it is within the capability of symbolic manipulation packages to do the expansion (46) into four real components, from which further analysis and graphical display may be performed.

5. Conclusions

A new system of analysis has been constructed for one hypercomplex variable of four dimensions. This analysis may be interpreted and applied as a vector analysis for four-dimensional vectors, or may be interpreted as a "complex-like" analysis of a four-component variable, including functions, derivatives, and integrals with all the classical properties. Complete representations are given in terms of $\mathbf{1}, \mathbf{i}, \mathbf{j}, \mathbf{k}$ unit vectors, $\mathbf{e}_1, \mathbf{e}_2$ unit vectors, 2×2 complex matrices, and 4×4 real matrices.

The basis algebra is a commutative ring with unity, $\mathbb{C}^2$, with elements (ξ, η) of complex pairs. However, the complex pairs are constructed from the four coordinates x, y, z, ct in such a way that they can be equal only at the origin $(0, 0, 0, 0)$, and that neither can be zero unless $x = \pm ct$, which lies in the (physically-unreachable) relativistic limit. We may freely use the new analysis in any real-world, physical problem. Just as for classical complex functions, isolated singularities and discontinuities may appear. A Clifford-algebra representation is given that embeds $\mathbb{C}^2$ in the 16-dimensional $\mathcal{G}(4, 0)$ algebra.

Functions and operations on the four-dimensional variable $\mathbf{Z}$ are defined as the same operation on pairs (ξ, η) of classical complex variables. Consequently, the four-dimensional functions obey all the same axioms and have the same properties as for the classical complex variables. This is remarkable, considering that all this complex behavior is representable in a 4×4 real matrix form.

The function theory admits a four-dimensional set of Cauchy-Riemann conditions. Of their many consequences, the one that is investigated in the present paper is that connecting partial derivatives to the full, ordinary derivative [see (58), above]. This relationship is a straightforward extension of the classical complex case, and implicitly embodies the C-R conditions. It provides a means to explicitly enforce continuity in the formulation of field problems in physics and engineering, with the bonus of changing PDEs into ODEs, in such a way that simplifies solution.

Use of the new analysis with modern symbolic manipulation software, such as Derive^{TM}, Maple^{TM}, MACSYMA^{TM}, or Mathematica^{TM}, is discussed. All operations and functions of a four-dimensional variable $\mathbf{Z}$ reduce to the same operations upon two classically-complex variables of the form

$$\xi = (x - ct) + i(y + z), \tag{95}$$

$$\eta = (x + ct) + i(y - z), \tag{96}$$

with the result that four-dimensional analysis can be done with existing symbolic manipulation software packages without having to modify the software.

References

M. J. Crowe: 1967, 'A History of Vector Analysis', University of Notre Dame Press, South Bend.

C. M. Davenport: 1978, 'An Extension of the Complex Calculus to Four Real Dimensions, with an Application to Special Relativity', M. S. Thesis, University of Tennessee, Knoxville.

C. M. Davenport: 1991, 'A Commutative Hypercomplex Calculus, with Applications to Special Relativity', Privately Published, Knoxville; ISBN 0-9623837-0-8.

W. R. Hamilton: 1844, 'On a New Species of Imaginary Quantities Connected with the Theory of Quaternions', *Proceedings of the Royal Irish Academy* **2**, pp. 424-434.

C. C. MacDuffee: 1946, 'The Theory of Matrices', Chelsea Publishing Company, New York.

G. B. Price: 1991, 'An Introduction to Multicomplex Spaces and Functions', Marcel Dekker, Inc., New York.

G. Scheffers: 1893, 'Verallgemeinerung def Grundlagen der gewhnlich complexen Functionen', *Berichte der Gesellschaft der Wissenschaften zu Leipzig* **45**, pp. 829-848.

G. Sobczyk: 1988, Private correspondence.

ON GENERALIZED CLIFFORD ALGEBRAS — RECENT APPLICATIONS

W. BAJGUZ
Institute of Physics
Warsaw University Campus Białystok
ul. Przytorowa 2 A
15-104 Białystok, Poland

A. K. KWAŚNIEWSKI
Institute of Mathematics
Technical University of Białystok
ul. Wiejska 45 A, room 133, Poland

and

Institute of Physics
Warsaw University Campus Białystok
ul. Przytorowa 2 A
15-104 Białystok, Poland
e-mail: Kwandr@cksr.ac.bialystok.pl

Abstract. Apart from well established applications of generalized Clifford algebras new recent ones are quoted and described.

Key words: Hurwitz Theorem, division algebras, quaternions, generalized Clifford algebras, finite dimensional quantum mechanics.

The growing area of applications of Clifford algebras and naturalness of their use in formulating problems for direct calculation entitles one to call them Clifford numbers (M. Riesz, D. Hestenes, G. Sobczyk, P. Lounesto, B. Jancewicz, Delanghe *et al.* [1]–[5]).

The purpose of this note is to indicate that elements of generalized Clifford algebras deserve to be called Clifford numbers – ..., too. The known extension of real numbers considered as division algebra over reals leads to complex numbers, quaternions and octonions via Cayley–Dickson procedure and Hurwitz theorem.

Another extension of the $\mathbb{R}$ & $\mathbb{C}$ fields and of quaternions – provides us with $Cl_{p,q}$ Clifford algebras [6]; these Clifford numbers we owe to the nineteenth century mathematics.

The way Clifford numbers are related to quadratic forms, generalized Clifford algebras are related to k–ubic forms. These were invented by Morinaga, Nono *et al.* [7]; (for systematic exposition see [8]), although their appearance is to be noted already in H. Weyl's famous book [14]. However seeds of the " k–th order idea" may be traced back to Weierstrass [9] who considered possible commutative extensions of complex numbers to the case of arbitrary number of real dimensions. This possibility being afterwards realized [10, 11, 12] provides us with quasi–number systems which form commutative subalgebras of generalized Clifford algebras – as indicated in [12]. These "special" Clifford numbers are perfectly suited for the development of generalized "hyberbolon & ellipton" trigonometries as exposed in [10] and [12]. Generalized Clifford algebras possess inherent $\mathbb{Z}_k \oplus \mathbb{Z}_k \oplus ... \oplus \mathbb{Z}$ grading (where number of summands equals to the number of generators & k-labels k-ubic form).

This makes generalized Clifford numbers a very convenient and efficient apparatus to deal with spin lattice systems (see [13] and references therein) as – in particular – the transfer matrix for $\mathbb{Z}_k$-Potts models is the product of two specially related to each other generalized Clifford numbers.

Another vast field of application of generalized Clifford numbers originates from Herman Weyl's example [14] of finite dimensional quantum mechanics. There – one degree of freedom is represented by a toroidal grid $\mathbb{Z}_k \times \mathbb{Z}_k$ i.e. classical phase space. At the same time the group $\mathbb{Z}_k \times \mathbb{Z}_k$ is the grading group for the resulting generalized Clifford algebra; see [14]–[17] and references therein. Generators of these generalized Clifford algebras serve as realization of Weyl relation interpreted quantum–mechanically via corresponding transitive system of imprimitivities (see [19]). Toroidal grid $\mathbb{Z}_k \times \mathbb{Z}_k$-phase space was treated in [18] as a stage for linear maps' quantization. The Hanney & Berry method of quantization leads one naturally to Fibonacci–like sequences (labeling trajectories of quantum states) and thus makes links between coding theory and generalized Clifford numbers – noticeable [19], as already announced in [12]. These links are now under study by the authors of this note.

Anyhow ideas involved are also based on the observation that convolution of sampled functions (which are then regarded periodic) leads to circulant matrices [20] i.e. to quasi–numbers i.e. to special generalized Clifford numbers. This is to be confronted with a plan to treat finite dimensional quantum mechanics as a theory of digital image processing (– for the physical motivation of the above attitude see [21]).

As generalized Clifford numbers are defined [8] via k-ubic form replacing quadratic one for ordinary construction of an appropriate ideal of tensor algebra – it appears to be not of extreme surprise that generators of generalized Clifford (or Pauli) algebra and the corresponding group of automorphisms preserving the very k-ubic form are strictly related to the Last Fermat Theorem (see [22] for recent review). Namely, as it was recently observed [23] the Last Fermat Theorem (LFT) is equivalent to the statement that all rational solutions of $x^k + y^k = 1$ equation ($k \geq 2$) are provided by an orbit of rationally parameterized subgroup of a group preserving k-ubic form used to define the algebra of generalized Clifford numbers.

Further links of generalized Clifford numbers and LFT as well as elliptic curves are now under study.

References

1. M. Riesz: 1958, *Clifford Numbers and Spinors*, University of Maryland, reprinted as facsimile by Kluwer, 1993 (E. F. Bolinder, P. Lounesto, eds.).
2. D. Hestenes, G. Sobczyk: 1984, 1987 *Clifford Algebra to Geometric Calculus*, Reidel, Dordrecht.
3. P. Lounesto: 1993, 'Clifford algebras and Hestenes spinors', *Foundations of Physics* **23 No. 9**, pp. 1203–1237.

4. B. Jancewicz: 1988, *Multivectors and Clifford Algebra in Electrodynamics*, Word–Scientific, Singapore.
5. R. Delanghe, F. Sommen, V. Souček: 1992, *Clifford Algebra and Spinor-Valued Functions*, Kluwer.
6. M. P. Atiyah, R. Bott and A. Shapiro: 1964, 'Clifford modules', *Topology* **3** (1), pp. 3–38.
7. K. Morinaga and T. Nono: 1952, 'On the linearization of a form of higher degree and its representation', *J. Sci. Hiroshima Univ., Ser. A, Math. Phys. Chem.* **16**, pp. 13–41.
    K. Yamazaki: 1964, 'On projective representations and ring extensions', *J. Fac. Sci. Univ. Tokyo*, Sect. 1, **10**, pp. 147–195.
8. A. K. Kwaśniewski: 1985, 'Clifford and Grassmann-like algebras – old and new', *J. Math. Phys.* **26** (9), pp. 2234–2238.
9. K. Weierstrass: 1884, *Zur Theorie der aus n Haupteinheiten Gebildeten Grössen*, Leipzig.
10. N. Fleury, M. Rauch de Traubenberg, R. M. Yamaleev: 1991, 'Commutative extended complex numbers and connected trigonometry', Université Louis Pasteur, Strasbourg, CRN–PHTH/91–07.
    N. Fleury, M. Rauch de Traubenberg: 1992, 'Linearization of polynomials', *J. Math. Phys.* **33** (10), pp. 3356–3366.
11. L. L. Silverman: 1953, 'Generalization of hyperbolic and trigonometric functions', *Riveon Lematematica* **6**, pp. 53–60; and literature there in.
12. A. K. Kwaśniewski, R. Czech: 1992, 'On quasi – numbers algebras', *Reports on Math. Phys.* **31 No. 3**, pp. 341–351.
13. A. K. Kwaśniewski: 1986, 'On the Onsager problem for Potts models', *J. Phys. Math. Gen.* A **19**, pp. 1469–1476.
14. H. Weyl: 1950, *Theory of Groups and Quantum Mechanics*, E. P. Dutton Co., New York, Sec. 4.14; first edition 1929 (in German).
    J. Schwinger: 1960, *Proceedings of the National Academy of Sciences (U.S.A.)*, **46**, p. 570.
15. T. S. Santhanam: 1982, 'Quantum mechanics in a finite number of dimensions', *Physica* **114 A**, pp. 445–447 and references therein.
16. R. Balian, C. Itzykson: 1986, 'Observations sur la mécanique quantique finie', *C. R. Acad. Sc. Paris* **303**, serie 1, **No. 16**, pp. 773–778.
17. D. Galetti, A. F. R. de Toledo Piza: 1992, 'Discrete quantum phase spaces and the mod N invariance', *Physica* **A, 186**, pp. 513–523 and references therein.
18. J. H. Hannay, M. V. Berry: 1980, 'Quantization of linear maps on a torus – Fresnel diffraction by a periodic grating', *Physica* **1D**, pp. 267–290.
19. W. Bajguz, A. K. Kwaśniewski: 1994, 'On quantum mechanics and Fibonacci sequences', *Advances in Applied Clifford Algebras* **Vol. 4** (1), pp. 73–88.
20. R. G. Gonzales, P. Wintz: 1984, *Digital Image Processing*, Addison–Wesley Publishing Company.
21. A. K. Kwaśniewski: 1996, 'Could we manage without micro_UFO's?', *Physics Essays*, Vol.. 9, No. 2, Ottawa, Canada.
22. W. Narkiewicz: 1993, *Wielkie Twierdzenie Fermata*, *Wiadomości Matematyczne* **XXX** 1, pp. 1–17 (in Polish).
23. A. K. Kwaśniewski: 'On the last Fermat Theorem and generalized Clifford algebras', submitted.

ORIENTED PROJECTIVE GEOMETRY WITH CLIFFORD ALGEBRA

RICHARD C. PAPPAS
Department of Mathematics
Widener University,
Chester, PA 19013
e-mail: pappas@kuratowski.math.widener.edu

Abstract. Classical projective geometry can be efficiently formulated using the Clifford–algebra based "geometric algebra" developed by Hestenes and Sobczyk. In this formulation, points, lines, planes, etc. in a projective space $\mathbf{P}^{n-1}$ are represented by equivalence classes of vectors, bivectors, trivectors, etc. in the Clifford algebra Cl_n. Two k-blades A and B are in the same equivalence class iff $AB = A \cdot B$; that is, iff A and B differ by a scalar factor. We show that if the notion of equivalence is restricted, so that two blades of the same grade are equivalent iff they differ by a *positive* scalar factor, then the geometric objects represented inherit an orientation and provide the building blocks for oriented projective spaces. Projective concepts such as meet, join, duality, and collineation can be extended to such spaces, while new features, such as relative orientation, give these spaces a much richer structure with many important applications. Examples of computations using CLICAL are provided.

Key words: Projective geometry, orientation.

1. Introduction

Projective geometry originated from the study of perspective by artists. Although its development as a practical science can be traced from antiquity, systematic study of the mathematics involved was initiated only in the seventeenth century by Desargues and Pascal. Their treatment was synthetic. Classical projective geometry can also be formulated analytically, and this is the form in which it is used in computer graphics. (See Penna and Patterson (1986).) However, the analytic approach introduces (homogeneous) coordinates, which have no intrinsic geometric significance; it has therefore occasionally been regarded as unsatisfactory, in spite of its computational usefulness.

Attempts to develop an algebraic formalism which would express geometric objects and relations in a coordinate-free manner may be traced back to Grassmann. (Forder (1941) gives an extensive exposition of Grassmann's ideas.) More recently, Rota and Stein (1976) have used Cayley algebras and Hestenes and Ziegler (1991) have used Clifford (or "geometric") algebra as the basis for projective geometry. The Clifford algebra formulation seems to have some distinct advantages, including a more satisfactory treatment of the important concept of duality.

From the point of view of computational geometry, however, the most serious deficiency of classical projective geometry is not adequately dealt with by either the analytic or algebraic formulations. As is well known, the projective plane is not orientable. Removing a line from the projective plane, for example, leaves a set of points topologically equivalent to a disk, and, consequently, one cannot give any meaning to the statement that two points are "on the same side" of a line. Similarly, one cannot tell if a point lies "between" two given points. Stolfi (1991) has therefore argued that what is needed for computational purposes is an *oriented* projective geometry (OPG) in which all points, lines, planes, ... , and hyperplanes are replaced with two oppositely oriented copies. The result of this double covering for the projective plane, for example, is thus an oriented spherical geometry.

Stolfi uses a version of the Cayley algebra developed by Rota and co-workers (Barnabei, Brini, and Rota (1985); Rota and Stein (1976)) for his theoretical development. We show that a simple modification of the work of Hestenes and Ziegler (1991) also provides a basis for the development of OPG and that certain key concepts can be implemented more naturally with Clifford algebra than with Cayley algebra.

We follow the notation of Hestenes and Ziegler (1991) with two exceptions: they denote the *reverse* of a multivector A by $A^\dagger$ and the *dual* by $\tilde{A}$, while we have interchanged these two notations.

As shown in Hestenes and Sobczyk (1984), the blades in $C\ell(V^n)$ determine an "algebra of subspaces" of V^n. In Section 2, we recall some of the basic ideas and introduce a series of definitions which express the concepts of orientation, meet, join, and duality in terms of algebraic operations in the geometric algebra $C\ell(V^n)$. Quite a few examples are given to show how computations may be made using geometric algebra. Section 3 illustrates in detail how some of the problems associated with the non orientability of the classical projective plane are overcome in the oriented framework. Section 4 defines and discusses projective transformations in oriented spaces. CLICAL has been used throughout in the numerical examples (Lounesto, *et al.* (1989)).

2. Definitions

2.1. ORIENTATION

Every non-zero r-blade in $C\ell(V^n)$ determines a (unique) r-dimensional subspace V^r. Thus if $A = \langle A \rangle_r$, then

$$V^r = \{x \in V^n | x \wedge A = 0\}.$$

Multiplication and addition of vectors in V^r generates a geometric algebra $C\ell(V^r)$ in which A is a pseudoscalar. We express the relationship by saying that "V^r is the *support* of A", or, "A is the pseudoscalar of V^r". Clearly the choice of pseudoscalar for V^r is unique only up to a scalar factor. The outer product of any set of r vectors in V^r gives an r-blade that is proportional to the unit pseudoscalar I:

$$a_1 \wedge a_2 \wedge \ldots \wedge a_r = \lambda I$$

where $\lambda = 0$ iff $\{a_i\}$ are linearly dependent. Thus the space $\bigwedge^r V^r$ is a real one-dimensional vector space, so $\bigwedge^r V^r \setminus \{0\}$ has exactly two connected components, represented by $\pm I$. We have the

Definition 2.1.1 *An* orientation *for a vector space V^r is a choice of one of the components of $\bigwedge^r V^r \setminus \{0\}$. Let this component be represented by I. Then a pseudoscalar $A \in \bigwedge^r V^r$ is called* positive *if $A = \lambda I$, $\lambda > 0$; and a basis $\mathcal{B} = \{e_i\}_{i=1,\dots,r}$ of V^r is called* positively oriented *iff*

$$(e_1 \wedge \dots \wedge e_r) \cdot \tilde{A} > 0$$

for any positive A.

We shall see below that it is useful to have a notation for the magnitude of a pseudoscalar relative to the unit pseudoscalar I.

Definition 2.1.2 *The magnitude of a pseudoscalar A relative to I is called the* bracket *of A, and we write*

$$[A] = AI^{-1}.$$

Thus, we may say that $\{e_i\}$ is a positively oriented basis of V^r
    iff

$$[e_1 \wedge e_2 \dots \wedge e_r] > 0,$$

which we will also write as $[e_1 e_2 \dots e_r] > 0$.

The concept of an oriented projective space may be introduced as follows:

Definition 2.1.3 *Let V be a vector space with an orientation, as in Def. 2.1.1. The (real)* oriented projective space *derived from V, denoted $\Pi(V)$, is the quotient $V \setminus \{0\}$ by the equivalence relation*

$$x \sim y \qquad \textit{iff} \qquad y = \lambda x$$

for some real $\lambda > 0$.

We will be considering projective spaces based on oriented $\mathbb{R}^n$, and will write Π^{n-1} for $\Pi(\mathbb{R}^n)$, since the dimension of $\Pi(\mathbb{R}^n)$ is $n-1$. The dimension can be justified, but we shall not do so, since it is quite plausible: $\Pi(\mathbb{R}^n)$, being a set of lines in $\mathbb{R}^n$, should have dimension one less than $\mathbb{R}^n$.

It can also be shown that Π^{n-1} is homeomorphic to S^{n-1}, the unit sphere in $\mathbb{R}^n$. We will take advantage of this below, especially in the case $n = 3$, as an aid in visualizing some of the algebraic results.

2.2. FLATS

Since we will associate a k-blade in the Clifford algebra $C\ell_n$ with a $(k-1)$-dimensional oriented subspace of Π^{n-1}, we will not use a separate notation for these two concepts — the subspace will be denoted by the blade. It is convenient, however, to have a separate term for the subspace.

Definition 2.2.1 *Oriented subspaces of Π^{n-1} are called* flats.

The fact that a flat A of grade r is a subspace of Π^{n-1} is expressed algebraically by the fact that A is a factor of the unit n-blade I; in other words, there exists a unique $(n-r)$-blade B such that $AB = I$. For every flat A in Π^{n-1}, there is an *opposite* flat, $-A$, which is of course distinct from A.

Example 2.2.2 *Zero-dimensional flats.*

A zero-dimensional flat consists of two points, with one designated as "positive". If p is a vector representing this point, the opposite point is $-p$. In general, the points of Π^{n-1} are its zero-dimensional flats. ◁

Example 2.2.3 *Lines in* Π^1.

The one-dimensional space Π^1 has exactly two lines, represented by a bivector $B \in C\ell_2$ and by $-B$. These may be visualized as oppositely oriented (unit) circles in $\mathbb{R}^2$. ◁

Example 2.2.4 *Lines in* Π^2.

A line in Π^2 is represented by a two-dimensional subspace of $\mathbb{R}^3$, and hence by a bivector $B \in C\ell_3$. This may be visualized as a plane through the origin of $\mathbb{R}^3$ with an orientation assigned to it. The intersection of the oriented plane with the unit sphere in $\mathbb{R}^3$ gives an oriented great circle, which is the line in the spherical model of Π^2. ◁

Example 2.2.5 *Planes in* Π^2.

The two-dimensional space Π^2 has exactly two planes, represented by the pseudoscalars $\pm I_2$. ◁

In general, k-dimensional flats in Π^{n-1} are represented by $(k+1)$-blades in $C\ell_n$, and blades which differ by a positive scalar factor represent the same flat.

2.3. Join

In classical projective geometry, the join of two subspaces A and B of a projective space is defined as the smallest space containing both A and B. (It is not just the union of the point sets A and B.) In OPG, the definition of join must of course specify how the resulting space is oriented. Let us consider some examples of how this can be done.

Example 2.3.1 *Join of two points.*

If p and q are points in Π^{n-1}, then the *join of p to q* is the oriented line determined by the bivector $p \wedge q$. This makes sense iff p and q are independent, i.e., $p \neq q$ and $p \neq -q$. The properties of the wedge product give

$$
\begin{aligned}
q \wedge p &= -(p \wedge q) \\
&= (-p) \wedge q \\
&= p \wedge (-q).
\end{aligned}
$$

The first equality shows that the join of q to p has the opposite orientation from the join of p to q, and hence is a different line. (See Figure 1.)

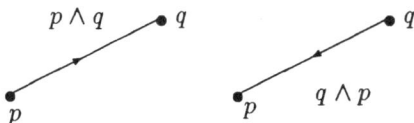

Fig. 1.   The two lines through p and q

Of course the same oriented line may be determined by different points on it. Clearly, $a \wedge b$ represents the same line as $p \wedge q$ iff

$$a \wedge p \wedge q = b \wedge p \wedge q = 0$$

i.e., a and b are on $p \wedge q$; and

$$(a \wedge b)(p \wedge q)^{-1} > 0,$$

i.e., $a \wedge b$ has the same orientation as $p \wedge q$. ◁

Example 2.3.2 *Join of a point and a line (See Figure 2).*
If L is a line and p is a point not on L, so that $p \wedge L \neq 0$, then the *join of p to L* is the plane in Π^{n-1} containing both p and L and which is positively oriented (Def. 2.1.1); that is,

$$(p \wedge L) \cdot \tilde{I} > 0,$$

or, equivalently

$$[pL] > 0.$$

Notice that, since L is a bivector, $p \wedge L = L \wedge p$ and

$$\begin{aligned} p \wedge (-L) &= (-p) \wedge L \\ &= -(p \wedge L), \end{aligned}$$

showing that the orientation of a point-line join depends on the orientation of both factors. ◁

We may generalize the preceding examples as follows.

Definition 2.3.3 *The join J of the flat A to the flat B is the outer product of A with B when this product does not vanish:*

$$J = A \wedge B.$$

Thus J is the "common dividend of lowest grade" of A and B (Hestenes and Ziegler (1991)), which corresponds to the geometric idea of the smallest flat containing both of the flats A and B. Definition 2.3.3 makes sense iff A and B have no common factor, that is, iff the corresponding flats have zero intersection. The support of the join is then the "lattice join" of the supports of A and B.

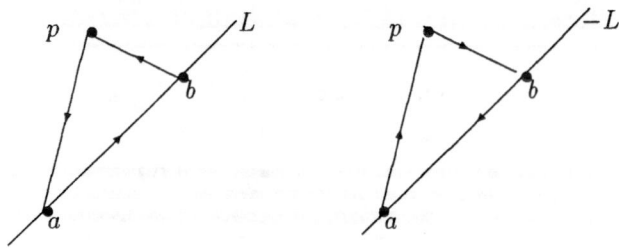

Fig. 2. $p \wedge (-L) = -(p \wedge L)$

2.4. DUALITY

The important classical concept of projective duality is illustrated by the situation in $\mathbf{P}^2$. For every theorem or axiom of the projective plane, there is a dual statement, obtained by interchanging "point" and "line". Such dual statements are always true. For example, the axiom "Any two distinct points are incident with exactly one line" has as its dual: "Any two distinct lines are incident with exactly one point." Geometric algebra implements the idea of duality as follows:

Definition 2.4.1 *The dual $A^\dagger$ of an r-blade A is the $(n-r)$-blade*

$$A^\dagger = AI^{-1} = A \cdot I^{-1} = (-1)^{r(n-r)} I^{-1} A.$$

For the "double dual", we have

$$(A^\dagger)^\dagger = (-1)^{\frac{n(n-1)}{2}} A.$$

Thus, geometrically, the dual of a point would be a hyperplane, for example, since the dual of a vector is an $(n-1)$-blade.

Observe that if grade$(A) = r$ and grade$(B) = s$, and if $r + s = n$, then

$$A \cdot B^\dagger = \langle AB^\dagger \rangle_0 = [A \wedge B] = [AB],$$

so the bracket (Def. 2.1.2) is a particular duality relation. It follows that

$$A^\dagger \cdot B = (-1)^{n-1}[AB].$$

2.5. MEET

Classically, the meet of two subspaces is their set intersection. In OPG, the intersection must also have an orientation. Geometric algebra provides a consistent way of assigning such an orientation (Hestenes and Sobczyk (1984), pp. 24–26).

Suppose that A and B are blades with a common factor. Specifically, if C is a common factor of maximal grade, then we can write

$$A = A' \wedge C, \qquad\qquad B = C^{-1} \wedge B',$$

and the part of AB with maximal grade is

$$\langle AB \rangle_{max} = A' \wedge B' = (A \cdot C) \wedge (C^{-1} \cdot B).$$

The intersection of the supports $C\ell^1(A)$ and $C\ell^1(B)$ is

$$C\ell^1(A) \cap C\ell^1(B) = C\ell^1(C),$$

and the sum is

$$C\ell^1(A) \cup C\ell^1(B) = C\ell^1(J_{AB}),$$

where

Definition 2.5.1 $J_{AB} \equiv C \wedge A' \wedge B'$.

J_{AB} is a blade of minimum grade containing both A and B. Note that it is indeed well-defined: B' has no common factors with C and is a factor of B; so it is not a factor of A, since by assumption C is the common factor of A and B with largest grade. Also, if A and B have *no* common factor, then $C = 0$, $\langle AB \rangle_{max} = A \wedge B$, and the definition of the sum is consistent with the discussion of the join above (Def. 2.3.3).

If we take duals with respect to J_{AB}, then, as shown in Hestenes and Sobczyk (1984), the following definition of the meet of A and B leads to $A \vee B = C$.

Definition 2.5.2 $A \vee B \equiv (A^\dagger \wedge B^\dagger) J_{AB} = (A^\dagger \wedge B^\dagger) \cdot J_{AB}$.

Multiplication by J_{AB}^{-1} gives $(A \vee B)^\dagger = A^\dagger \wedge B^\dagger$. Thus, "the dual of the meet is the join of the duals". In the important case where $\text{grade}(A^\dagger \wedge B^\dagger) \leq n = \text{grade}(J_{AB})$, we have a useful theorem.

Theorem 2.5.3 $A \vee B = A^\dagger \cdot B = (-1)^{s(n-s)} A \cdot B^\dagger$, *where s is the grade of B.*

Proof: From Def. 2.5.2,

$$\begin{aligned}
A \vee B &= (A^\dagger \wedge B^\dagger) \cdot J \\
&= A^\dagger \cdot (B^\dagger \cdot J) \\
&= A^\dagger \cdot (BJ^{-1}J) \\
&= A^\dagger \cdot B,
\end{aligned}$$

where we have used Eq. (1–1.25b) of Hestenes and Sobczyk (1984) to get the second equality and have omitted subscripts on J for brevity. Continuing, using the definition of the dual and Eq. (1–1.25c) from the same source, we have

$$\begin{aligned}
A^\dagger \cdot B &= (A \cdot J^{-1}) \cdot B \\
&= A \cdot (J^{-1} \cdot B) \\
&= (-1)^{s(s-n)} A \cdot (B \cdot J^{-1}) \\
&= (-1)^{s(s-n)} A \cdot B^\dagger.
\end{aligned}$$

This shows clearly that once the orientation for J_{AB} has been chosen, $A \vee B$ is unique, since the grade of J_{AB} is fixed and all blades of a given grade are in one of two equivalence classes according to sign. An opposite choice of sign for J_{AB} would lead to an opposite orientation for the meet of A and B. Specifically, using an obvious notation,

Corollary 2.5.4 $A \vee_{(-J)} B = -(A \vee_J B)$.

If A and B have no common factor, then $A \vee B$ is undefined. Also, $A \vee B$ is undefined if $A = B$ or $A = -B$. Let us illustrate these definitions by computing the meet for some fundamental configurations in Π^2 and Π^3.

Example 2.5.5 *The meet of two lines in Π^2.*
As is evident from the spherical model of Π^2, two distinct lines in general intersect in a pair of (antipodal) points. Let us see how Def. 2.5.1 allows us to select one of the two points as the meet of the two lines. First, let $A = a_1 \wedge a_2$, $B = b_1 \wedge b_2$ and take I to be the unit pseudoscalar of grade 3. Then we have

$$
\begin{aligned}
A \vee B &= A^\dagger \cdot B \\
&= (A^\dagger \cdot b_1)b_2 - (A^\dagger \cdot b_2)b_1
\end{aligned}
$$

since $\mathrm{grade}(A^\dagger) = 1$. But

$$
\begin{aligned}
A^\dagger \cdot b_i &= ((a_1 \wedge a_2) \cdot \tilde{I}) \cdot b_i \\
&= b_i \cdot ((a_1 \wedge a_2) \cdot \tilde{I}) \\
&= (b_i \wedge a_1 \wedge a_2) \cdot \tilde{I} \\
&= [b_i a_1 a_2] \\
&= [a_1 a_2 b_i]
\end{aligned}
$$

using the bracket notation from Def. 2.1.2. Thus,

$$
A \vee B = [a_1 a_2 b_1]b_2 - [a_1 a_2 b_2]b_1,
$$

which shows that two distinct lines must intersect, since $b_1 \neq 0$, $b_2 \neq 0$, and in general $[a_1 a_2 b_i]$ cannot both vanish. (If they did, then either $a_1 \wedge a_2 = 0$, which is not true, or both b_1 and b_2 are linear combinations of a_1 and a_2. However, in that case, $b_1 \wedge b_2$ would be a multiple of $a_1 \wedge a_2$, so the lines would not be distinct.)

Next, to see how to select a unique point of intersection, let us identify the common factor of A and B as p and rewrite them as $A = a \wedge p$, $B = p \wedge b$. The preceding formula then gives

$$
A \vee B = -[apb]\,p.
$$

The bracket here, however, is taken with respect to I, not J_{AB}. To compute J_{AB}, we first find

$$
\begin{aligned}
\langle AB \rangle_{max} &= (A \cdot p) \wedge (p^{-1} \cdot B) \\
&= -(a \cdot p)\,p \wedge b + p^2\, a \wedge b - (p \cdot b)\, a \wedge p.
\end{aligned}
$$

Then, by Def. 2.5.1,

$$\begin{aligned} J_{AB} &= p \wedge \langle AB \rangle_{max} \\ &= -p^2 \, (a \wedge p \wedge b). \end{aligned}$$

But since J_{AB} is determined only up to a positive scalar factor, we have $J_{AB} = -[apb]I$. Taking duals with respect to J_{AB} therefore selects p, not $-p$, as the meet:

$$A \vee_{J_{AB}} B = p.$$

If we compute the duals using $-J_{AB} = [apb]\,I$, then $A \vee_{(-J_{AB})} B = -p$, so, as mentioned above, the meet of two flats (in this case, two lines) depends on the orientation selected for the flat containing them (in this case, the whole plane Π^2).

Finally, note that $B \vee_{J_{AB}} A = -p$; in other words, in Π^2, *if the line A meets the line B at p, then B meets A at $-p$.* ◁

Example 2.5.6 *Meet of two lines in Π^3.*
We saw in Example 2.5.5 that a choice of orientation for the space Π^2 determined unambiguously the meet of two lines. Algebraically, this is because grade$(J_{AB}) = $ grade(I_3), and therefore $[J_{AB}] = J_{AB} \cdot I_3^{-1}$ has a definite sign. But in Π^3 a choice of pseudoscalar I_4 for the whole space does not fix the sign of the pseudoscalar J_{AB} for the plane containing the lines A and B, since in this case grade$(J_{AB}) = 3 \neq$ grade(I_4). Hence, although for $A = a \wedge p$ and $B = p \wedge b$ we find as before that

$$A \vee B = -[apb]\,p,$$

the sign of $[apb]$ is not determined. In general, the oriented meet of two coplanar lines in Π^m, $m \geq 3$, cannot be defined absolutely, but only relative to the orientation of the flat of minimum dimension which contains the lines. (See Stolfi (1991) for a more geometrical analysis of this configuration.) ◁

Example 2.5.7 *Meet of a plane and line in Π^3 (See Figure 3).*
Let $A = a_1 \wedge a_2 \wedge a_3$ be the plane and $B = b_1 \wedge b_2$ the line. Take the duals with respect to I_4. By a similar calculation to the one in Example 2.5.5 we get the formula

$$A \vee B = [b_1 b_2 a_2 a_3]\,a_1 + [b_1 b_2 a_3 a_1]\,a_2 + [b_1 b_2 a_1 a_2]\,a_3.$$

If B is parallel to or contained in A, then each of the brackets is zero and of course there is no point of intersection. If we now identify the point of intersection by putting $a_3 = b_1 = p$, we get

$$\begin{aligned} A \vee B &= (a_1 \wedge a_2 \wedge p) \vee (p \wedge b_2) \\ &= [pb_2 a_1 a_2]\,p. \end{aligned}$$

Computation of J_{AB} gives

$$\begin{aligned} J_{AB} &= p^2 \, a_1 \wedge a_2 \wedge p \wedge b_2 \\ &\equiv a_1 \wedge a_2 \wedge p \wedge b_2 \\ &= [a_1 a_2 p b_2]\,I_4 \\ &= [pb_2 a_1 a_2]\,I_4. \end{aligned}$$

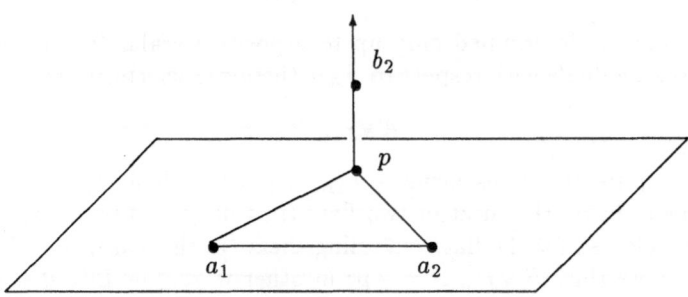

Fig. 3.   The meet of $a_1 \wedge a_2 \wedge p$ and $p \wedge b_2$ in Π^3.

So if the duals are with respect to J_{AB}, then $A \vee B = p$. Note that $A \vee B = B \vee A$.
◁

Example 2.5.8 *Meet of two planes in* Π^3.
Let the planes be $A = a_1 \wedge a_2 \wedge a_3$ and $B = b_1 \wedge b_2 \wedge b_3$ and take the duals with respect to I_4. One obtains

$$A \vee B = - [a_1 a_2 a_3 b_3]\, b_1 \wedge b_2 + [a_1 a_2 a_3 b_2]\, b_1 \wedge b_2 - [a_1 a_2 a_3 b_1]\, b_2 \wedge b_3.$$

Thus, the two planes must intersect in a line; the argument is similar to the one in Example 2.5.1. Now set $a_1 = a$, $b_3 = b$, and $a_2 = b_1 = p$, $a_3 = b_2 = q$. Then

$$A \vee B = - [apqb]\, p \wedge q.$$

But we find that

$$\begin{aligned}
J_{AB} &= p \wedge q \wedge \langle AB \rangle_{max} \\
&= p \wedge q \wedge b \wedge a \\
&= - [apqb]\, I_4.
\end{aligned}$$

Hence, taking duals with respect to J_{AB}, we have

$$A \vee B = p \wedge q$$

as required. Note that $A \vee B = -B \vee A$. ◁

2.6. General properties of meet and join

We conclude this section of definitions by briefly demonstrating some general properties of the meet and join which are useful in more involved computations.

Theorem 2.6.1 *The join and meet in* Π^{n-1} *satisfy the following commutativity and associativity rules:*

(a) $A \wedge B = (-1)^{rs} B \wedge A$

(b) $A \vee B = (-1)^{(n-r)(n-s)} B \vee A$

(c) $A \wedge (B \wedge C) = (A \wedge B) \wedge C$

(d) $A \vee (B \vee C) = (A \vee B) \vee C$

(e) $A \vee I = I \vee A = A$

where $r = \mathrm{grade}(A)$, $s = \mathrm{grade}(B)$, *and* $r + s \geq n$ *is assumed in (b) and (d).*

Proof: (a) and (d) are simply the well-known properties of the outer product. To prove (b), we have from Def. 2.5.2

$$A \vee B = (A^\dagger \wedge B^\dagger) \cdot J.$$

But $\mathrm{grade}(A^\dagger) = n - r$, $\mathrm{grade}(B^\dagger) = n - s$, so

$$B \vee A = (-1)^{(n-r)(n-s)} A \vee B$$

using (a). To prove (d), we again use Def. 2.5.2 to write

$$
\begin{aligned}
A \vee (B \vee C) &= (A^\dagger \wedge (B \vee C)^\dagger) J \\
&= (A^\dagger \wedge (B^\dagger \wedge C^\dagger)) J.
\end{aligned}
$$

Thus, (d) follows from (c). Property (e) follows from Theorem 2.5.3 and the definition of duality (Def. 2.4.1). The first equality in (e) follows from (b). ∎

3. Relative orientation

Recall that the following are among the difficulties associated with the non orientability of $\mathbf{P}^2$ which make computational geometry more complicated than in $\mathbb{R}^2$: (1) Given three points a, b, and p on a line in $\mathbf{P}^2$, it is not possible to say, for example, that "p lies between a and b", because a and b divide the line into *two* simple segments which cannot be consistently distinguished. (2) Given two points p, q and a line L in $\mathbf{P}^2$, it is not possible to say that "p and q are on the same side of L". The homeomorphism between $\mathbf{P}^2$ and the unit sphere with antipodal points identified should clarify the nature of these difficulties.

In the following subsection we shall see how the oriented structure of Π^2, as represented by elements of the geometric algebra $C\ell_3$, allows us to resolve the first difficulty. In 3.2 we discuss "two-sided lines" and generalize the idea of relative orientation to flats of higher dimension; and in 3.3 we state a "separation theorem" for oriented projective spaces. Interestingly enough, not only can flats be given a relative orientation with respect to flats of complementary grade, but they can be uniquely characterized by an "orientation function".

We illustrate some of the constructions by numerical examples, using CLICAL. Although, as we have shown in Section 2, the basic definitions of OPG can be formulated in a coordinate-free way using geometric algebra, for numerical illustrations it is of course convenient to use homogeneous coordinates. (See Penna and Patterson

(1986), Chapter 2, Section 1, for example.) Geometric algebra allows this to be done in the usual way. We need, however, to recall that our oriented spaces are double coverings of the classical projective spaces due to our restriction on what constitutes an equivalence class of points.

Thus, for the oriented projective line Π^1 we use elements $\langle p_1, p_2 \rangle$ of $\mathbb{R}^2 \setminus \{0\}$ to represent the point p_1/p_2, which is on the "front" or the "back" copy of the line depending on whether $p_2 > 0$ or $p_2 < 0$, respectively. So $\langle 1, 2 \rangle$ represents $1/2$ on the front side of Π^1 and $\langle -1, -2 \rangle$ represents $1/2$ on the back copy of the line. $\langle 1, 0 \rangle$ and $\langle -1, 0 \rangle$ are identified with $+\infty$ and $-\infty$, respectively.

Similarly, we coordinatize the oriented projective plane Π^2 by letting $\langle p_1, p_2, p_3 \rangle$ in $\mathbb{R}^3 \setminus \{0\}$ correspond to $(p_1/p_3, p_2/p_3)$. Again, Π^2 is double-sided, so that, for example, $\langle 1, 2, 3 \rangle$ corresponds to $(1/3, 2/3)$ on the front side and $\langle -1, -2, -3 \rangle$ corresponds to $(1/3, 2/3)$ on the back.

3.1. SIGNATURE OF A POINT

Consider the segment $\overrightarrow{ab}$ on the line $A = a \wedge b$ in Π^2. In the spherical model, $\overrightarrow{ab}$ is the shortest arc of a great circle through a and b, directed from a to b. The positions of a third point p on A are, assuming $p \neq a$ and $p \neq b$, as shown in the figure:

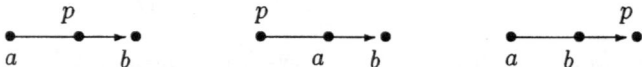

Fig. 4.   The signatures of p are $(+,+)$, $(-,+)$, and $(+,-)$, respectively.

The positions can be distinguished algebraically by examining the signs of $a \wedge p$ and $p \wedge b$. In case (a) for example, $a \wedge p$ and $p \wedge b$ have the same sign as $a \wedge b$. Thus,

$$(a \wedge p)(a \wedge b)^{-1} > 0, \qquad\qquad (p \wedge b)(a \wedge b)^{-1} > 0.$$

This suggests the following:

Definition 3.1.1 *The* signature *of a point* p *with respect to a line segment from* a *to* b *is*

$$\sigma_p(ab) \equiv (\text{sign}\,(a \wedge p)(a \wedge b)^{-1},\ \text{sign}\,(p \wedge b)(a \wedge b)^{-1}).$$

The three configurations in the figure therefore have the signatures $(+,+)$, $(-,+)$, $(+,-)$, respectively. A signature $(0,+)$ or $(+,0)$ indicates that p coincides with a or b, respectively.

The definition of signature may be extended in two ways. First, the same computations can obviously be made for a line segment in Π^{n-1}. Second, the location of a point may be specified with respect to a *simplex* of higher dimension. (See Stolfi (1991) for details.)

Example 3.1.2 *Numerical example using CLICAL.*
"Determine the location of the meet p of $A = a_1 \wedge a_2$ and $B = b_1 \wedge b_2$ with respect to the segments $\overrightarrow{a_1 a_2}$ and $\overrightarrow{b_1 b_2}$ if $a_1 = \langle 2, 1, -1 \rangle$, $a_2 = \langle -1, 1, 1 \rangle$, $b_1 = \langle 2, -1, 1 \rangle$, and $b_2 = \langle -1, 0, 1 \rangle$ give the homogeneous coordinates of the points in Π^2."

Solution: Using CLICAL and Example 2.5.5, we find that $A = a_1 \wedge a_2 = 3e_{12} + e_{13} + 2e_{23}$ and $B = b_1 \wedge b_2 = -e_{12} + 3e_{13} - e_{23}$. The dual of A is computed to be $A^\dagger = \langle 2, -1, 3 \rangle$ so we get for the meet

$$p = A^\dagger \cdot B = \langle -10, 1, 7 \rangle$$

with the signatures

$$\sigma_p(a_1 a_2) = (+, -), \qquad \sigma_p(b_1 b_2) = (+, -).$$

Hence, p is located beyond the terminal points of both segments. ◁

3.2. RELATIVE ORIENTATION

Let us next consider the second difficulty mentioned above, that of specifying which side of a line a point lies on. Considering Π^2 again for simplicity, we note that if p is a point not on the line $L = q \wedge r$, then $p \wedge q \wedge r$ is a pseudoscalar, and the discussion of orientation in Def. 2.1.1 shows that either $[pqr] > 0$ or $[pqr] < 0$. In the former case, we shall say that, by definition, the point p is on the *left* (or *positive*) side of L, and in the latter case, p is on the *right* (or *negative*) side of L.

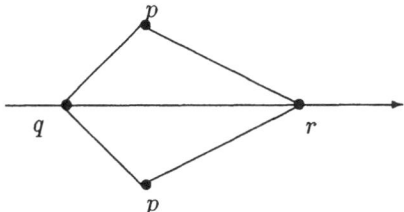

Fig. 5.   Relative orientation of p with respect to $q \wedge r$ in Π^2.

Generalizing this idea in an obvious way, we formulate the following definition:

Definition 3.2.1 *Let A and B be flats in Π^{n-1} such that* grade(A) + grade(B) = n. *(A and B are then said to have* complementary grade.) *The flat A is* positively *(negatively) oriented with respect to B iff* $[AB] > 0$ (< 0).

Suppose that one of the flats in Def. 3.2.1 is a point, p; the other is then a hyperplane, represented by a blade N of grade $n - 1$ (i.e., a pseudovector). Then of course we want the statement "p is on the positive side of N" to mean that $[pN] > 0$. But note that some care is necessary in this case. If n is even, then N has an odd number of vector factors and

$$p \wedge N = -N \wedge p,$$

so consistency in locating p requires that we make the convention that the point must stand to the left of the pseudovector in the bracket to identify the positive side of the hyperplane.

Example 3.2.2 *Numerical example: Point and plane in* Π^3.

Let $a_1 = \langle 2, -1, 0, -7 \rangle$, $a_2 = \langle 1, 1, -1, 0 \rangle$, $a_3 = \langle -1, 0, 0, 3 \rangle$ be points in Π^3. Then

$$
\begin{aligned}
N &= a_1 \wedge a_2 \wedge a_3 \\
&= -e_{123} + 2e_{124} + e_{134} + 3e_{234}
\end{aligned}
$$

is a plane in Π^3. To find the relative locations of the points $p = \langle -2, -3, -4, 1 \rangle$ and $q = \langle 2, -3, 4, 1 \rangle$ with respect to N, we use CLICAL to compute $[pN] = -10$ and $[qN] = 18$. Hence, p is on the *right* side of N and q is on the *left* side of N.
◁

Example 3.2.3 *Numerical example: Relative orientation of two lines in* Π^3.

Consider the following lines in Π^3: $L = a \wedge b$ and $M = p \wedge q$, where $a = \langle 2, 1, -2, -1 \rangle$, $b = \langle 1, 3, 4, 2 \rangle$, $p = \langle 1, 2, 1, 3 \rangle$, $q = \langle -3, -1, 1, 2 \rangle$. A computation shows that $[LM] > 0$, which means that L is oriented with respect to M "according to the right-hand rule". (See Figure 6.) ◁

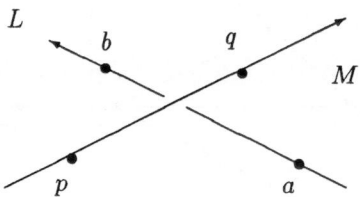

Fig. 6.   Relative orientation of L and M is positive.

3.3. THE SEPARATION THEOREM

Following Stolfi (1991), we can go a bit further and use the idea of the relative orientation to classify flats, at least in principle. Consider first Π^2.

Definition 3.3.1 *Two flats A and B in Π^2 are said to be* separated *by a point* y *iff*

$$
[Ay] = -[By] = 1.
$$

Then it is easy to prove that

Theorem 3.3.2 *Two lines in Π^2 are distinct iff they are separated by a point.*

Proof: Of course if there is a point y such that $[Ay] = -[By] = 1$, then A and B are necessarily distinct.

Conversely, let $A = a \wedge p$ and $B = p \wedge b$, with $a \neq b$, so that A and B are distinct. (Recall from Example 2.5.5 that there is no loss of generality in writing the two lines in this form.) Then

$$A \wedge b = a \wedge p \wedge b \neq 0$$

and we may in fact take $A \wedge b = I_3$. Now let

$$y = \lambda a + b$$

where λ is a scalar that we may choose for our convenience. We have

$$A \wedge y = I_3$$

and

$$B \wedge y = \lambda I_3,$$

so by taking $\lambda = -1$ we have constructed a point $y = b - a$ that separates A and B. ∎

The dual of Theorem 3.3.2 is of course

Theorem 3.3.3 *Two points in Π^2 are distinct iff they are separated by a line.*

In general the dual of a theorem in OPG is true, just as in classical projective geometry. To show this, one would need to axiomatize OPG and show that the duals of the axioms were true. This is straightforward if the axioms are first translated into the language of geometric algebra. We will not, however, attempt that here. We simply remark that by taking the duals of the lines A, B and the point y in Theorem 3.3.2, we find that $[A^\dagger y^\dagger] = -[B^\dagger y^\dagger] = 1$, so we can say that the line $y^\dagger$ separates the points $A^\dagger$ and $B^\dagger$.

These results can be generalized:

Theorem 3.3.4 *Two flats A and B of the same grade in Π^{n-1} are distinct iff there is a flat Y of complementary grade such that $[AY] = -[BY] = 1$.*

We have not succeeded in using geometric algebra to improve on the demonstration of this theorem in Stolfi (1991), so we refer the reader to that source. As is pointed out there, Theorem 3.3.4 implies immediately that a flat in an oriented projective space is uniquely characterized by an "orientation function" which in our notation is simply $[AX]$. As X varies over all flats with complementary grade to A, this function takes on the values $\{1,0,-1\}$ according to the location of A with respect to those flats.

4. Projective transformations

4.1. DEFINITIONS

In the classical projective geometry of $\mathbf{P}^2$, a *perspective transformation* (or *perspectivity*) between two lines is a point transformation in which each pair of corresponding points is collinear with a fixed point, the *center of perspectivity*. (See Figure 7.) A *projective transformation* (or projectivity) is a composition of perspectivities. One can show that: a transformation is a projectivity iff it preserves the cross ratio of four points; a transformation preserves cross ratio iff it is linear; and, therefore, a transformation is a projectivity iff it is linear. The axioms of the projective plane require that projectivities be one-to-one; hence, they are represented by non singular linear transformations. Similar considerations establish the identification of projectivities and linear transformations in higher dimensions.

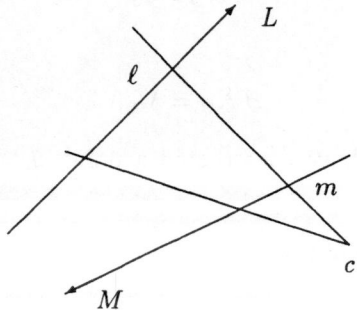

Fig. 7.   Perspectivity with center c.

If f is a linear transformation mapping vectors to vectors in V^n, then it can be extended uniquely to a linear transformation $\underline{f}$ on multivectors in $C\ell(V^n)$ called the *outermorphism* of f, which acts on blades as follows:

$$\underline{f}(a_1 \wedge a_2 \wedge \ldots \wedge a_r) = f(a_1) \wedge f(a_2) \wedge \ldots \wedge f(a_r).$$

Thus, $\underline{f}$ is grade-preserving. If I is the unit pseudoscalar of $C\ell(V^n)$, then $\underline{f}$ can only transform I by a scalar factor, which in fact is the determinant of f:

$$\underline{f}(I) = (\det f)\, I.$$

Clearly, $\underline{f}$ will preserve orientations iff $\det f > 0$, and we can define a projective transformation in OPG as follows:

Definition 4.1.1 *A* projective transformation *of an oriented projective space into itself is a linear transformation with a positive determinant. A transformation with a negative determinant will be said to be* orientation reversing.

Two projective transformations f and g are the same iff $f = \lambda g$ for some $\lambda > 0$. We note the important fact that a projective transformation defined this way satisfies the essential geometric requirement of mapping points to points, lines to lines, planes to planes, etc., since the outermorphism is grade-preserving. One can also show (Hestenes (1991)) that the meet and join as defined above are preserved by projective transformations.

4.2. SINGULAR REPRESENTATION OF A PERSPECTIVITY

For computational purposes, it is convenient to classify projective transformations in more detail than simply as compositions of perspectivities. One can distinguish those transformations which are extensions to the projective context of transformations of Euclidean spaces (rotations, translations, reflections, shearings, and scalings) from perspectivities. Canonical forms for the Euclidean type using geometric algebra are well-known (Hestenes and Sobczyk (1984), Chapter 3). We will consider here only perspectivities in Π^2.

Let L and M be two lines in Π^2. If ℓ is a point on L, then its image m under a perspectivity f with center c is given by (see Figure 7):

$$
\begin{aligned}
f(\ell) &= (\ell \wedge c) \vee M \\
&= (\ell \wedge c) \cdot M^\dagger \\
&= (c \cdot M^\dagger)\ell - (\ell \cdot M^\dagger)c \\
&= [cM]\ell - [\ell M]c.
\end{aligned}
$$

First, note that the representation of the perspectivity is actually *singular*. Indeed, considered as a map of Π^2, it is clearly not one-to-one since all points on $\ell \wedge c$ get mapped to m, for example. Formally, using the outermorphism $\underline{f}$, we have

$$
\begin{aligned}
\underline{f}(I_3) &= f(\mathbf{e}_1) \wedge f(\mathbf{e}_2) \wedge f(\mathbf{e}_3) \\
&= ([cM]\mathbf{e}_1 - [\mathbf{e}_1 M]c) \wedge ([cM]\mathbf{e}_2 - [\mathbf{e}_2 M]c) \wedge ([cM]\mathbf{e}_3 - [\mathbf{e}_3 M]c)
\end{aligned}
$$

which is easily shown to be identically zero. Hence, $\det f = 0$. Although not strictly a projective transformation according to Def. 4.1.1, this so-called extrinsic representation of a perspectivity is nevertheless useful.

Example 4.2.1 *Numerical example.*
If in Figure 7 we take the center to be $c = \langle 1, 2, 1 \rangle$ and the lines to be $L = -\mathbf{e}_{12} + \mathbf{e}_{13} + \mathbf{e}_{23}$, $M = -\mathbf{e}_{12} + \mathbf{e}_{13} + 2\mathbf{e}_{23}$, then the image of the point $\ell = \langle 1, 1, 0 \rangle$ is

$$
\begin{aligned}
m &= (c \cdot M^\dagger)\ell - (\ell \cdot M^\dagger)c \\
&= \langle -2, -3, -1 \rangle
\end{aligned}
$$

which is easily verified to be on M. ◁

Observe that if m is on M, we should have $f(m) = m$, that is $f^2 = f$, since f should be idempotent. This is not guaranteed, however. Consider

$$
f(m) = (c \cdot M^\dagger)m - (m \cdot M^\dagger)c
$$

$$= (c \cdot M^\dagger)m$$
$$= [cM]m$$

where, since m is on M, $m \cdot M^\dagger = 0$. We see that we need $[cM] > 0$ for f to be idempotent. Recalling the discussion of relative orientation in Section 3.2 above, we must make the convention that the center of perspectivity is on the left (positive) side of M. Similar considerations apply if L and M are planes in Π^3.

References

M. Barnabei, A. Brini, and G.-C. Rota: 1985, "On the exterior calculus of invariant theory," *Journal of Algebra* **96**, pp. 120 – 160.

H. G. Forder: 1941, *The Calculus of Extension*, Cambridge University Press, Cambridge [Reprint: Chelsea, New York, 1960].

D. Hestenes: 1991, "The design of linear algebra and geometry," *Acta Applicandae Mathematicae* **23**, pp. 65 – 93.

D. Hestenes and G. Sobczyk: 1984, *Clifford Algebra to Geometric Calculus*, Reidel, Dordrecht.

D. Hestenes and R. Ziegler: 1991, "Projective geometry with Clifford algebra," *Acta Applicandae Mathematicae* **23**, pp. 25 – 63.

P. Lounesto, R. Mikkola, and V. Vierros: 1989, *CLICAL – Complex Number, Spinor and Clifford Algebra Calculations with a Personal Computer*, Version 3, Helsinki University of Technology.

M. A. Penna and R. R. Patterson: 1986, *Projective Geometry and its Applications to Computer Graphics*, Prentice-Hall, Englewood Cliffs.

G.-C. Rota and J. Stein: 1976, "Applications of Cayley algebras," *Atti dei Convegni Lincei* **17**, pp. 71 – 97.

J. Stolfi: 1991, *Oriented Projective Geometry*, Academic Press, San Diego.

THE APPLICATIONS OF CLIFFORD ALGEBRAS TO CRYSTALLOGRAPHY USING *MATHEMATICA*

A. GÓMEZ, J. L. ARAGÓN and O. CABALLERO
Instituto de Física
Universidad Nacional Autónoma de México
Apartado Postal 20-364
01000 México, Distrito Federal, México
e-mail: aragon@ifunam.ifisicacu.unam.mx

and

F. DÁVILA
Departamento de Matematicas
Escuela Superior de Física y Matemáticas-I.P.N.
U.P. Adolfo López Mateos, Edificio 9,
07300 México, Distrito Federal, México

Abstract. The discovery of materials such as incommensurate structures and quasicrystals makes necessary to extend crystallography to more than three dimensions. In this work, we explore the use of Clifford algebras to describe n-dimensional lattices. This point of view allows to phrase the geometrical crystallography in a concise language valid in any dimension. Applications to the problems of faceting and phason degrees of freedom in quasicrystals are presented and solved using a *Mathematica* package which is also presented in the work.

Key words: Clifford algebras, crystallography, quasicrystals, computer algebra.

1. Introduction

The idea of using algebra to study geometry is not new. Descartes himself with his discovery (invention?) of analytic geometry initiated this chain of thought that continues today. By the end of the last century and at the beginning of the 20-th century there was a sort of competition between rival models for a geometric algebra. The winner was Gibbs with his vector algebra that included the famous "cross" product. The losers were Hamilton (with his quaternions), Grassmann (with his exterior algebra) and Clifford (the father of Clifford algebras). The time now seems ripe for a "revenge". From the standpoint of crystallography, reasons for this are manifold:

1. The recent discovery of quasicrystals indicates that it is necessary to do crystallography in spaces of dimension higher than three.

2. The structures of metallic glasses and of Fullerenes (C_{60} molecules and related compounds) make it clear that one has to do crystallography in non-Euclidean spaces, in particular in hyperbolic spaces.

Thus what is needed is an algebra capable of expressing geometric relationships in various spaces; and recall that it is only in $\mathbb{R}^3$ and $\mathbb{R}^7$ where the cross product is defined. In the past people would argue that to use a Clifford algebra was cumbersome: in $\mathbb{R}^3$ the algebra is of dimension 8. However, with recent developments in numerical software (such as Matlab) and in symbolic algebra software (Maple, *Mathematica*) this excuse is no longer valid.

The purpose of this paper is to explore the use of Clifford algebras in crystallography. Here we do various crystallographic calculations in spaces of various dimensions including $\mathbb{R}^2$ where there is no cross product (here people often play a dirty trick of embedding it in $\mathbb{R}^3$). In Section 2 a brief summary of useful concepts of geometric crystallography (some of them expressed in terms of Clifford algebra) is given. In Section 3 we show how to do crystallography in two dimensions and how it can be done advantageously with a Clifford algebra: reciprocal lattices in two-dimensional spaces are delightfully simple in Clifford algebra terms. Since the main application of higher-dimensional crystallography is in the field of quasicrystals, a brief introduction to this topic is given in Section 4. In preparation for the applications of Clifford algebra to the crystallography of quasicrystals, a *Mathematica* package for doing Clifford algebra is introduced in Section 5. In Section 6, the package is applied to the problem of faceting (shapes) in quasicrystals. In Section 7 we indicate how Clifford algebras are relevant in problems related to phason degrees of freedom in quasicrystals.

2. Some concepts of geometric crystallography

2.1. LATTICES IN $\mathbb{R}^n$

A *lattice* spanned by a linearly independent set $\{\mathbf{a}_1, \mathbf{a}_2, \ldots, \mathbf{a}_n\} \subset \mathbb{R}^n$ is defined to be the set L given by

$$L = \left\{ \mathbf{x} \in \mathbb{R}^n \mid \mathbf{x} = \sum_{i=1}^{n} h_i \mathbf{a}_i; \ h_i \in \mathbf{Z} \right\}.$$

2.2. UNIT AND PRIMITIVE CELLS

Given $\{\mathbf{a}_1, \mathbf{a}_2, \ldots, \mathbf{a}_n\} \subset \mathbb{R}^n$, the parallelotope

$$\Gamma_n = \left\{ \mathbf{x} \in \mathbb{R}^n \mid \mathbf{x} = \sum_{i=1}^{n} h_i \mathbf{a}_i; \ 0 < h_i < 1 \right\}$$

is a fundamental domain of the translation group of L and it is called the *primitive unit cell* of L. When this cell is translated through all vectors in L, it just fills entire space without either overlapping itself or leaving gaps.

The space can also filled with non-primitive unit cells known simply as *unit cells*. In this case, a unit cell fills space without any overlapping when translated through some subset of the vectors of L.

In terms of Grassmann product, the volume of the primitive unit cell of L is given by

$$V = |\mathbf{a}_1 \wedge \mathbf{a}_2 \wedge \ldots \wedge \mathbf{a}_n|. \tag{1}$$

2.3. RECIPROCAL LATTICES

Given a basis $\{\mathbf{a}_1, \mathbf{a}_2, \ldots, \mathbf{a}_n\}$, the existence of the *reciprocal lattice* $\{\mathbf{a}^1, \mathbf{a}^2, \ldots, \mathbf{a}^n\}$ $(\mathbf{a}_i \cdot \mathbf{a}^j = \delta_{ij})$ can be seen most readily using the outer product. In fact, it turns out that (Hestenes and Sobczyk, 1984):

$$\mathbf{a}^j = -(-1)^i (\mathbf{a}_1 \wedge \mathbf{a}_2 \wedge \ldots \wedge \widehat{\mathbf{a}_i} \wedge \ldots \wedge \mathbf{a}_n)(\mathbf{a}_1 \wedge \mathbf{a}_2 \wedge \ldots \wedge \mathbf{a}_n)^{-1},$$

where $\widehat{\mathbf{a}_i}$ means that the i-th term is omitted.

2.4. LATTICE DIRECTIONS

In general, a lattice direction is defined as the direction of a line that passes through at least two lattice points. Since lattice points have integer coordinates in the $\{\mathbf{a}_1, \mathbf{a}_2, \ldots, \mathbf{a}_n\}$ basis, it follows that the lattice directions are parallel to the lattice vectors.

For a lattice vector $\mathbf{x} \in L$, the shorthand

$$[h_1, h_2, \ldots, h_n] = \mathbf{x} = \sum_{i=1}^{n} h_i \mathbf{a}_i$$

can be used whenever it is clear what basis is used. The lattice directions are, therefore, also labelled by $[h_1, h_2, \ldots, h_n]$ where the integers will be divided by their greatest common divisor since a multiplicative factor does not affect directions.

2.5. PLANES AND HYPERPLANES

The number of degrees of freedom of a p-flat (p-dimensional manifold) embedded in a n-dimensional space is given by

$$f = (n - p)(p + 1)$$

whereas the number of degrees of freedom of a p-flat in an n-dimensional space constrained to pass through a given r-flat is (Sommerville, 1958)

$$f = (n - p)(p - r).$$

2.5.1. *(n-1)-flats in* $\mathbb{R}^n$

The $(n-1)$-lattice-flats in $\mathbb{R}^n$, $((n-1)$-dimensional lattice hyper-planes), have the following characteristics:

1. They can be described by means of an $(n-1)$-vector M and the equation for the flat is

$$(\mathbf{x} - \mathbf{x}_0) \wedge M = 0$$

   where $\mathbf{x}_0$ is a point in the manifold (flat) and M has integer coordinates in the standard basis of the space of grade $n - 1$.

2. They can be described by means of intersections with n coordinate axes, and/or by the corresponding Miller indices (provided, of course, that the flat does not go through the origin).

3. They can be described as manifolds with normal $\mathbf{n}$, where $\mathbf{n}$ lies in the reciprocal lattice and the equation for the flat is

$$(\mathbf{x} - \mathbf{x}_0) \cdot \mathbf{n} = 0.$$

It is important to realize that in this case the number f of degrees of freedom for the flat is

$$f = (n - p)(p + 1) = n$$

in agreement with the fact that the intersections with the n crystal axes determine the flat uniquely and completely.

2.5.2. p-flats in $\mathbb{R}^n$ $(p \neq n - 1)$

In the general case $p \neq n - 1$ we observe the following general features:

1. Since the number of degrees of freedom f of the flat is now

$$f = (n - p)(p + 1) \neq n,$$

the flat cannot be specified in terms of intersection with the axes (or in terms of Miller indices).

2. However, it is still true that the flat through lattice points can be described by a p-vector M and the equation of the flat becomes

$$(\mathbf{x} - \mathbf{x}_0) \wedge M = 0$$

where M has integer coordinates in the standard basis of the space of grade p.

3. It is no longer true that the flat can be described by a normal that lies in the reciprocal lattice. This is because the set of vectors perpendicular to the given reciprocal lattice vector $\mathbf{n}$ forms an $(n - 1)$-dimensional space and not a p-dimensional one $(p \neq n - 1)$.

Consequently, in the general case the description of the lattice planes must be done in terms of the multivectors M with integer coordinates.

By the way, the coordinates of a multivector M that describe a flat constitute what is known since long in the mathematical literature as *Plücker coordinates*. So, given a lattice flat we can find its Plücker coordinates, which, as we have seen are integer. There is a partial converse to this: given $\binom{n}{p}$ integers, they are the components (indices!) of a multivector (i.e. they are Plücker coordinates) describing a lattice p-flat provided that the so-called Plücker condition (to be discussed below) holds.

2.5.3. *Plücker's condition*

A multivector M that can be written as a product

$$M = \mathbf{v}_1 \mathbf{v}_2 \cdots \mathbf{v}_p,$$

of linearly independent vectors $\{\mathbf{v}_1, \mathbf{v}_2, \ldots, \mathbf{v}_p\}$ is called *simple*. *Multivectors describing flats must be simple*. If a multivector M is simple then $M\tilde{M} \in \mathbb{R}$. This equation is known as *Plücker's condition* and it provides a necessary condition for the multivector M to represent a p-flat (a lattice p-flat if the coordinates are integer).

2.6. GENERALIZED MILLER INDICES

On the basis of the preceding discussion it can be said that a general lattice p-flat can be described by means of a p-vector M with integer coefficients [in the standard lattice basis for the space of p-vectors] and which satisfies Plücker's condition. In this case the equation for the hyperplane is

$$(\mathbf{x} - \mathbf{x}_0) \wedge M = 0$$

where $\mathbf{x}_0$ is any vector lying in the hyperplane.

We propose a generalization of the concept of Miller indices. The obvious candidate is the set of the integer coordinates of M : if they are $h_1, h_2, \ldots$ then the hyperplanes can be labelled as $(h_1, h_2, \ldots, h_m)$ where $m = \binom{n}{p}$.

However, by virtue of Plücker's condition the indices are not all independent and, consequently, cannot be all specified independently.

3. Two-dimensional crystallography

Let $\{\mathbf{a}, \mathbf{b}\} \subset \mathbb{R}^2$ be a linearly independent set and let

$$L = \{h\mathbf{a} + k\mathbf{b} \mid h, k \in \mathbf{Z}\}$$

be the lattice generated by $\{\mathbf{a}, \mathbf{b}\}$. A *reciprocal lattice* to L is the lattice L^* defined as

$$L^* = \{h\mathbf{a}^* + k\mathbf{b}^* \mid h, k \in \mathbf{Z}\}$$

where vectors $\mathbf{a}^*$ and $\mathbf{b}^*$ belong to the space spanned by $\mathbf{a}$ and $\mathbf{b}$. They are defined by the relations

$$\mathbf{a}^* \cdot \mathbf{a} = 1, \quad \mathbf{b}^* \cdot \mathbf{b} = 1, \quad \mathbf{a}^* \cdot \mathbf{b} = 0, \quad \mathbf{b}^* \cdot \mathbf{a} = 0.$$

Explicit expressions for the reciprocal lattice vectors can be found easily in various ways. One of them is to embed $\mathbb{R}^2$ in $\mathbb{R}^3$ and to use the cross product in $\mathbb{R}^3$. Here we present a simple approach using Geometric Algebra that does not require any embedding.

Form the product

$$\mathbf{a}^* (\mathbf{a} \wedge \mathbf{b}) = \mathbf{a}^* \cdot (\mathbf{a} \wedge \mathbf{b}) + \mathbf{a}^* \wedge (\mathbf{a} \wedge \mathbf{b}) = \mathbf{a}^* \cdot (\mathbf{a} \wedge \mathbf{b}) = (\mathbf{a}^* \cdot \mathbf{a}) \mathbf{b} - (\mathbf{a}^* \cdot \mathbf{b}) \mathbf{a} = \mathbf{b}$$

from which it can be inferred that

$$\mathbf{a}^* = \mathbf{b} (\mathbf{a} \wedge \mathbf{b})^{-1}$$

and

$$\mathbf{b}^* = -\mathbf{a}(\mathbf{a} \wedge \mathbf{b})^{-1}.$$

From these expressions, the following formulae usually found in Gibb's calculus can be derived:

$$\mathbf{a}^* = \frac{|\mathbf{b}|^2}{|\mathbf{a}|^2 |\mathbf{b}|^2 - (\mathbf{b} \cdot \mathbf{a})^2} \mathbf{a} + \frac{(-\mathbf{b} \cdot \mathbf{a})}{|\mathbf{a}|^2 |\mathbf{b}|^2 - (\mathbf{b} \cdot \mathbf{a})^2} \mathbf{b},$$

$$\mathbf{b}^* = \frac{-(\mathbf{b} \cdot \mathbf{a})}{|\mathbf{a}|^2 |\mathbf{b}|^2 - (\mathbf{b} \cdot \mathbf{a})^2} \mathbf{a} + \frac{|\mathbf{a}|^2}{|\mathbf{a}|^2 |\mathbf{b}|^2 - (\mathbf{b} \cdot \mathbf{a})^2} \mathbf{b}.$$

The Clifford algebra expressions are easier to manipulate.

4. Quasicrystals

Quasicrystals are metallic alloys whose diffraction patterns exhibit sharp spots but non-crystallographic symmetry. The sharp spots in the diffraction pattern mean that the structure has long-range order but the forbidden symmetry implies that the lattice underlying the atomic structure cannot be periodic. The lattice is quasiperiodic so it will be referred to as a *quasilattice* (Janot, 1992). The first step to understand quasiperiodic order is to develop a mathematical method to generate infinite quasilattices. In what follows, we shall discuss two popular and elegant methods: the so-called Cut and Projection Method and the Cut Method.

4.1. THE CUT AND PROJECTION METHOD

The Cut and Projection Method (Katz and Duneau, 1986), (Kalugin *et al.*, 1985) regards d-dimensional quasilattices as projections of a subset of a lattice in an n-dimensional space $\mathbb{R}^n$, $(n > d)$. For simplicity, we shall consider the case of quasilattices with icosahedral symmetry which are obtained by projecting a cubic lattice in the six-dimensional space.

Let us consider the cubic lattice $L_P = \mathbf{Z}^6 \subset \mathbb{R}^6$ equipped with an orthonormal basis $\{e_1, e_2, e_3, e_4, e_5, e_6\}$ and let us denote by Γ_6 the unit hypercube. Now, let $E^{\parallel}$ be the three-dimensional "physical space" of the quasilattice and assume that it does not contain any points of the lattice L_P. A "strip" S in L_P is generated by shifting Γ_6 along $E^{\parallel}$, i.e., $S = \Gamma_6 \oplus E^{\parallel}$. If $P^{\parallel} : L_P \to E^{\parallel}$ is the projection from the lattice L_P onto the physical space, a quasilattice Q_P is generated in $E^{\parallel}$ by projecting all the points of L_P that fall inside the strip S. That is

$$Q_P = P^{\parallel}(L_P \cap S).$$

Recall that $\mathbb{R}^6$ can be decomposed as $\mathbb{R}^6 = E^{\parallel} \oplus E^{\perp}$ where $E^{\perp}$ is the orthogonal complement to $E^{\parallel}$ that is often called "phason space". If $P^{\perp} : L_P \to E^{\perp}$ is the projection onto $E^{\perp}$, then the projection of all points inside the strip onto $E^{\perp}$ defines an "acceptance domain" $K = P^{\perp}(L_P \cap S)$. The name comes from the fact that a point $\mathbf{x} \in L_P$ is inside the strip S if $P^{\perp}(\mathbf{x}) \in K$. In this case, K is a rhombic triacontahedron.

The physical space $E^{\parallel}$ is fixed by the condition that

$$P^{\parallel}(e_1) = \alpha_i, \quad i = 1, \ldots, 6,$$

where $\{\alpha_1, \alpha_2, \alpha_3, \alpha_4, \alpha_5, \alpha_6\}$ are vectors pointing to the upper vertices of an icosahedron (they line up with the six five-fold symmetry axes). Relative to the orthonormal basis, the projection matrix is (Katz and Duneau, 1986):

$$P^{\parallel} = \begin{bmatrix} b & a & -a & -a & a & a \\ a & b & a & -a & -a & a \\ -a & a & b & a & -a & a \\ -a & -a & a & b & a & a \\ a & -a & -a & a & b & a \\ a & a & a & a & a & b \end{bmatrix}, \tag{2}$$

where $a = 1/\sqrt{20}$ and $b = 1/2$. The projector onto $E^{\perp}$ is given by $P^{\perp} = \mathbf{1} - P^{\parallel}$ where $\mathbf{1}$ is the 6×6 identity matrix.

4.2. THE CUT METHOD

Quasiperiodic structures can also be obtained as "cuts" of periodic structures in higher dimensions. Here, one needs a family s_i of surfaces in $\mathbb{R}^6$ with the periodicity of the lattice L. The intersections of s_i's with $E^\parallel$ give the positions of the atoms. Physical assumptions lead to restrictions on the shape and properties of the surfaces in $\mathbb{R}^6$ and will be discussed in Section 7.

The Cut Method is completely equivalent to the Cut and Projection Method provided that surfaces have only components in $E^\perp$, that is $\dim(s_i) = \dim E^\perp$ (Oguey *et al.*, 1988). For instance, the quasilattice obtained from $\mathbb{R}^6$ by the Cut and Projection Method, as described previously, can be obtained with the Cut Method by using rhombic triacontahedra as hypersurfaces in the vertices of a primitive six-dimensional lattice.

When surfaces have components along $E^\parallel$, the Cut and the Cut and Projection Methods are no longer equivalent. The Cut Method is therefore more general.

5. The Clifford package

In this section a *Mathematica* package for general calculations with the Clifford Algebra of $\mathbb{R}^{p,q}$ is briefly described. The package contains 31 functions and allows symbolic operations with no restriction on the dimension $n\ (= p + q)$. The only limitation is that a basis is needed since the package was designed for n-dimensional crystallographic calculations where a coordinate system is initially given.

All that we need in order to develop a package for calculations with Clifford Algebra is to define two basic operations: geometric product and grade operator. In the first case, a simple algorithm for the computation of the geometric product between multivectors can be devised by noticing that a general multivector in $C\ell_{p,q}$ is a linear combination of terms of the form

$$\mathrm{e}_1^{m_1}\mathrm{e}_2^{m_2}\cdots\mathrm{e}_n^{m_n}, \tag{3}$$

where $m_i = 1$ or 0, $i = 1,\ldots,n$. The geometric product of two n-blades is:

$$(\mathrm{e}_1^{m_1}\mathrm{e}_2^{m_2}\cdots\mathrm{e}_n^{m_n})(\mathrm{e}_1^{r_1}\mathrm{e}_2^{r_2}\cdots\mathrm{e}_n^{r_n}) = (-1)^s\,\mathrm{e}_1^{m_1+r_1}\mathrm{e}_2^{m_2+r_2}\cdots\mathrm{e}_n^{m_n+r_n},$$

where

$$s = \sum_{1\leq i<j\leq n} r_i m_j.$$

If $m_i + r_i = 2$ then, in order to have the right hand side in the form (3), $\mathrm{e}_i^{m_i+r_i}$ will be replaced with $<\mathrm{e}_i,\mathrm{e}_i>\mathrm{e}_i^0$, and in this case we will have

$$\mathrm{e}_1^{m_1+r_1}\cdots\mathrm{e}_i^{m_i+r_i}\cdots\mathrm{e}_n^{m_n+r_n} = <\mathrm{e}_i,\mathrm{e}_i>\mathrm{e}_1^{m_1+r_1}\cdots\mathrm{e}_i^0\cdots\mathrm{e}_n^{m_n+r_n}.$$

Equation (3) enables us to establish an isomorphism between blades and n-tuples $(m_1, m_2,\ldots,m_n)$ of zeros and ones that can be more easily handled from a computational point of view. The grade of a blade such as the one in (3) is simply $m_1 + m_2 + \cdots + m_n$.

The package `Clifford.m` uses this isomorphism and multivectors are translated into linear combinations of n-tuples. So, for instance, the multivector $A = 15 + 7e_1e_2e_4$ is represented internally as $15\,(0,0,0,0) + 7\,(1,1,0,1)$.

In a session with `Clifford.m`, basis vectors e_i are denoted by `e[i]`, so that the multivector A of the previous example is written as

```
A = 15 + 7 e[1]e[2]e[4]
```

Note that only for display purposes the blade $e_1e_2e_4$ is written as `e[1]e[2]e[4]` that involves the commutative product of *Mathematica*. This provides us with a simple notation since the proper way to put it should be `Gp[ e[1], e[2],e[4] ]`. This simplification is only applied to the geometric product of basis vectors and can be used provided that care is taken in preserving the canonical order of the basis vectors since *Mathematica* rearranges factors before operating on expressions which are then automatically rewritten in the canonical order. The use of the function `GeometricProduct` (alias `Gp`) is recommended in order to avoid mistakes. For example, the blade $B = e_1e_3e_2$ should be entered as

```
In[1]:= B = Gp[ e[1], e[3],e[2] ]
Out[1]:= -e[1]e[2]e[3]
```

but, as a shortcut, we can type it in directly as

```
In[1]:= B = -e[1] e[2] e[3]
Out[1]:= -e[1] e[2] e[3]
```

This is obviously not valid any longer for arbitrary vectors or multivectors. For instance, the product between vectors $\mathbf{u} = (a, 2, 0)$ and $\mathbf{v} = (b^2, 0, 1)$ is calculated with the function `Gp`:

```
In[1]:= Gp[ a e[1] + 2 e[2], b^2 e[1] + e[3] ]
              2       2
Out[1]:= a b  - 2 b  e[1] e[2] + a e[1] e[3] + 2 e[2] e[3]
```

The function `Gp` is linear and it accepts symbolic, constant and numeric coefficients. The signature of the bilinear form $<\mathbf{x},\mathbf{y}>$ can be set by using `$SetSignature=p`. If no value is specified at the beginning of the session, the default is `p=20`.

The description of some other functions of the package will be given through a particular application in the field of quasicrystals which is the topic of the next section.

6. The morphology of quasicrystals

It is an experimental fact that crystals frequently grow with polyhedral external shapes (habits). This fact can be understood on the basis that crystals consist of identical units that are repeated according to a lattice. The lattice periodicity also explains why the various facets (flat faces of the polyhedra) of crystals form always the same angles among themselves (law of constancy of angles) and why the directions of the normals to the facets can be indexed (Miller indices) with small integers (law of rational indexes) (Vainshtein, 1981). Quasicrystals, although non-crystalline, behave in a similar way. They also present polyhedral habits, the law of constancy of angles is obeyed, and they can be indexed, in Fourier module, with small integers (Janssen *et al.*, 1989). The observed morphology of icosahedral quasicrystals depends of the type of the six-dimensional lattice associated with the quasilattice. In six dimensions, there exists three Bravais lattices invariant under the icosahedral group (Martinals, 1987): simple cubic (P), face centered (F) and body centered (I). Defined as in (Conway and Sloane, 1988)

$$L_P = \left\{ \sum_{i=1}^{6} n_i a_i \mid n_i \in \mathbf{Z} \right\},$$

$$L_F = \left\{ \sum_{i=1}^{6} n_i a_i \mid \sum_{i=1}^{6} n_i = 0 \ (\text{mod } 2) \right\},$$

$$L_I = \left\{ \sum_{i=1}^{6} \frac{n_i}{2} a_i \mid n_i = n_j \ (\text{mod } 2) \right\}.$$

Quasicrystals belonging to two of these lattices have been identified experimentally. P icosahedral quasicrystals (such as Al_2CuLi_3) show the polyhedral shape of a rhombic triacontahedron (Dubost *et al.*, 1986) while the more likely shape of F quasicrystals (such as $Al_{63}Cu_{25}Fe_{12}$ and $Al_{70}Mn_9Pd_{21}$) is a pentagonal dodecahedron (Tsai *et al.*, 1987), (Beele and Nissen, 1993). Quasicrystals of type I have no yet been observed experimentally.

Bravais' Law of reticular density states that the important faces of a crystal should be those parallel to the densest lattice planes (which at the same time are those with the largest distances among themselves). It has been surprisingly successful in predicting the morphology of crystals of many compositions (Vainshtein, 1981). This idea can be extended to quasicrystals in a direct manner using the Cut and Projection Method described in Section 4.1. The assumption is made that the important quasilattice planes in quasicrystals are also those with the highest density and largest separation. Thus, a hypothesis can bee formulated that relevant planes in $\mathbb{R}^3$ are those that correspond (via projection) to the densest lattice hyperplanes in $\mathbb{R}^6$. We solved this problem for the six-dimensional lattices P, F and I using standard methods of linear algebra (Aragón *et al.*, 1995) and it is the aim of this section to show how this problem can be formulated and solved in a more transparent fashion using Clifford Algebra.

Our procedure can be summarized this way:

1. Consider a plane p of the quasilattice characterized by the 2-blade B_p.

2. Find the sublattice $L_W \subset L_p$ whose points project onto the plane p. The density ρ of points in p is directly related to the density of L_W.

3. Find the average density $\bar{\rho}$ of the family of planes parallel to p. The more dense this family of planes is, the more likely it is that a facet will appear and develop.

We present in detail only the case of planes invariant under the five-fold rotations (the five-fold planes) in a P quasilattice. Similar approach ma be applied to the other cases including quasilattices of types F and I.

A unitary P six-dimensional lattice is spanned by the set $\{a_1, a_2, \ldots, a_6\}$ where $a_i = e_i$, $i = 1, \ldots, 6$.

For the numbering of the vectors pointing to the vertices of an icosahedron, we will stick to the convention that e_6 is surrounded in cyclic order by $e_1, e_2, \ldots, e_5$ (see for instance Fig. 1 in (Aragón *et al.*, 1995)). According to this convention, a five-fold plane p is characterized by a 2-blade

$$B_p = (\alpha_3 - \alpha_2) \wedge (\alpha_1 - \alpha_2), \tag{4}$$

where

$$\alpha_3 - \alpha_2 = P^{\parallel}(e_3) - P^{\parallel}(e_2), \qquad \alpha_1 - \alpha_2 = P^{\parallel}(e_1) - P^{\parallel}(e_2).$$

Now consider the subspace $W \subset \mathbb{R}^6$ defined as

$$W = \text{span} \left\{ \alpha_3 - \alpha_2, \alpha_1 - \alpha_2, \alpha_3^{\perp} - \alpha_2^{\perp}, \alpha_1^{\perp} - \alpha_2^{\perp}, \alpha_6^{\perp} \right\},$$

where α_1, α_2 are as before and $\alpha_i^{\perp} = P^{\perp}(e_i)$, $i = 1, 2, 3, 6$. Since $P^{\parallel}(\alpha_i) = \alpha_i$, and $P^{\parallel}(\alpha_i^{\perp}) = 0$, all vectors in W project onto the plane characterized by (4). Then the five-dimensional subspace W projects onto the five-fold plane of the quasilattice. The sublattice $L_W \subset L_p$ is therefore

$$L_W = W \cap L_P.$$

A basis for L_W can be obtained by noticing that a vector $\mathbf{x} = \sum_{i=1}^{6} x_i e_i$ lies in L_W if and only if x_i are integers and if

$$B_W \wedge \mathbf{x} = 0, \tag{5}$$

where $B_W = (\alpha_3 - \alpha_2) \wedge (\alpha_1 - \alpha_2) \wedge \left(\alpha_3^{\perp} - \alpha_2^{\perp} \right) \wedge \left(\alpha_1^{\perp} - \alpha_2^{\perp} \right) \wedge \alpha_6^{\perp}$.

With respect to the orthonormal basis $\{e_i\}_{i=1}^{6}$, the components of vectors α_i and $\alpha_i^{\perp}$ form the i-th column of the projection matrix $P^{\parallel}$ and $P^{\perp}$, respectively (see Eq. 2). Therefore, simple but tedious calculations are needed to solve (5). We shall use `Clifford.m` to achieve this.

Here are the projection matrices $P^{\parallel}$ and $P^{\perp}$:

```
In[1] := a = 1/Sqrt[20];
In[2] := b = 1/2;
In[3] := para = { {b,a,-a,-a,a,a},{a,b,a,-a,-a,a},{-a,a,b,a,-a,a},
                  {-a,-a,a,b,a,a},{a,-a,-a,a,b,a},{a,a,a,a,a,b} };
In[4] := perp = IdentityMatrix[6] - para.
```

This defines the 5-blade B_W :

```
In[5]  := <<'clifford'
In[6]  := t1 = ToBasis[ Transpose[para][[3]]-Transpose[para][[2]] ];
In[7]  := t2 = ToBasis[ Transpose[para][[1]]-Transpose[para][[2]] ];
In[8]  := t3 = ToBasis[ Transpose[perp][[3]]-Transpose[perp][[2]] ];
In[9]  := t4 = ToBasis[ Transpose[perp][[1]]-Transpose[perp][[2]] ];
In[10]:= t5 = ToBasis[ Transpose[perp][[6]]];
In[11]:= Bw = Oup[ t1,t2,t3,t4,t5]
             -( e[1] e[2] e[3] e[4] e[5] )   e[1] e[2] e[3] e[4] e[6]
Out[11] := --------------------------- + --------------------------- -
                      2 Sqrt[5]                        10
             e[1] e[2] e[3] e[5] e[6]   e[1] e[2] e[4] e[5] e[6]
             ----------------------- + ----------------------- -
                        10                        10
             e[1] e[3] e[4] e[5] e[6]   e[2] e[3] e[4] e[5] e[6]
             ----------------------- + -----------------------
                        10                        10
```

The function ToBasis transforms a vector from the list notation used by *Mathematica* into a linear combination of the vectors e[i], and Oup is an alias for OuterProduct.

The left hand side of (5) is

```
In[12]:= Oup[Bw,ToBasis[{x1,x2,x3,x4,x5,x6}]] // Simplify
Out[12]:= -((x1 + x2 + x3 + x4 + x5 + Sqrt[5] x6) e[1] e[2]
             e[3]e[4]e[5] e[6]) / 10
```

Consequently, (5) is fulfilled if $x_1 + x_2 + x_3 + x_4 + x_5 + \frac{1}{\sqrt{5}}x_6 = 0$, i.e., $x_6 = 0$ and $x_1 + x_2 + x_3 + x_4 + x_5 = 0$. So, a basis for L_W is given by

$$\{(1,0,0,0,-1,0),(0,1,0,0,-1,0),(0,0,1,0,-1,0),(0,0,0,1,-1,0)\}$$

or

$$\{e_1 - e_5, e_2 - e_5, e_3 - e_5, e_4 - e_5\}. \tag{6}$$

As we can see, the lattice L_W that projects onto the five-fold plane characterized by B_p is four-dimensional. One can obtain the pattern of projected points onto this five-fold plane by the Cut and Projection Method in L_W (Katz and Duneau, 1986), (Aragón *et al.*, 1995).

There is a two-dimensional lattice $L_{W\perp}$ orthogonal to L_W which projects onto a one-dimensional space orthogonal to the quasilattice plane, i.e., it projects onto a space spanned by the normal to the five-fold plane. By considering the 4-blade

$$B_W = (e_1 - e_5) \wedge (e_2 - e_5) \wedge (e_3 - e_5) \wedge (e_4 - e_5),$$

a 2-blade $B_{W\perp}$ that characterizes a two-dimensional space containing $L_{W\perp}$ can be found. It must satisfy the condition $B_W B_{W\perp} = \lambda e_{12\ldots 6}$ where $\lambda \in \mathbb{R}$. Therefore

$$B_{W\perp} = \frac{\lambda e_{12\ldots 6}}{B_W}$$

and it can be calculated as follows:

```
In[14]:= Bw = Oup[ e[1]-e[5], e[2]-e[5], e[3]-e[5], e[4]-e[5] ] ;
In[14]:= BwPerp = lambda Gp[ e[1]e[2]e[3]e[4]e[5]e[6], Mvi[Bw] ]
                 e[1] e[6]    e[2] e[6]    e[3] e[6]    e[4] e[6]
Out[14]:= lambda ( --------- + --------- + --------- + --------- +
                      5            5            5            5
                 e[5] e[6]
                 --------- )
                    5
```

where **Mvi** is the alias for **MultivectorInverse**. The last result can be rewritten as

$$B_{W\perp} = \frac{\lambda}{5}\,(e_1 + e_2 + e_3 + e_4 + e_5)\,e_6 = \frac{\lambda}{5}\,(e_1 + e_2 + e_3 + e_4 + e_5) \wedge e_6,$$

so a basis for $L_{W\perp}$ is

$$\{e_1 + e_2 + e_3 + e_4 + e_5, e_6\}. \tag{7}$$

The Cut and Projection Method in $L_{W\perp}$ gives the pattern of vertices along the normal to the five-fold plane that represents the sequence of separations between planes. For our purposes, all that matters is the lattice L_W generated by (6), whose points project onto the five-fold plane and which contains information about the density of vertices in this plane.

The procedure described above was applied also to the six-dimensional lattices of types F and I. Calculations were made to determine morphological importance of high symmetry planes, i.e., five-fold, three-fold and two-fold facets. Results for the three type of the six-dimensional lattices are summarized in Table 1 which gives generators of three different quasilattice planes in three dimensions and a basis for the four-dimensional lattices L_W associated with each plane in terms of the orthonormal basis.

For a given type of lattice (L_P, L_F or L_I) and for a given type of facet (two-fold, three-fold, five-fold or any other), the average occupation density (number of vertices per unit area) of the family of planes parallel to the facet is inversely proportional to the volume $V(L_W)$ of the primitive cell of the four-dimensional lattice L_W that projects onto one plane of the family [1]:

$$\bar{\rho} \propto \frac{1}{V(L_W)}. \tag{8}$$

From the generators of each L_W in Table 1, the volume of the unit cell in each case can be calculated using Eq. 1. Table 2 shows volumes of the unit cells calculated for the three families of planes of the three quasilattices L_P, L_F and L_I. The average density $\bar{\rho}$ has been calculated using the approximation (8) and is given in the last column.

From Table 2, the following predictions about the ideal shape of the icosahedral quasilattices can be made:

---

[1] A formal derivation of the exact expression of $\bar{\rho}$, taking into account the Cut Method in L_W, is presented in (Aragón *et al.*, 1995)

TABLE I

Generators of the primitive cells of lattices L_W for the three possible lattices in $\mathbb{R}^6$.

Lattice	Direction	Generators in 3D	Generators of Λ_W
P	2-fold	$\{\alpha_1, \alpha_2\}$	$\{e_1, e_2, e_3 - e_5, e_4 - e_6\}$
	3-fold	$\{\alpha_1 + \alpha_2, \alpha_2 + \alpha_3\}$	$\{e_1 - e_3, e_2 + e_3, e_4 - e_6, e_5 - e_6\}$
	5-fold	$\{\alpha_3 - \alpha_2, \alpha_1 - \alpha_2\}$	$\{e_1 - e_5, e_2 - e_5, e_3 - e_5, e_4 - e_5\}$
F	2-fold	$\{-\alpha_1 - \alpha_2, \alpha_1 - \alpha_2\}$	$\{-e_1 - e_2, e_1 - e_2,$ $e_3 - e_4 - e_5 + e_6, e_4 - e_6\}$
	3-fold	$\{-\alpha_1 - \alpha_2, \alpha_5 - \alpha_6\}$	$\{-e_1 - e_2, e_1 - e_3, e_4 - e_5, e_5 - e_6\}$
	5-fold	$\{\alpha_1 - \alpha_2, \alpha_2 - \alpha_3\}$	$\{e_1 - e_2, e_2 - e_3, e_3 - e_4, e_4 - e_5\}$
I	2-fold	$\{\alpha_1, \alpha_2\}$	$\{e_1, e_2, e_3 - e_5,$ $(-e_1 - e_2 - e_3 + e_5 + e_5 - e_6)/2\}$
	3-fold	$\{\alpha_1 + \alpha_2, \alpha_2 + \alpha_3\}$	$\{-e_2 - e_3 - e_4 + 2e_5 - e_6,$ $e_1 + 2e_2 + e_3 + e_4 - 2e_5 + e_6,$ $-e_1 - e_2 - e_4 + 2e_5 - e_6, e_4 - e_5\}$
	5-fold	$\{\alpha_3 - \alpha_2, \alpha_1 - \alpha_2\}$	$\{e_1 - e_5, e_2 - e_5, e_3 - e_5, e_4 - e_5\}$

TABLE II

Volumes of the primitive cells of lattices L_W and average density of planes for each case.

Lattice	Direction	$V(\Lambda_W)$	$\bar{\rho}$
P	2-fold	2	0.500
	3-fold	3	0.333
	5-fold	$\sqrt{5}$	0.447
F	2-fold	4	0.250
	3-fold	3	0.333
	5-fold	$\sqrt{5}$	0.447
I	2-fold	1	1.000
	3-fold	3	0.333
	5-fold	$\sqrt{5}$	0.447

— In the case P, the largest occupation density is carried by the planes with a normal along the two-fold axis of the icosahedron. The polyhedral shape should have big two-fold facets resembling a rhombic triacontahedron.

— In the case F, the densest planes are those of the five-fold family. In this case we have a polyhedron with big five-fold facets. The ideal shape is therefore a pentagonal dodecahedron.

— In the yet unobserved case I we have the same result as in the P case but with larger two-fold facets. This resembles also a rhombic triacontahedron.

All these prediction are in agreement with experimental observations (Dubost *et al.*,

1986), (Tsai *et al.*, 1987), (Beele and Nissen, 1993).

7. Transversality of atomic surfaces

Let us consider again quasicrystals as obtained by the Cut Method. As discussed in 4.2, physical assumptions impose restrictions on the shape and properties of the atomic surfaces. Here we will be concerned with such restrictions.

It is assumed that each atomic surface has a one-to-one projection on the phason space $E^\perp$. It is also required that there exists a constant $h > 0$ such that for any two atoms their distance is greater than h. We will also restrict our attention to the case where the atomic surfaces are continuous.

Under these circumstances it can be shown that the atomic surfaces are plane-like, i.e. they stay within finite distance from an hyper-plane (manifold) of dimension 3.

7.1. ROLE OF SYMMETRY

If the quasicrystal is invariant under a space group G, then given an atomic surface s it may happen that the new surface s' (resulting from s by acting with an element $g \in G$), cuts or intersects s. This leads to unphysical results since then two atoms could approach arbitrarily. Hence the surfaces cannot intersect, a restriction that receives formal expression in terms of the concept of transversality.

7.2. TRANSVERSALITY

Let E be a vector space and V, W, subspaces. For $\mathbf{x}_0 \in V$ and $\mathbf{y}_0 \in V$ define manifolds

$$A = \mathbf{x}_0 + V = \{\mathbf{p} \in E \mid \mathbf{p} = \mathbf{x}_0 + \mathbf{v}, \, \mathbf{v} \in V\},$$

$$A = \mathbf{y}_0 + W = \{\mathbf{p} \in E \mid \mathbf{p} = \mathbf{y}_0 + \mathbf{w}, \, \mathbf{w} \in W\}.$$

The manifolds are said to be *transverse* if $\forall \mathbf{x}_0, \mathbf{y}_0 \in E$, $A \cap B \neq \phi$. Then A and B are transverse if and only if $\forall \mathbf{x}_0, \, \mathbf{y}_0 \in E$ there exist $\mathbf{v} \in V$ and $\mathbf{u} \in W$ such that

$$\mathbf{x}_0 - \mathbf{y}_0 = -\mathbf{v} + \mathbf{u},$$

or, equivalently,

$$E = V + W.$$

In the applications we are interested in, only the case where $\dim V + \dim W = \dim E$ matters; so, A and B are not transverse if and only if $\dim(V \cap W) > 0$ or $\dim(V + W) < \dim E$.

If we now call ξ the volume element of E, v the volume element of V and ω the volume element of W, the manifolds A and B are not transverse if and only if

$$v \wedge \omega = 0. \tag{9}$$

It should be remarked that in the present case it is always true that

$$v \wedge \omega = \lambda \xi,$$

for some scalar λ. λ is zero in the non-transverse case and it is not zero in the general, transverse case.

For "physical" quasicrystals it is always required that their approximating manifolds be non-transverse and Equation 9 must hold.

7.3. AN EXAMPLE

Frenkel (Frenkel and Gómez, 1991) has considered a simple case of a two-dimensional quasicrystal with octagonal symmetry that is a cut from a periodic structure in $\mathbb{R}^4$. In that case, the physical space $E^{\parallel}$ is spanned by

$$\mathbf{T}_1 = (1, 2^{\frac{1}{2}}, 1, 0), \qquad \mathbf{T}_2 = (-1, 0, 1, 2^{\frac{1}{2}}),$$

and the phason space by

$$\mathbf{n}_1 = (2^{\frac{1}{2}}, -1, 0, 1), \qquad \mathbf{n}_2 = (0, 1, -2^{\frac{1}{2}}, 1).$$

In this case the group G that leaves both spaces invariant is generated by

$$R = \begin{bmatrix} 0 & 1 & 0 & 0 \\ 0 & 0 & 1 & 0 \\ 0 & 0 & 0 & 1 \\ -1 & 0 & 0 & 0 \end{bmatrix}.$$

The condition that the manifolds spanned by $\{\mathbf{t}_1, \mathbf{t}_2\}$ and $\{R^k(\mathbf{t}_1), R^k(\mathbf{t}_2)\}$ are not transverse becomes

$$\mathbf{t}_1 \wedge \mathbf{t}_2 \wedge R^k(\mathbf{t}_1) \wedge R^k(\mathbf{t}_2) = 0.$$

Here $\mathbf{t}_1, \mathbf{t}_2$ are arbitrary vectors and k can be $1, 2, \ldots, 8$ since R generates a cyclic group of order 8.

As explained by Frenkel (1991) in this case only $k = 1$ and $k = 2$ give rise to independent equations so the non-transversality amounts to

$$\mathbf{t}_1 \wedge \mathbf{t}_2 \wedge R(\mathbf{t}_1) \wedge R(\mathbf{t}_2) = 0 \quad \text{and} \quad \mathbf{t}_1 \wedge \mathbf{t}_2 \wedge R^2(\mathbf{t}_1) \wedge R^2(\mathbf{t}_2) = 0.$$

When the various quantities are substituted into the above equation, one obtains exactly the same equations other authors have obtained by other means. The important point to stress here is that the problem can be expressed and solved completely in Clifford algebra terms. Actually the various quantities that appear in the standard treatments are nothing but components of bivectors and various multivectors. In the literature one finds constantly "Plücker coordinates" (which in our example are the components of $\mathbf{t}_1 \wedge \mathbf{t}_2$) in disguise. A Clifford algebra treatment is clearer in that the geometric relevance of all quantities involved is never lost (the equations above have only solutions in which $\{\mathbf{t}_1, \mathbf{t}_2\} = \{\mathbf{T}_1, \mathbf{T}_2\}$ or $\{\mathbf{t}_1, \mathbf{t}_2\} = \{\mathbf{n}_1, \mathbf{n}_2\}$, the relevance of this to quasicrystallography is that there are no rational solutions).

8. Conclusions

Many crystallographic calculations in spaces of dimension other than three can be conveniently performed using a Clifford algebra. Even in two-dimensional spaces there are some advantages of using such an algebra.

Besides ease of manipulation, Clifford algebras are useful in crystallography because they are ideally suited to the task of exhibiting geometrical relationships.

As a matter of fact, some calculations already in the literature are Clifford algebra manipulations in disguise with certain determinants in place of wedge products and homogeneous coordinates (Plücker coordinates) instead of multivectors.

A *Mathematica* package for performing all the required computations has been developed. Among other things, such a package is relevant because calculations even in $\mathbb{R}^3$ are cumbersome.

Acknowledgments. This work is supported by CONACYT (México) under grants 3348-E93, 5165-E and 1759-E.

References

J.L. Aragón, F. Dávila and A. Gómez: 1995, 'Prediction of the external shape of ideal icosahedral quasicrystals', *Phys. Rev.* **B51**, pp. 857 – 863.

C. Beeli and H.-U. Nissen: 1993, 'Growth morphology of isosahedral Al–Mn–Pd single quasicrystals', *Phil. Mag.* **B68**, pp. 487 – 512.

J.H. Conway and N.A. Sloane: 1988, *Sphere Packings, Lattices and Groups.* Springer-Verlag, New York.

B. Dubost, J.M. Lang, M. Tanaka, P. Sainfort and M. Audier: 1986, 'Large Al–Cu–Li single quasicrystals with triacontahedral solidification morphology', *Nature* **324**, pp. 48 – 50.

D.M. Frenkel and A. Gómez: 1991, 'Impossibility of continuous deterministic phasons in octagonal quasicrystals,' *Phys. Rev.* **B43**, pp. 10751 – 10754.

D. Hestenes and G. Sobczyk: 1984, *Clifford Algebra to Geometric Calculus.* D. Reidel Publishing Co., Holland.

C. Janot: 1992, *Quasicrystals. A Primer.* Clarendon Press, Oxford.

T. Jansen, A. Janner, P. Bennema: 1989, 'On the morphology of quasicrystals,' *Phil. Mag.* **B59**, pp. 233 – 242.

P.A. Kalugin, A.Y. Kitayev and L.S. Levitov: 1985, '$Al_{0.86}Mn_{0.14}$: a six-dimensional crystal', *JETP Lett.* **41**, pp. 145 – 149.

A. Katz and M. Duneau: 1986, 'Quasiperiodic patterns and icosahedral symmetry', *J. Phys. France* **47**, pp. 181 – 186.

D. Martinais: 1987, *C.R. Acad. Sci.* **305**, p. 509.

C. Oguey, M. Duneau and A. Katz: 1988, 'A geometrical approach to quasiperiodic tilings', *Commun. Math. Phys.* **118**, pp. 99 – 118.

D.M.Y. Sommerville: 1958, *An Introduction to the Geometry of N Dimensions.* Dover, New York.

A.P. Tsai, A. Inoue and T. Masumoto: 1987, 'A stable quasicrystal in Al–Cu–Fe system', *Jap. J. Appl. Phys.* **26**, pp. L1505 – L1507.

B.K. Vainshtein: 1981, *Modern Crystallography I. Symmetry of Crystals. Methods of structural crystallography*, Springer Series in Solid-State Sciences. Springer-Verlag, Berlin, Heidelberg, New York.

4

Numerical Methods in Clifford Algebras

ORTHONORMAL BASIS SETS IN CLIFFORD ALGEBRAS

G. BERGDOLT
Centre de Recherches Nucleaires, C.N.R.S.
B.P.28
F-67037 Strasbourg Cedex-2, France
e-mail: BERGDOLT@frcpn11.in2p3.fr

Abstract. Orthonormal basis sets define isomorphisms and automorphisms in Clifford algebras. Orthonormal basis sets (ONB) are defined as sets of multivectors satisfying scalar product relations. A FORTRAN program determining ONBs is described. It is shown that any simple Clifford algebra is isomorphic to the tensor product of a Clifford algebra $Cl_{m,m}$ and a Clifford algebra isomorphic to $\mathbb{R}$, $\mathbb{C}$ or $\mathbb{H}$. From the construction of matrix algebras isomorphic to $Cl_{m,m}$ given by the second FORTRAN program, matrix algebras with entries in $\mathbb{R}$, $\mathbb{C}$ or $\mathbb{H}$ can be used to construct isomorphisms to all simple Clifford algebras.

Key words: Isomorphisms, automorphisms, representations, classification of real Clifford algebras.

1. Introduction

Let $\mathbb{R}^{p,q}$ be a real vector space with a metric of signature (p, q) and A an associative algebra with unit. A Clifford map $\varphi : \mathbb{R}^{p,q} \to A$ is a linear homomorphism such that

$$(\varphi(X))^2 = g(X, X)I \tag{1}$$

where g is the metric tensor and $X \in \mathbb{R}^{p,q}$. From the point of view of category theory, Clifford algebra $Cl_{p,q}$ is the universal object with respect to Clifford maps (Atiyah *et al.*, 1964). The Clifford algebra is the unique algebra with an injective Clifford map $i : \mathbb{R}^{p,q} \to Cl_{p,q}$ such that an algebra homomorphism $\psi : Cl_{p,q} \to A$ exists for any algebra with a Clifford map. The mappings satisfy:

$$
\begin{array}{ccc}
 & i & \\
\mathbb{R}^{p,q} & \to & Cl_{p,q} \\
\varphi & \searrow & \downarrow \psi \\
 & A &
\end{array}
$$

The word *unique* in the definition of universal objects should be understood to mean *up to isomorphic algebras*. Isomorphism is a category dependent concept; in the case of algebras it means that the mapping should preserve the additive and multiplicative composition laws and commute with multiplication by a scalar. According to the classification in (Lounesto, 1981), real Clifford algebras can be assigned to five classes of algebras isomorphic to matrix algebras. In Section 2 the classification and factorization properties of real Clifford algebras are put in a form suitable for our purpose.

Projectors are defined as elements P_{ij} satisfying

$$P_{ij} P_{kl} = \delta_{jk} P_{il}. \tag{2}$$

The set of projectors generates a matrix algebra. In Section 2, a set of projectors is constructed for real Clifford algebras of neutral signature $Cl_{m,m}$.

Given a basis $\mathbf{e}_i$ $(i = 1, \ldots, n)$ in $\mathbb{R}^{p,q}$, the following relations hold:

$$\varphi(\mathbf{e}_i)\varphi(\mathbf{e}_j) + \varphi(\mathbf{e}_j)\varphi(\mathbf{e}_i) = 2g_{ij}\mathbf{1}. \tag{3}$$

In (3), 1 is the unit element of $Cl_{p,q}$. Relations (3) are referred to as *scalar product relations*. Assume now that the basis has been orthonormalized so that $g = \mathrm{diag}(+1, +1, \ldots, -1, -1, \ldots)$ with p times $+1$ and q times -1 (signature (p, q)). If we can find a set of elements w_i $(i = 1, \ldots, n)$ in algebra A such that

$$w_i w_j + w_j w_i = 2g_{ij}, \tag{4}$$

we have a Clifford map and hence an algebra homomorphism. If $A = Cl_{p',q'}$ is another Clifford algebra, scalar product relations can be realized by multivectors. We use the name multivector in the traditional sense of exterior product of vectors. Following (Dimakis, 1989), a set of multivectors w_i $(i = 1, \ldots, n)$ satisfying the scalar product relations (4) is called *an orthonormal set*. If in addition the w_i generate the algebra $Cl_{p',q'}$, the set is called *an orthonormal basis set* (abbreviated ONB). In the case of an ONB, the homomorphism of algebras resulting from a Clifford map is an isomorphism. The factorization theorems of Clifford algebras as tensor products of lower dimensional Clifford algebras follow from the construction of commuting sets of ONB's (Dimakis, 1989).

In Section 3 all possible orthonormal basis sets for Clifford algebras with 4 dimensional vector spaces are determined by a FORTRAN program.

Conclusions are drawn in Section 4.

2. Classification of real Clifford algebras

Let w_i $(i = 1, \ldots, n)$ be an orthonormal basis set. The w_i generate a finite group defined by:

$$\begin{aligned}
w_i w_j &= -w_j w_i \\
w_i^2 &= +1 \text{ for } i = 1, \ldots, p \\
w_j^2 &= -1 \text{ for } j = p+1, \ldots, p+q.
\end{aligned} \tag{5}$$

Since opposite multivectors are distinct elements of the group, the group is of order 2^{n+1}. These groups have been studied by (Salingaros, 1981) who found a connection with extra-special p-groups, and by (Braden, 1985) who used the factorization properties of extra-special p-groups to establish the classification of real Clifford algebras. Since no definite name has emerged we call these groups *multivector groups*. A classification of multivector groups can be derived from the theory of extra-special p-groups. A p-group is a finite group of order p^k where p is a prime and $k > 1$ is an integer. A p-group is *extra-special* if its center is of order p and it coincides with the commutator subgroup. The main result is that extra-special p-groups can be factored as central products of non abelian p-groups of order p^3. In central products the centers are amalgamated.

From relations (5) it can be seen that the multivector group of a simple Clifford algebra is an extra-special 2-group. The factorization properties of real Clifford algebras correspond to the factorization properties of extra-special 2-groups. Clifford algebras can be considered as group algebras provided the definition of a group algebra is extended as follows: we require that in a conventional group algebra the group unit e and the field unit 1 are identified. To obtain a Clifford algebra the elements e and $-e$ of the extra-special 2-group are identified with $+1$ and -1 of $\mathbb{R}$. The elements $\{e, -e\}$ and $\{1, -1\}$ define isomorphic central subgroups in $G^{p,q}$ and $\mathbb{R}^*$ and the identification of these subgroups corresponds to amalgamated group products (Serre, 1983).

Classification Theorem: Real Clifford algebras $C\ell_{p,q}$ belong to 5 classes of algebras labelled by an index $\beta = 0, 1, 2, 3, 4$ related to the signature (p, q) by

$$p - q - 1 = 8k \pm \beta \tag{6}$$

These classes of algebras are isomorphic to the matrix algebras $2^\sigma \mathrm{Mat}(d, K_\eta)$ where

- $\sigma = 0, 1$ is a simplicity parameter such that the algebra is the direct sum of 2^σ simple algebras: $\sigma = 1$ if $\beta = 0, 4$ and $\sigma = 0$ otherwise,
- η is a field parameter such that K_η ($\eta = 0, 1, 2$) denotes $\mathbb{R}, \mathbb{C}, \mathbb{H}$; $\eta = \beta - 1$ if $\sigma = 0$, and $\eta = \beta/2$ if $\sigma = 1$,
- $d = 2^\xi$, where $\xi = (n - \eta - \sigma)/2$, is the dimension of the matrices. ∎

The -1 in the left hand side of (6) indicates a shift of the symmetry axis of the algebra structure with respect to $p - q$. The occurrence of this term is due to the fact that the field $\mathbb{R}$ is part of the Clifford algebra and constitutes a one dimensional vector space not taken in account in (p, q). Let $\alpha \in \mathbb{R}$ and $x \in V$. Then, the square of $X = \alpha + x \in C\ell_{p,q}$ is

$$X^2 = (\alpha + x)^2 = \alpha^2 + 2\alpha x + g(x, x) = 2\alpha X + g(x, x) - \alpha^2.$$

Since the scalar part of X^2 is $\beta = g(x, x) - \alpha^2 \in \mathbb{R}$, the vector space $\mathbb{R}$ is to be counted as a negative square. If we set $\alpha' = 2\alpha$ in the relation above then $X^2 = \alpha' X + \beta$ is the defining relation of quadratic algebras. Note that not all elements of a Clifford algebra satisfy the quadratic relation.

Table I.
Factorization of real Clifford algebras

β	$C\ell_{p,q}$	$G^{p,q}$
0	$C\ell_{1,1} \otimes \cdots \otimes C\ell_{1,1} \otimes C\ell_{1,0}$	$D_4 \circ \cdots \circ D_4 \circ (Z_2 \oplus Z_2)$
1	$C\ell_{1,1} \otimes \cdots \otimes C\ell_{1,1}$	$D_4 \circ \cdots \circ D_4$
2	$C\ell_{1,1} \otimes \cdots \otimes C\ell_{1,1} \otimes C\ell_{0,1}$	$D_4 \circ \cdots \circ D_4 \circ Z_4$
3	$C\ell_{1,1} \otimes \cdots \otimes C\ell_{1,1} \otimes C\ell_{0,2}$	$D_4 \circ \cdots \circ D_4 \circ Q_4$
4	$C\ell_{1,1} \otimes \cdots \otimes C\ell_{1,1} \otimes C\ell_{0,2} \otimes C\ell_{1,0}$	$D_4 \circ \cdots \circ D_4 \circ Q_4 \circ (Z_2 \oplus Z_2)$

Table I shows tensor product factorizations of real Clifford algebras and central product factorization of multivector groups $G^{p,q}$. The number of factors in the dotted products is $\xi = (n - \eta - \sigma)/2$.

The dihedral group D_4 is generated by $\{a, b, e\}$ with relations $a^4 = b^2 = (ab)^2 = e$. It is not difficult to see that it is the multivector group of $Cl_{1,1}$.

The quaternion group Q_4 is generated by $a^4 = b^4 = (ab)^4 = e$ and it is the multivector group of $Cl_{0,2}$.

The group Z_4 is the group of integers under addition modulo 4 and it is obviously isomorphic to a cyclic group generated by i. It is the multivector group of $Cl_{0,1}$.

Recall that $\mathbb{R}$, $\mathbb{C}$ and $\mathbb{H}$ are Clifford algebras $Cl_{0,0}$, $Cl_{0,1}$ and $Cl_{0,2}$, respectively. The factorization in Table I is derived from (Dimakis, 1989) and (Braden, 1985).

In the next section we describe a construction of matrix algebras isomorphic to real Clifford algebras $Cl_{m,m}$ with neutral signature. Since

$$Cl_{1,1} \otimes \cdots \otimes Cl_{1,1} \simeq Cl_{\xi,\xi} \simeq \text{Mat}(d, \mathbb{R})$$

we have the following well known result: simple Clifford algebras are isomorphic to the real tensor product $\text{Mat}(d, \mathbb{R}) \otimes K_\eta$, i.e., to matrix algebras with entries in $\mathbb{R}$, $\mathbb{C}$ or $\mathbb{H}$. Note that the program IMTV described in Appendix A computes matrices of $\text{Mat}(d, \mathbb{R}) \simeq Cl_{m,m}$.

For comparison, we recall the groups studied in connection with Clifford algebras:

a) the group of invertible elements called *the group of units*,

b) the subgroup of the group of units whose elements preserve vectors under the morphism $x \to gx\hat{g}^{-1}$ called *Lipschitz group*,

c) the **Pin** group whose elements have spinor norm ± 1,

d) the **Spin** group intersection of the **Pin** group with the even Clifford subalgebra.

These groups have been extended to degenerate Clifford algebras by (Ablamowicz, 1986). The multivector group is obviously a subgroup of the group of units.

2.1. PROJECTORS OF $Cl_{m,m}$ ALGEBRAS

It has been pointed out in (Schönberg, 1957) that a simple or full matrix algebra is generated by a set of elements P_{ij} $(i, j = 1, \ldots, m)$ called projectors satisfying the relations:

$$P_{ij} P_{kl} = \delta_{jk} P_{il} \tag{7}$$

These projectors have the following properties:

a) the m^2 projectors form a basis of the algebra,

b) the projectors P_{ij}, $j = 1, \ldots, m$ and i fixed, span a right ideal,

c) the projectors P_{ij}, $i = 1, \ldots, m$ and j fixed, span a left ideal.

The properties follow from the product relations (7). In the case of a real vector space with signature (m, m), let $\{e_i, e_i'\}$, $(i = 1, \ldots, m)$ be basis vectors satisfying $e_i'^2 = -1$ and $e_i^2 = 1$. A set of projectors can be obtained by a construction due to (Schönberg, 1957). The projectors are Clifford products of isotropic vectors $I_i^+ = (e_i + e_i')/2$ and $I_i^- = (e_i - e_i')/2$. The $\{I_i^+\}$ (resp. $\{I_i^-\}$) span maximal completely isotropic subspaces. Projectors are defined by:

$$P_{i_1 \ldots i_r, j_1 \ldots j_s} = I_{i_r \ldots i_1}^- I_{1 \ldots m}^+ I_{m \ldots 1}^- I_{j_1 \ldots j_s}^+ \tag{8}$$

where $I_{i_r \ldots i_1}^-$ denotes the multivector $I_{i_r}^- \cdots I_{i_1}^-$. The projectors defined by (8) satisfy the product relations (7) since the following relations hold

$$I_{m \ldots 1}^- I_{j_1 \ldots j_s}^+ I_{i_r \ldots i_1}^- I_{1 \ldots m}^- = \delta_{rs} \delta_{j_1 i_1} \cdots \delta_{j_s i_s} I_{m \ldots 1}^- I_{1 \ldots m}^+$$

and

$$I_{m \ldots 1}^- I_{1 \ldots m}^+ I_{m \ldots 1}^- I_{1 \ldots m}^+ = I_{m \ldots 1}^- I_{1 \ldots m}^+.$$

The following expansion of projectors can be derived (Bergdolt, 1988):

$$P_{i_r \ldots i_1, j_1 \ldots j_s} = \sum (-1)^t I_{i_r \ldots i_1}^- I_{k_t \ldots k_1}^- I_{k_1 \ldots k_t}^+ I_{j_1 \ldots j_s}^+. \tag{9}$$

The summation is over sets of indices $\{k_1, \ldots, k_t\}$ restricted by

$$\{k_1, \ldots, k_t\} \cap \{i_1, \ldots, i_r\} = \{0\}, \{k_1, \ldots, k_t\} \cap \{j_1, \ldots, j_s\} = \{0\}.$$

The isotropic vectors can be permuted to standard order, i.e., decreasing for the I^- vectors and increasing for the I^+ vectors so that projectors can be expressed as linear combinations of these ordered products. The projectors for the Clifford algebra $C\ell_{2,2}$ expressed as linear combination of isotropic multivectors are given in (Schönberg, 1957).

The isotropic multivectors $I_{i_r \ldots i_1}^-$ or $I_{j_1 \ldots j_s}^+$ can be represented by an m-tuple of binary numbers $\alpha = \{\alpha_0, \ldots, \alpha_{m-1}\}$ where $\alpha_i = 1$ if I_i is present in the multivector and $\alpha_i = 0$ otherwise. If the multi-indices in the relations above are replaced by the m-tuples, relation (9) can be written as

$$P_{\beta\gamma} = \sum_\alpha (-1)^{d(\alpha) + s(\alpha,\beta) + s(\alpha,\gamma)} I_{\alpha \cup \beta}^- I_{\alpha \cup \gamma}^+. \tag{10}$$

The sum is over the m-tuples α such that $\alpha \cap \beta = 0$ and $\alpha \cap \gamma = 0$. The set theoretic $\cup$ is used to denote the logical 'exclusive or'. This function is implemented in computers by the bit handling function IEOR. The set theoretic intersection $\cap$ is implemented by the function IAND. The degree $d(\alpha)$ of a multivector is equal to $\sum_i \alpha_i$. The function $s(\alpha, \beta) = 0, 1$ is a sign function due to the rearrangement of vectors in normal order. It can be evaluated to be $s(\alpha, \beta) = \sum_{i > j} \alpha_i \beta_j$.

The multivector products $I_\beta^- I_\gamma^+$ form a basis of the algebra. Hence the relations (10) can be inverted, the result is (Bergdolt, 1988)

$$I_\beta^- I_\gamma^+ = \sum_\alpha P_{\alpha \cup \beta, \alpha \cup \gamma} (-1)^{s(\alpha,\beta) + s(\alpha,\gamma)}. \tag{11}$$

The sum is over the values of α such that $\alpha \cap \beta = 0$ and $\alpha \cap \gamma = 0$.

Example 1: For $m = 1$ we have one isotropic vector I^- and one vector I^+. For $\beta = \{1\}, \gamma = \{0\}$ we have

$$I^- = (I^-)^1 (I^+)^0 = P_{1,0} = \begin{pmatrix} 0 & 0 \\ 1 & 0 \end{pmatrix}.$$

For $\beta = \{0\}, \gamma = \{1\}$ we obtain

$$I^+ = (I^-)^0 (I^+)^1 = P_{0,1} = \begin{pmatrix} 0 & 1 \\ 0 & 0 \end{pmatrix}.$$

The computation of the matrices of isotropic multivector products for a given m is implemented in the FORTRAN program IMTV listed in Appendix A.

Example 2: We obtain from the program IMTV the 4 isotropic vectors of $C\ell_{2,2}$:

$$I_1^- = P_{1,0} - P_{3,2} = \begin{pmatrix} 0 & 0 & 0 & 0 \\ 1 & 0 & 0 & 0 \\ 0 & 0 & 0 & 0 \\ 0 & 0 & -1 & 0 \end{pmatrix}, \quad I_2^- = P_{2,0} + P_{3,1} = \begin{pmatrix} 0 & 0 & 0 & 0 \\ 0 & 0 & 0 & 0 \\ 1 & 0 & 0 & 0 \\ 0 & 1 & 0 & 0 \end{pmatrix},$$

$$I_1^+ = P_{0,1} - P_{2,3} = \begin{pmatrix} 0 & 1 & 0 & 0 \\ 0 & 0 & 0 & 0 \\ 0 & 0 & 0 & -1 \\ 0 & 0 & 0 & 0 \end{pmatrix}, \quad I_2^+ = P_{0,2} + P_{1,3} = \begin{pmatrix} 0 & 0 & 1 & 0 \\ 0 & 0 & 0 & 1 \\ 0 & 0 & 0 & 0 \\ 0 & 0 & 0 & 0 \end{pmatrix}.$$

The matrix calculus of any computer algebra system can then be used to check that the scalar product relations of the isotropic vectors are satisfied:

$$I_i^\sigma I_j^\tau + I_j^\tau I_i^\sigma = \delta_{ij} \delta(\sigma + \tau).$$

Here $\sigma, \tau = \pm$ and $\delta(\sigma + \tau) = 1$ if the two signs are opposite.

3. Orthonormal basis sets

Orthonormal sets (Dimakis, 1989) are defined as sets of n elements w_i, ($i = 1, \ldots, n$) satisfying
$$w_i w_j + w_j w_i = 2g_{ij}$$
where g is a diagonal matrix with p entries $+1$ and q entries -1. The set is a basis set if w_i generate $C\ell_{p,q}$. Let e_i, $i = 0, \ldots, n - 1$, be an orthonormalized vector basis. Then, w_i can be multivectors defined by an n-tuple of binary numbers:

$$w^\alpha = (e_0)^{\alpha_0} \cdots (e_{n-1})^{\alpha_{n-1}}.$$

The indexing starts from 0 in agreement with computer practice. We work out the sign factor in

$$w^\alpha w^\beta = (-1)^\sigma w^\beta w^\alpha.$$

and obtain:

$$\sigma(\alpha, \beta) = \sum_i \alpha_i \sum_i \beta_i - \sum_k \alpha_k \beta_k. \tag{12}$$

The result is independent of the signature of the basis vectors e_i. In order to determine the orthogonal sets of multivectors, program ONBS was written. The program tests all sets of n multivectors for mutual anticommutation using (12). The n-tuples corresponding to an ONB can be arranged as a matrix of binary numbers and it is easily seen that the condition (12) remains valid if rows or columns of the matrix are permuted. This corresponds to permutation of the multivectors or basis vectors. The program tests all possible matrices up to permutations of rows and/or columns. An orthogonal set is a basis set if the basis vectors can be recovered by products of multivectors. The existence of an n-tuple $\{\beta_0, \ldots, \beta_{n-1}\}$ such that

$$(w^{\alpha^0})^{\beta_0}(w^{\alpha^1})^{\beta_1} \cdots (w^{\alpha^{n-1}})^{\beta_{n-1}} = e_i \tag{13}$$

where e_i is a basis vector is sufficient for the set $\{w^{\alpha^0}, \ldots, w^{\alpha^{n-1}}\}$ to be a basis set. Hence we obtain:

An orthogonal set defined by matrix α is a basis set if α is invertible.

More precisely, the above result is valid if a matrix β exists such that $\beta\alpha$ is a permutation matrix. A permutation matrix is a matrix with a unique 1 in each row and column. In our program all multivectors are tested for condition (13) until a set is found such that $\beta\alpha$ is a permutation matrix.

The square of a multivector $w = e_{i_1} \cdots e_{i_k}$ is

$$w^2 = (-1)^{k(k-1)/2}(e_{i_1})^2 \cdots (e_{i_k})^2$$

and it obviously depends on the squares of the basis vectors. The signs of the squared multivectors for all combinations of signs for the squares of basis vectors are computed in the program ONBS. For Clifford algebras $Cl_{p,q}$ with $p + q = 4$ the output is given in Table II. The matrices to the left define the multivector sets as explained above. The 11 solutions cannot by reduced to a smaller set by permutation of rows and/or columns. The signs of the squared multivectors are listed as a function of the squares of basis vectors given in item 1. It can be seen that any of the 16 combinations of signs can be found for each ONB.

A consequence of the classification theorem is that $|p - q - 1|$ (mod 8) evaluated for a vertical quadruplet of signs is constant in each column. The signatures $(0,4)$, $(4,0)$ and $(1,3)$ correspond to the class $\beta = 3$. According to Table I these algebras can be factored and we have $Cl_{0,4} \simeq Cl_{4,0} \simeq Cl_{1,3} \simeq Cl_{1,1} \otimes Cl_{0,2}$. The conditions for a set of multivectors to span a factorization are different from the conditions to constitute an ONB. In the case at hand the multivectors should split into two commuting ONBs of signatures $(1,1)$ and $(0,2)$. A FORTRAN program to determine these sets can be written similar to ONBS. The signatures $(3,1)$ and $(2,2)$ belong to the class $\beta = 1$ and admit a real matrix representation.

Table II
Orthogonal basis sets in 4 dimensions.

```
1
1 0 0 0    - + - + - + - + - + - + - + - +
0 1 0 0    - - + + - - + + - - + + - - + +
0 0 1 0    - - - - + + + + - - - - + + + +
0 0 0 1    - - - - - - - - + + + + + + + +
2
1 0 0 0    - + - + - + - + - + - + - + - +
0 1 0 0    - - + + - - + + - - + + - - + +
0 0 1 0    - - - - + + + + - - - - + + + +
1 1 1 1    + - - + - + + - - + + - + - - +
3
1 0 0 0    - + - + - + - + - + - + - + - +
1 1 0 0    - + + - - + + - - + + - - + + -
1 0 1 0    - + - + + - + - - + - + + - + -
1 0 0 1    - + - + - + - + + - + - + - + -
4
1 0 0 0    - + - + - + - + - + - + - + - +
1 1 0 0    - + + - - + + - - + + - - + + -
1 0 1 0    - + - + + - + - - + - + + - + -
0 1 1 1    + + - - - - + + - - + + + + - -
5
1 1 0 0    - + + - - + + - - + + - - + + -
1 0 1 0    - + - + + - + - - + - + + - + -
1 0 0 1    - + - + - + - + + - + - + - + -
0 1 1 1    + + - - - - + + - - + + + + - -
6
0 0 1 0    - - - - + + + + - - - - + + + +
1 0 1 0    - + - + + - + - - + - + + - + -
1 1 0 1    + - - + + - - + - + + - - + + -
0 0 1 1    - - - - + + + + + + + + - - - -
7
1 0 1 0    - + - + + - + - - + - + + - + -
0 1 1 0    - - + + + + - - - - + + + + - -
1 1 0 1    + - - + + - - + - + + - - + + -
0 0 1 1    - - - - + + + + + + + + - - - -
8
1 1 1 0    + - - + - + + - + - - + - + + -
0 0 0 1    - - - - - - - - + + + + + + + +
1 0 0 1    - + - + - + - + + - + - + - + -
0 1 0 1    - - + + - - + + + + - - + + - -
9
1 1 1 0    + - - + - + + - + - - + - + + -
1 0 0 1    - + - + - + - + + - + - + - + -
0 1 0 1    - - + + - - + + + + - - + + - -
```

```
0 0 1 1    - - - - + + + + + + + + + - - - -
10
1 1 1 0    + - - + - + + - + - - + - + + -
1 1 0 1    + - - + + - - + - + + - - + + -
1 0 1 1    + - + - - + - + - + - + + - + -
0 1 1 1    + + - - - - + + - - + + + + - -
11
1 1 1 0    + - - + - + + - + - - + - + + -
1 1 0 1    + - - + + - - + - + + - - + + -
1 0 1 1    + - + - - + - + - + - + + - + -
1 1 1 1    + - - + - + + - - + + - + - - +
```

Table II can be used to obtain a matrix representation by the following procedure: for $(3, 1)$ locate this signature in item 1 (this is found in column 8). In the same column we find signature $(2, 2)$ in item 3. Hence Table II provides the following data: for $e_1^2 = e_2^2 = e_3^2 = 1$, $e_4^2 = -1$ we have $(e_1)^2 = 1, (e_1 e_2)^2 = (e_1 e_3)^2 = -1, (e_1 e_4)^2 = 1$. Since $I_i^- - I_i^+$ has a negative square and $I_i^- + I_i^+$ has a positive square, we can make the following assignments to the multivectors:

$$e_1 = I_1^- + I_1^+, \; e_1 e_2 = I_1^- - I_1^+, \; e_1 e_3 = I_2^- - I_2^+, \; e_1 e_4 = I_2^- + I_2^+.$$

Multiplying by e_1 we obtain the basis vectors:

$$e_2 = 1 - 2 I_1^- I_1^+,$$
$$e_3 = -I_2^- I_1^- - I_1^- I_2^+ - I_2^- I_1^+ - I_1^+ I_2^+,$$
$$e_4 = -I_2^- I_1^- + I_1^- I_2^+ - I_2^- I_1^+ + I_1^+ I_2^+.$$

The corresponding matrices can be obtained from the program IMTV or by multiplying the matrices given in Example 2. The result is

$$e_1 = \begin{pmatrix} 0 & 1 & 0 & 0 \\ 1 & 0 & 0 & 0 \\ 0 & 0 & 0 & -1 \\ 0 & 0 & -1 & 0 \end{pmatrix}, \quad e_2 = \begin{pmatrix} 1 & 0 & 0 & 0 \\ 0 & -1 & 0 & 0 \\ 0 & 0 & 1 & 0 \\ 0 & 0 & 0 & -1 \end{pmatrix},$$

$$e_3 = \begin{pmatrix} 0 & 0 & 0 & -1 \\ 0 & 0 & -1 & 0 \\ 0 & -1 & 0 & 0 \\ -1 & 0 & 0 & 0 \end{pmatrix}, \quad e_4 = \begin{pmatrix} 0 & 0 & 0 & 1 \\ 0 & 0 & 1 & 0 \\ 0 & -1 & 0 & 0 \\ -1 & 0 & 0 & 0 \end{pmatrix}.$$

Again, the matrix calculus facility of a computer algebra system can be used to check that the matrices satisfy scalar product relations in the signature $(3, 1)$. The consistency of the approach described above can be checked by a computer algebra as follows:

— compute 16 ordered products

$$e_{i_1} \cdots e_{i_k} \; (i_1 < i_2 \cdots < i_k, \; k = 1, \ldots, n).$$

Since the matrix elements can be considered as coefficients of the expansion into projectors, this gives the elements of the multivector basis as linear combinations of the projectors;

- using matrix inversion or solution of linear systems we may express the projectors as linear combinations of the multivectors;
- using the implementation of Clifford algebra we may check that the product relations $P_{ij}P_{kl} = \delta_{jk}P_{il}$ are satisfied.

4. Conclusions

Orthogonal basis sets define isomorphisms between Clifford algebras of different signatures. Automorphisms are obtained in the same way. In Table II take a quadruplet of a given signature in item 1 and look for the same signature in the column below. To each vector a multivector with the same square can be assigned. Together with orthogonality this defines an automorphism of the algebra.

From Table I it follows that the structure of a Clifford algebra is determined by the structure of its multivector group rather than the signature of the vector space $\mathbb{R}^{p,q}$. Moreover the relation (6) assigns the same classification parameter to different signatures. Conversely a given Clifford algebra defined by its isomorphic matrix algebra contains real vector spaces of different signatures. This could have some effect in applications to physics.

Appendix

A. Appendix A

The program IMTV computes the expansion of isotropic multivectors defined by

$$II(IA, IB) = (I_{m-1}^-)^{\alpha_{m-1}} \cdots (I_0^-)^{\alpha_0}(I_0^+)^{\beta_0} \cdots (I_{m-1}^+)^{\beta_{m-1}}$$

in projectors (matrix units). The m-tuple of binary numbers $\alpha_0, \ldots, \alpha_{m-1}$ is the binary representation of IA, the same is valid for IB. The isotropic vectors I_i^+ and I_i^- are defined in Section 2. The program input is MD half the dimension of the vector space and can be changed by modifying the DATA instruction.

```
C     MATRICES OF ISOTROPIC MULTIVECTOR PRODUCTS
CHARACTER*8 PR(8),PX
DATA IMP/32/
C     INPUT DATA MD HALF DIMENSION
DATA MD/2/
LN=0
ND=ISHFT(MD,1)
MM=0
MM=IBSET(MM,MD)
MN=MM-1
DO 4 MY=0,MN
DO 4 MZ=0,MN
WRITE(IMP,104)MY,MZ
104 FORMAT(1X,'II',2I2,' = ')
MYS=ISHFT(MY,MD)
```

```
MZS=ISHFT(MZ,MD)
DO 6 MX=0,MN
IF(IAND(MX,MY).NE.0)GO TO 6
IF(IAND(MX,MZ).NE.0)GO TO 6
C     SIGN OF REARRANGEMENT IN MX*MY AND MX*MZ
M=MX
K=0
DO 8 N=1,MD
M=ISHFT(M,1)
MA=IAND(M,MYS)
MB=IAND(M,MZS)
DO 9 I=MD,ND
IF(BTEST(MA,I))K=K+1
IF(BTEST(MB,I))K=K+1
9 CONTINUE
8 CONTINUE
MU=IEOR(MX,MY)
MV=IEOR(MX,MZ)
IF(.NOT.BTEST(K,0))WRITE(PX,100)MU,MV
100 FORMAT(2X,'+P',2I2)
IF(BTEST(K,0))WRITE(PX,101)MU,MV
101 FORMAT(2X,'-P',2I2)
LN=LN+1
PR(LN)=PX
IF(LN.LT.8)GO TO 6
WRITE(IMP,103)(PR(L),L=1,8)
103 FORMAT(5X,8A8)
LN=0
6 CONTINUE
IF(LN.EQ.0)GO TO 4
WRITE(IMP,103)(PR(L),L=1,LN)
LN=0
4 CONTINUE
END
```

B. Appendix B

The program ONBS determines orthonormal basis sets for Clifford algebras with underlying vectors space of dimension ND. The program eliminates ONBs corresponding to permutations of multivectors and/or basis vectors. The value of ND can be changed by modifying the DATA instruction. The inverse of the ONB matrices is printed if the flag KFA is set to 1. The sign combinations of the squared multivectors as functions of the squares of basis vectors are printed if the flag KFB is set to 1. The number of multivectors for n dimensional vector spaces is $m = 2^n - 1$ and the number of sets of n multivectors is to be tested is $m!/[n!(m-n)!]$. The computing time becomes prohibitive for $n > 6$.

```
C  ONB DETERMINATION
CHARACTER*4 LN(16),PL,MN
COMMON/DCM/NBC,LI(0:31)
DIMENSION MH(0:31),MV(0:31),MT(0:31,256)
DIMENSION MO(0:31),KA(0:31)
C    KFA =1 PRINT OF INVERSE OF ONB
DATA KFA/0/
C    KFB =1 PRINT OF SIGNS OF SQUARES
DATA KFB/1/
DATA IMP/32/
DATA PL,MN/'   +',' -'/
C   INPUT DATA ND DIMENSION OF VEC SPACE
DATA ND/4/
MA=0
C   MA=2**ND
MA = IBSET(MA,ND)
MB = MA-1
NB=ND-1
NBC=NB
131 FORMAT()
LT=0
LX=0
N=0
MH(0)=1
200 N=N+1
MH(N)=MH(N-1)+1
IF(N.LT.NB)GO TO 200
201 CONTINUE
C   MH(0:NB) ARE THE MULTIVECTORS TO BE TESTED
DO 210 JX=0,NB
NX=MH(JX)
LNX=LNGT(NX)
DO 210 JY=JX+1,NB
NY=MH(JY)
LNY=LNGT(NY)
K=LNX*LNY-LSCL(NX,NY)
IF(.NOT.BTEST(K,0))GO TO 202
210 CONTINUE
C   COLUMN TEST MVI COLUMN MULTIVECTORS
DO 301 I=0,NB
MV(I)=0
301 CONTINUE
DO 302 I=0,NB
DO 302 J=0,NB
IF(BTEST(MH(J),I))MV(I)=IBSET(MV(I),J)
302 CONTINUE
```

```
C    COMPARISON WITH TESTED MULTIVECTORS IN MT
DO 305 L=1,LT
DO 306 I=0,NB
DO 307 J=0,NB
IF(MV(I).EQ.MT(J,L))GO TO 306
307 CONTINUE
GO TO 305
306 CONTINUE
GO TO 202
305 CONTINUE
LT=LT+1
IF(LT.GT.256)STOP 'ARRAY MT INSUFFICIENT'
DO 310 I=0,NB
MT(I,LT)=MV(I)
310 CONTINUE
C    INVERSE OF ONB
DO 21 NX=0,NB
KA(NX)=0
21 CONTINUE
J=0
DO 20 M=1,MB
IF(J.GT.NB)GO TO 26
L=0
DO 22 NX=0,NB
K=0
DO 24 I=0,NB
IF(.NOT.BTEST(M,I))GO TO 24
IF(BTEST(MH(I),NX))K=K+1
24 CONTINUE
IF(.NOT.BTEST(K,0))GO TO 22
L=L+1
NY=NX
22 CONTINUE
IF(L.NE.1)GO TO 20
C     IF(KA(NY).NE.0)GO TO 20
MO(J)=M
J=J+1
KA(NY)=1
20 CONTINUE
26 CONTINUE
DO 28 NX=0,NB
IF(KA(NX).EQ.1)GO TO 28
WRITE(IMP,121)
121 FORMAT('   DETERMINATION OF INVERSE FAILED')
GO TO 202
28 CONTINUE
```

```
LX=LX+1
WRITE(IMP,124)LX
124 FORMAT(2X,I4)
DO 311 I=0,NB
CALL BITS(MH(I))
WRITE(IMP,122)(LI(K),K=0,NB)
122 FORMAT(6X,16I4)
311 CONTINUE
IF(KFA.EQ.0)GO TO 204
WRITE(IMP,131)
WRITE(IMP,125)
125 FORMAT('     INVERSE')
WRITE(IMP,131)
DO 406 I=0,NB
CALL BITS(MO(I))
WRITE(IMP,122)(LI(K),K=0,NB)
406 CONTINUE
WRITE(IMP,131)
204 CONTINUE
IF(KFB.EQ.0)GO TO 202
C    SIGNS OF SQUARES
DO 404 L=0,MB
DO 403 J=0,NB
LG=LNGT(MH(J))
LG=LG*(LG+1)
K=0
IF(BTEST(LG,1))K=1
DO 405 I=0,NB
IF(.NOT.BTEST(MH(J),I))GO TO 405
IF(BTEST(L,I))K=K+1
405 CONTINUE
LL=J+1
IF(.NOT.BTEST(K,0))LN(LL)=PL
IF(BTEST(K,0))LN(LL)=MN
403 CONTINUE
WRITE(IMP,128)L,(LN(LL),LL=1,ND)
128 FORMAT(2X,I4,16A4)
404 CONTINUE
WRITE(IMP,131)
202 CONTINUE
C    NEXT SET OF MULTIVECTORS
MH(N)=MH(N)+1
IF(MH(N).LE.MB)GO TO 201
MX=MB
203 N=N-1
IF(N.LT.0)STOP
```

```
MX=MX-1
IF(MH(N).EQ.MX)GO TO 203
MH(N)=MH(N)+1
GO TO 200
END
INTEGER FUNCTION LNGT(NX)
COMMON/DCM/NB,LI(0:31)
K=0
DO 10 I=0,NB
IF(BTEST(NX,I))K = K + 1
10 CONTINUE
LNGT=K
RETURN
END
INTEGER FUNCTION LSCL(NX,NY)
COMMON/DCM/NB,LI(0:31)
NU = IAND(NX,NY)
K=0
DO 12 I = 0,NB
IF(BTEST(NU,I))K = K+1
12 CONTINUE
LSCL=K
RETURN
END
SUBROUTINE BITS(NX)
COMMON/DCM/NB,LI(0:31)
DO 10 I=0,NB
LI(I)=0
IF(BTEST(NX,I))LI(I)=1
10 CONTINUE
RETURN
END
```

References

R. Ablamowicz: (1986) "Deformation and contraction in Clifford algebras", *J. Math. Phys.* **27**, pp. 423–427.

M. F. Atiyah, R. Bott and A. Shapiro: (1964) "Clifford modules", *Topology* **3** Suppl.1, pp. 3–38.

G. Bergdolt: (1988) "Projector bases and algebraic spinors", *J. Math. Phys.* **29**, pp. 2519–2522.

G. Bergdolt: (1993) "The complete reduction of Clifford algebras", *J. of Math. Phys.* **Vol. 34**, pp. 5924–5934.

H. W. Braden: (1985) "Dimensional spinors; their properties in terms of finite groups", *J. Math. Phys.* **26**, pp. 613–620.

A. Dimakis: (1989) "A new representation of Clifford algebras", *J. Phys. A. Math. Gen.* **22**, pp. 3171–3193.

P. Lounesto: (1981) "Scalar products of spinors and an extension of Brauer-Weil groups", *Found. of Phys.* **11**, pp. 721–740.

N. Salingaros: (1981) "Realization, extension and classification of certain physically important groups", *J. Math. Phys.* **22**, pp. 226–232.

J. P. Serre: (1983) "Arbres, amalgames, SL2", *Asterisque* **46**, Société mathematique de France, pp. 1–188.

M. Schönberg: (1957) "Quantum kinematics and geometry", *Nuovo Cimento Suppl.* **6**, pp. 356–380.

COMPLEX CONJUGATION – RELATIVE TO WHAT?

ALEXANDER M. SOIGUINE
191011 St. Petersburg, Fontanka 53 #37
Russia

Abstract. Some initial, technically simple but fundamentally important statements concerning the very origin of the notion of a complex number are formulated in terms of the Clifford (Geometric) algebra generated by vectors in some geometrically and physically sensitive dimensions. A new insight into the sense of geometrical product is given. It is shown that it makes no sense to speak about complex numbers without identifying a corresponding two-dimensional plane. This is particularly important if the given physical situation is set in higher dimensions. Because of great importance of these questions in education and because of increasing use of graphical computer programs in mathematical education and research, some components of a computer program implementing the Geometric Algebra approach are outlined in terms of classes of the object-oriented computer language C++.

Key words: Cognitive process – imaginary unit – CLICAL

1. Introduction

Teaching of mathematics has improved significantly particularly where mathematical formalism can be represented by a computer program able to manipulate symbols according to certain rules. The improvement is most effective if the involved mathematical objects have an explicit and unambiguous geometric interpretation transformed into computer visual images. In such case the cognitive process is enforced radically by including in it human visual system – our most powerful sensor system for communication with the outer world.

Clifford (or, maybe, better to say "Geometric") algebra is one of the best examples of such a field in mathematics, the more so as its mathematical formalism is tending to become a unified language for theoretical physics (Hestenes and Sobczyk, 1987). Geometric algebra is superbly convenient to be supplied with adequate computer representation programs. It is extremely important today when we have, from one side, a tendency among students of mathematics or physics for a more intuitively obvious and even visualizable basis for mathematical constructions and, from another side, we have available computer facilities, in hardware and software, to meet such desires.

For several years now we have had CLICAL, a well known and widely recognized computer program developed by professor Pertti Lounesto and his colleagues (Lounesto *et al.*, 1987). This program allows a student to make all algebraic calculations in the field of Clifford algebras – from complex numbers to the special relativity theory. At the same time it should be said that CLICAL code is written (and, sure, couldn't be written in another way at those days) in the text mode. While the program can only manipulate numbers and symbols, objects of the Geometric Algebras, at least in low, visualizable dimensions, have direct geometrical interpretation. So today we have a challenge to transform CLICAL into a graphical interface or, really, to develop a *Clicalian* computer program able to appropriately manipulate and visualize corresponding geometrical objects. We have now good computer displays and we have adequate and powerful computer languages to deal with graphical images namely object-oriented languages such as, first of all, C++. Development and implementation of such program could have far reaching educational consequences and may be considered as the next generation of Lounesto's CLICAL.

Before delving into a discussion of C++ classes representing geometric objects and their behaviors, we clarify some basic and well known geometric facts about Clifford algebras to make them absolutely unambiguous in their geometrical interpretation hence more suitable for computer implementation.

Hestenes said that "physicists quickly become impatient with any discussion of elementary concept" (Hestenes, 1985). This statement is not applicable to the author because his goal here is not to make a show on super complicated mathematical wisdom but to reveal, at a maximal degree, initial geometrical sense of some basic operations in the Clifford algebra and track possible relations with corresponding C++ units. From this point of view the current work has a methodological and educational content. So, we shall start with recollection of some ideas.

2. Geometric algebra of the plane

The arena of the performance will now be a two-dimensional plane denoted here as E_2. At the very beginning I emphasize that in no way E_2 is viewed here as a two-dimensional linear space, that is as a set of linear combinations $\alpha^1 e_1 + \alpha^2 e_2$ where e_1, e_2 are two arbitrary linearly independent vectors from E_2 and α^1, α^2 are scalars.[1] For our purposes E_2 should be thought of as a geometrical object, an abstraction arising from such real things as a table-desk or a sheet of paper. I won't make any attempt to give a formal definition of E_2, I am just considering it as an intuitively obvious geometrical object.

Certainly, elements of the form $\alpha^1 e_1 + \alpha^2 e_2$ are in E_2. These are directed segments of straight lines in the plane, or arrows. We shall call them (two-dimensional) vectors. A vector may be identified by its length (value) and direction (orientation).

However, there exists at least one another sort of geometrical objects inhabiting E_2 which are of the greatest importance for in Geometric Algebra considerations. I mean closed oriented curves without self-intersections lying in E_2.

Closed oriented curves are identified, similarly to usual vectors, by value and orientation. By the value of such a geometrical object we mean a non-negative number which is equal to the area inside the curve. By the orientation we mean one of the two possible directions of movement along the closed curve. One will be called positive, another negative. For shortness, we will call the oriented closed curves on E_2 *bivectors* as is the common practice.

Ordinary free vectors in the plane are defined up to parallel movements in E_2.

---

[1] I do not say *real* scalars since, I hope, it would be evident soon that it has a little sense to speak about complex-valued scalars. If we would follow that practice, then elements of any field, over which a linear space is considered, should be called scalars.

Bivectors are defined up to arbitrary movements and curve deformations which don't change the area inside the curve.

Let us make some agreements about notation. Usual vectors will be denoted by small Latin letters: $a, b, \ldots$. Bivectors will be denoted by capital Latin letters: $A, B, \ldots$. For scalars we will use Greek alphabet: $\alpha, \beta, \ldots$. Values of vectors and bivectors are denoted, along with the absolute values of scalars, as $|a|, |A|, \ldots$. Orientation of a bivector A will be denoted as $\mathcal{O}_A$ and if it is given then the two possible values are denoted as $\mathcal{O}_A^+$ and $\mathcal{O}_A^-$.

The set of vectors in E_2 is a linear space with the known geometrical sense of the operations. Now we have to supply the set of bivectors in E_2 with the structure of a linear space.

Given two bivectors, A, B, their sum is a bivector $C \equiv A + B$, the value of which is equal to $|A| + |B|$ in the case of coinciding orientations of A and B, $\mathcal{O}_A = \mathcal{O}_B$, and $||A| - |B||$ in the case of opposite orientations, $\mathcal{O}_A = -\mathcal{O}_B$. The orientation of C is the same as the common orientation of A and B in the first case and is equal to the orientation of bivector with the greater value in the second case.

Multiplication by a scalar is defined in an obvious way. Given a bivector A and a scalar λ, the product λA is a bivector with $|\lambda A| = |\lambda||A|$ and $\mathcal{O}_{\lambda A} = \mathcal{O}_A$ if $\lambda > 0$, and $\mathcal{O}_{\lambda A} = -\mathcal{O}_A$, if $\lambda < 0$.

The zero element for addition is the equivalence class of oriented curves with zero inside area.

The unit positive oriented bivector will be denoted by i_{E_2}, or simply i, because we have only one plane E_2 at our disposal for the moment. Generally, an "imaginary unit" must be supplied with a subscript denoting the plane in which the unit bivector lies.

All axioms of a linear space are verified immediately for bivectors. The linear space of bivectors in E_2 will be denoted by $\mathbf{B}$, the linear space of vectors will be denoted by $\mathbf{V}$, and the linear space of scalars will be denoted here by $\mathbf{S}$.

Now we can form outer direct sums of these linear spaces in various combinations: $\mathbf{S} \oplus \mathbf{V}, \mathbf{S} \oplus \mathbf{B}, \mathbf{V} \oplus \mathbf{B}$ and $\mathbf{S} \oplus \mathbf{V} \oplus \mathbf{B}$. Recall that the direct sum of linear spaces is a linear space of all sequences of elements belonging to the corresponding spaces. We may consider direct sum as the set of all given linear spaces "put in one bag", where they don't mix up, similar to distinct purchases in a bag. Linearity is checked easily, for example, addition in $\mathbf{V} \oplus \mathbf{B}$ is defined as:

$$\mathbf{V} \oplus \mathbf{B} \ni M_1, M_2 : M_1 + M_2 = \left((M_1)_{\mathbf{V}} + (M_2)_{\mathbf{V}}, (M_1)_{\mathbf{B}} + (M_2)_{\mathbf{B}}\right) \in \mathbf{V} \oplus \mathbf{B}$$

where $(M_i)_{\mathbf{V}}$ and $(M_i)_{\mathbf{B}}$ are, respectively, the vector and the bivector components of $M_i \in \mathbf{V} + \mathbf{B}$.

The linear space $\mathbf{C} \equiv \mathbf{S} \oplus \mathbf{B} \equiv \mathbf{S} \oplus i\mathbf{B}$ is the set of complex numbers in E_2. It is a subalgebra of $\mathbf{G}_2 \equiv \mathbf{S} \oplus \mathbf{V} \oplus \mathbf{B}$, the Geometric (Clifford) Algebra of the plane E_2. To explain the usage of the term "algebra" we shall define multiplication in $\mathbf{G}_2$.

Let's begin with a definition of vector multiplication since multiplication by scalars is trivial. Given two vectors $a, b \in \mathbf{V}$ we define their product as a map

$$\mathbf{V} \times \mathbf{V} \ni (a, b) \rightarrow ab = (a \cdot b, a \wedge b) \in \mathbf{S} \oplus \mathbf{B}$$

where $a \cdot b$ is a scalar, equal to the scalar product of vectors a and b computed as

$|a||b|\cos\widehat{(a,b)}$ and $a \wedge b$ is a bivector with the value $|a||b||\sin\widehat{(a,b)}|$ and orientation defined by rotation of a to b through the angle which is less than π.

If we identify in a natural way elements of each space, comprising the direct sum, with elements of the form $(\alpha, 0_{\mathbf{V}}, 0_{\mathbf{B}})$, $(0_{\mathbf{S}}, a, 0_{\mathbf{B}})$ and $(0_{\mathbf{S}}, 0_{\mathbf{V}}, A)$ and denote the addition in $\mathbf{S} \oplus \mathbf{V} \oplus \mathbf{B}$ by the usual "plus", then an arbitrary $M \in \mathbf{S} \oplus \mathbf{V} \oplus \mathbf{B}$ is written as $M = \alpha + a + A$. In the same way $\mathbf{S} \oplus \mathbf{B}$ is identified with the subalgebra of $\mathbf{G}_2$ of elements of the form $\alpha + A$, so $ab = a \cdot b + i_{E_2}|a \wedge b|$.

It remains to define product of a vector with a bivector and product of two bivectors.

For $a \in \mathbf{V}$ and $A \equiv i_{E_2}\alpha \in \mathbf{B}$, multiplication of a by A from the right, aA, is vector of value $|\alpha||a|$, rotated in the direction of the positive orientation in E_2 by the angle $\frac{\pi}{2}$, with the opposite direction in the case of negative α. If the operands have opposite order of multiplication, Aa, the rotation is made in the direction of negative orientation on E_2. Clearly, $aA = -Aa$.

To calculate $i_{E_2}{}^2$ we first check the associativity of the geometric product of vectors. If the distributivity of multiplication of vectors relatively to their addition is valid then it suffices to consider products of three basis vectors. The proof of distributivity is a tedious exercise in trigonometry, and we take it for granted. Then we have for basis vectors $\mathbf{e}_1(\mathbf{e}_1\mathbf{e}_2) = \mathbf{e}_1 i_{E_2} = \mathbf{e}_2$ and $(\mathbf{e}_1\mathbf{e}_1)\mathbf{e}_2 = \mathbf{e}_2$, thus associativity holds. Hence, together with anticommutativity of orthogonal vectors, it follows that $i_{E_2}{}^2 = \mathbf{e}_1\mathbf{e}_2\mathbf{e}_1\mathbf{e}_2 = -1$.

Now let's turn to complex conjugation. In conventional geometric interpretation complex numbers are represented as vectors in some given orthogonal reference frame on the plane which is called the *complex plane*. This counterclockwise frame has the first axis as the *real* one, and the second axis, rotated by $\frac{\pi}{2}$ in positive direction, as the *imaginary* axis. A complex number (vector) has the first projection called the *real* part, and the second projection called the *imaginary* part. Conventionally, when written algebraically as $a + ib$, where a and b are usual (real) numbers and i is formally introduced as "imaginary unit" with the crucial property $i^2 = -1$, the complex numbers possess the operation of "complex conjugation": $a + ib \rightarrow a - ib$, which will be seen to be basis dependent.

At the same time, unambiguous and the only correct representation of complex numbers as elements of $\mathbf{S} \oplus \mathbf{B}$ and having the form $\alpha + i_{E_2}\beta$ has no such deficiency. The map $\alpha + i_{E_2}\beta \rightarrow \alpha - i_{E_2}\beta$ doesn't depend on a basis in E_2 and means that orientation of the bivector part in a pair $(\alpha, i_{E_2}\beta)$ is changed to the opposite.

Then, what to do with the conventional complex conjugation considered as the reflection of the vector relative to the real axis? First, we remember that there exists a correspondence between elements $\alpha + i_{E_2}\beta$ of $\mathbf{S} \oplus \mathbf{B}$ considered as complex numbers in a strict sense and vectors from $\mathbf{V}$.

Suppose, we've chosen a unit vector $\mathbf{e}_1$ on E_2, the latter is by assumption counterclockwise positive oriented. Given a complex number $\alpha + i_{E_2}\beta$, multiply it by $\mathbf{e}_1$ from the left. Then we get the vector $\alpha\mathbf{e}_1 + \beta\mathbf{e}_2$, where $\mathbf{e}_2$ is the unit vector rotated from $\mathbf{e}_1$ by $\frac{\pi}{2}$ in the positive direction. So, we get the conventional geometric representation of a complex number as a vector. Arbitrarily chosen unit vector $\mathbf{e}_1$ fixes the real axis, $\mathbf{e}_2$ – the imaginary one. Complex conjugation in these terms is the reflection of $\alpha\mathbf{e}_1 + \beta\mathbf{e}_2$ relative to $\mathbf{e}_2$.

We can call $\alpha e_1 + \beta e_2$ a *relative vector representation* of a complex number $\alpha + i_{E_2}\beta$ defined by the vector e_1.

If we choose another unit vector, say f_1, in E_2, we get the same "vectorial" picture together with complex conjugation as the reflection relative to the direction orthogonal to f_1 but the picture is rotated in E_2 by the angle between e_1 and f_1. Here we have maybe the simplest example of a gauge transformation connecting different relative representations of geometrically invariant object, a complex number in this case. It is worth mentioning that a rotation in E_2 is generated by multiplication by elements $\alpha + i_{E_2}\beta$ where $\alpha^2 + \beta^2 = 1$. For an explanation of this well known fact see (Gull *et al.*, 1993). Because of the above condition on α and β, bivector generating a rotation may be written as $\cos\phi + i_{E_2}\sin\phi$, where ϕ is the angle of rotation. Since the rules for addition of the angle arguments in multiplication of such bivectors coincide with those in multiplication of exponents, rotation generating bivectors are symbolically written as $e^{i_{E_2}\phi}$, taking into account the Euler's formulas and accepting the fact that the function sin *feels* orientation, whereas cos doesn't.

Having reviewed briefly some facts about the two-dimensional plane, which we consider as insufficiently clarified in the works on Geometric Algebra, we pass to even more fragmental discussion on higher dimensions.

3. Higher dimensions

Now we'll touch briefly the cases of higher physical dimensions. All results are principally known and only the interpretation, transparent geometrical sense and unambiguous relations between different representations are important to us. Some additional interesting details may be found in (Soiguine, 1990).

Suppose we are in the physical three dimensions. The Geometrical product of any two vectors is the same as in the plane, but here the plane of the bivector component is defined as the plane spanned by vector multipliers. Orientation of that plane is determined by given orientation of the three-dimensional space, which may be left or right by definition.

We are to define a geometrical product of a vector and bivector when the former is not in the bivector plane. To do this, we expand vector in two components, parallel and perpendicular to the bivector plane, $a = a_{\parallel} + a_{\perp}$. The product of the parallel component $a_{\parallel}$ with given bivector B is calculated by the rule of the previous section. It gives vector $a_{\parallel}i_B|B|$ or $i_B a_{\parallel}|B|$ depending on the order of the multipliers. The product $a_{\perp}B$ gives oriented volume equal to $|a_{\perp}||B|$, the orientation being determined by a screw representing both rotation in the bivector B plane in the direction of B and movement along $a_{\perp}$. It is very convenient to imagine an oriented volume as a cylinder of the given volume with a string wound around it in one of the two possible ways. The oriented volumes, or trivectors, may be written as αi_3, where i_3 is a positively oriented unit volume.

Geometric product of a vector a with a trivector αi_3 gives, independently of the order, a bivector of value $|\alpha||a|$ lying in the plane orthogonal to a and oriented in such way that its orientation and the direction of a restore the orientation of αi_3.

This rule actually gives the correspondence of duality between vectors and bivectors in three dimensions. If we have $A = i_3a$ then A is called *dual* to a.

If A is a bivector then $i_3A = Ai_3$ is vector of the value $|A|$, orthogonal to the plane of A and having orientation which, together with the orientation of A, gives three-dimensional orientation opposite to that of i_3.

If $a = i_3A \equiv Ai_3$ then a is called dual to A.

The inversion of the resulting common orientation in duality between vector and bivector, and bivector and vector follows from the fact that double duality changes the sign of a geometrical object. For example, if $i_3a = A$ then multiplying by i_3 from the left gives $-a = i_3A$. Here we use the fact that $i_3^2 = -1$, which may be checked from possible representation $i_3 = e_1e_2e_3$ where $\{e_i\}$ – unit orthogonal basis of given orientation in three dimensions.

Now, a general element of the Geometrical algebra in the three physical dimensions, $\mathbf{G_3} \equiv \mathbf{S} \oplus \mathbf{V} \oplus \mathbf{B} \oplus \mathbf{T}$, has the form:

$$\mathbf{G_3} \ni M = \alpha + a + i_3b + i_3\beta.$$

What is the complex conjugation in this case? For example, the subset of elements of the form $\alpha + i_3b$ comprise a subalgebra, if we mean the algebraic form of elements. At the same time any such element determines its own "complex plane", defined by the direction of vector b in three dimensions and orientation of i_3. In multiplication two such elements give an element of the same algebraic form but with its new complex plane. If $i_3b = B$ then always we can write $\alpha + i_3b = \alpha + i_B\beta$, where $|\beta| = |b| = |B|$. So, we can conclude that quaternions are really complex numbers differing from the convention in a single aspect that a quaternion "complex plane" is arbitrarily imbedded in the three dimensions.

At the end of this section I would like to suggest an interesting question to the reader: Is it possible to adjoin to the indefinite metric in the special relativity a "complex conjugation" remembering that the latter is nothing else but an inversion of orientation?

4. C++ classes implementation of Geometric Algebra in the plane

Now we consider possible computer C++ language implementation of geometrically obvious operations in Geometric algebra in the plane restricting ourselves to the simplest case of complex numbers. The C++ language is adequate in its **class** type variable structure to describe real physical things, and geometric objects in the plane in particular.

As it was mentioned earlier, the forthcoming computer program should be written in a graphical mode and because of that user input operations and screen output operations will become more complicated. The program is menu-controlled. Input may be either digital or graphical while complex numbers are input as vectors with the mouse manipulations.

The **class** type variable representing complex numbers together with operations on them may be defined as:

```
typedef class tagCOMPLEXNUMBER {
double RealPart;
```

```
    double ImagPart;
    int ScrRealPart;
    int ScrImagPart;
    char StrRealPart[];
    char StrImagPart[];
    int xScreenOrigine;
    int yScreenOrigine;
    int RSEntered;
    int ISEntered;
    int GetStrReal(int,int,char*);
    int GetStrImag(int,int,char*);
    int WriteNumber(int);
    void EraseNumberPicture(void);
    void EnlargeNumberPicture(double times);
    void ConvertRealToString(void);
    void ConvertImagToString(void);
    void ConvertStringToReal(void);
    void ConvertStringToImag(void);

      public:

    friend tagCOMPLEXNUMBER operator+(tagCOMPLEXNUMBER,
tagCOMPLEXNUMBER);

    friend tagCOMPLEXNUMBER operator-(tagCOMPLEXNUMBER,
tagCOMPLEXNUMBER);

    friend tagCOMPLEXNUMBER operator*(tagCOMPLEXNUMBER,
tagCOMPLEXNUMBER);

    friend tagCOMPLEXNUMBER operator/(tagCOMPLEXNUMBER,
tagCOMPLEXNUMBER);

    void SetRealPart(double value) {RealPart = value;}
    double GetRealPart(void) {return RealPart;}
    void SetImagPart(double value) {ImagPart = value;}
    double GetImagPart(void) {return ImagPart;}
    void SetScrRealPart(int value) {ScrRealPart = value;}
    int GetScrRealPart(void) {return ScrRealPart;}
    void SetScrImagPart(int value) {ScrImagPart = value;}
    int GetScrImagPart(void) {return ScrImagPart;}
    int EnterNumber(int);
    void DrawNumber(void);
        } COMPLEXNUMBER;
```

Only a part of the type definition is written here, consisting of comprehensive class variables and functions.

Partition between private and public functions in the **COMPLEXNUMBER** class depends on concrete realization. For example, one of the main public functions, **EnterNumber**, has the C++ code:

```
int tagCOMPLEXNUMBER::EnterNumber(int number) {
int result=0;
do {
switch(mode % 2) {
case 0:
if (number==1)
{
if
(WriteNumber((strlen("operand")+2)*charwidth))
result=1;
else result=0; } else
if (number==2)
{
if
(WriteNumber((strlen("operand")+4)*charwidth+X_ max/2))
result=1;
else result=0; }
break;
case 1:  break;
} } while(!result);
return result;
}
```

The private function **WriteNumber**, in turn, has the code:

```
int tagCOMPLEXNUMBER::WriteNumber(int leftboundary) {
static int RResult=0;
static int IResult=0;
char* invfigptr = 0;
if (!(mode % 2))
do {
if (leftboundary<X_ max/2)
WashRect(leftboundary,Ymax+1,X_ max/2-2,Ystatusline-17);
else if(leftboundary>=X_ max/2)
WashRect(leftboundary,Ymax+1,X_ max-1,Ystatusline-17);
GetStrReal(leftboundary+4,Ymax+11,"Real part = ");
if (RSEntered) RResult=1;
SetRealPart(strtod(StrRealPart,& invfigptr));
if(*invfigptr!=NULL) Beep();
}
while ((*invfigptr!=NULL)&& (!(mode % 2)));
else { WashRect(1,Ymax+1,X_ max-1,Ystatusline-1); }
```

```
invfigptr = 0;
if (!(mode % 2))
do {
if (leftboundary<X_ max/2)
WashRect(leftboundary,Ymax+17,X_ max/2-2,Ystatusline-1);
else if(leftboundary>=X_ max/2)
WashRect(leftboundary,Ymax+17,X_ max-1,Ystatusline-1);
GetStrImag(leftboundary+4,Ystatusline-5,"Imaginary part = ");
if (ISEntered) IResult=1;
SetImagPart(strtod(StrImagPart,& invfigptr));
if(*invfigptr!=NULL) Beep();
}
while ((*invfigptr!=NULL)&& (!(mode % 2)));
else { WashRect(1,Ymax+1,X_ max-1,Ystatusline-1); }
if (RResult&& IResult) return 1; else return 0;
}
```

Here, for example, the function GetStrReal, implementing the input of a string containing the value of the real (scalar) part of complex number, has the code:

```
int tagCOMPLEXNUMBER::GetStrReal(int X,int Y,char* mes)
{ RSEntered = 0;
int len=strlen(mes), i=0, j, maxlen, k;
char ch, s2[2];
maxlen = charwidth*(len+12); s2[1] = '\0';
outtextxy(X,Y,mes);
while(1)
{ j=(len+i)*charwidth;
if(j>=maxlen) break;
TxtCursor(X+j, Y-6);
ch=getch();
if (!ch) ch=getch();
TxtCursor(X+j, Y-6);
if(ch==ESC || ch==BACKSPACE || ch==ENTER)
{ if (ch==ESC)
{
if(i>0) {k=(len+i)*charwidth; i=0;
WashRect(X+len*charwidth,Y-charheight,X+k+charwidth,Y); }
} else
if (ch==BACKSPACE)
{ if(i>0) i--;
k=(len+i)*charwidth;
WashRect(X+k,Y-charheight,X+k+charwidth,Y);
}
else
if (ch==ENTER) {RSEntered=1;break; }
```

```
}
else
{ StrRealPart[i]=s2[0]=ch;
outtextxy(X+j,Y,s2);
  i++; } } if (i==0) {StrRealPart[i]=s2[0]='0';
outtextxy(X+j,Y,s2);i++; }
  StrRealPart[i]='\0';
  return *StrRealPart;
}
```

The above portions of C++ source code illustrate work which is in progress now and, I hope, will soon result in a valuable computer program implementing Geometric algebra visual interpretation.

5. Conclusions

The author's intentions here were mainly pedagogical and methodological. The point is that Geometric algebra of the plane or the three-dimensional physical space has unambiguous and transparent geometric interpretations not realized yet in all details. This interpretation may be realized in computer images in terms of adequate object-oriented language C++. The work is in progress and is expected to yield next graphical version of the well known CLICAL.

References

S. Gull, A. Lasenby, Ch. Doran: 1993, 'Imaginary numbers are not real – the Geometric Algebra of spacetime', *Found. Phys.* **23**, pp. 1175–1201.

D. Hestenes, G. Sobczyk, *Clifford Algebra to Geometric Calculus*, Reidel, Dordrecht.

D. Hestenes: 1985, 'Clifford algebra and interpretation of quantum mechanics', *Proc. NATO and SERC Workshop on Clifford Algebras and Their Applications in Mathematical Physics*, Canterbury, pp. 321–346.

P. Lounesto, R. Mikkola, V. Vierros: 1987, *CLICAL, Program and User Manual*, Helsinki University of Technology, Helsinki.

A.M. Soiguine. *Vector Algebra in Applied Problems*, Naval Academy Publ., St. Petersburg, 1990. (in Russian)

OBJECT-ORIENTED IMPLEMENTATIONS OF CLIFFORD ALGEBRAS IN C++: A PROTOTYPE

ARVIND RAJA
Institute of Mathematics
Helsinki University of Technology (HUT)
Otakaari 1, FIN-02150 Espoo
Finland
e-mail: Arvind.Raja@hut.fi

Abstract. This paper describes an evolving library of data types and algorithms for numerical Clifford algebra computations. C++ classes for representing elements of a Clifford algebra $Cl_{p,q}$ are presented. Examples illustrating use of the library are included. These implementations will form the computation engine of a new interactive program, which is also under development. The library illustrates how "high level" mathematical descriptions can be implemented, elegantly and efficiently, in a "low level" programming language. This is achieved by using the object-oriented features supported by C++ (data abstraction, operator overloading and many others) At the time of writing, fundamental operations with elements of a Clifford algebra $Cl_{p,q}$ can be performed. Routines for evaluating transcendental functions of Clifford elements, via matrix functions, are evolving.

Key words: Clifford algebras, object-oriented programming; C++.

1. Introduction

Broadly speaking, interactive software such as CLICAL, Matlab and Maple, consists of a front end (user interface) and a computation engine (kernel). Amongst other things, the front end is responsible for accepting user input and transforming it into specific data objects stored within the computer system. The computation engine works with the data objects and carries out the actual calculations. Typically the computation engine is a set of highly optimized routines, often written and compiled in C/C++ or FORTRAN. Code in the form of say CLICAL or Maple *scripts* is interpreted (not compiled) as it is read (from a file) or entered (from the keyboard), allowing the software to be used interactively.

This paper focuses on a growing library of data types and algorithms for numerical Clifford algebra computations. The implementation language is C++, since it is particularly well suited for creating complex user-defined data types. Moreover, the user-defined data types can be made to work in a manner similar to the built-in types. The library can be used in C++ application programs. The main purpose of the library is to use it as the computation engine of a new interactive program for Clifford algebra calculations—also being developed by the author.

The reader is expected to be familiar with the C++ programming language and technical terms associated with it. For a complete description of the C++ language, the reader is referred to (Stroustrup, 1991) —the definitive guide to C++ written by the designer and implementor of the language. For a step-by-step approach to learning the language, from the ground up, the reader is referred to the excellent book by Lafore (Lafore, 1991), even though it doesn't cover *templates* and *exception handling*. Before embarking on a major software development project in C++, the book by Meyers (Meyers, 1992) is a "must" read.

At present the implementation forms a framework onto which increasing functionality is continuously being added. Thus the library is a prototype and far from complete. Fundamental data structures have been implemented and a comfortable **class clifford** is supported. Basic operations for elements of a Clifford algebra $Cl_{p,q}$ can be performed with ease and flexibility—in a syntactic style similar to hand-written expressions. Routines for evaluating transcendental functions of Clifford elements are evolving. Transcendental functions and inverses of Clifford elements are being implemented through matrix representations. Classes for vectors, matrices and complex numbers have also been made, but they are not presented here.

A sizable number of test problems were run. The results agreed with those produced by CLICAL. The library has been tested on the following systems:

Hardware	Operating System	Compiler
IBM PC compatible 486/33Mhz	MS-DOS 5.0	Borland C++ 3.1
IBM PC compatible 486/33Mhz	MS-DOS 5.0	djgpp (gcc 2.6.0)
IBM RS/6000	AIX 3.2.5	GNU gcc 2.6.3
Silicon Graphics Workstation	IRIX 6.0.1	GNU gcc 2.6.3
Hewlett-Packard Workstation	HP-UX	HP CC

Section 2 of this paper contains example calculations, corresponding program listings and explanations. The examples serve to exhibit the salient features of the library.

Section 3 describes the main aspects of the implementation, accompanied by a collection of remarks and comments.

Section 4 is a reference section. Some of the *global, member* and *friend* functions are documented there in a terse format. Standard C++ and mathematical notations are used there.

2. Using the library

In this section, I present some working examples to illustrate the main features of the library. Many of the examples are drawn from (Lounesto *et al.*, 1987) and (Lounesto, 1992). The explanations accompanying the examples are general so as to hide implementation details. Important functions are documented in Section 4 where more details can be found. In the program listings, line numbers appear for reference only and are *not* part of the code.

Example 1 Let OPQ be a triangle, where $P = (4, 3)$, $Q = (2, 3)$ and O is the origin $(0, 0)$. Calculate the length r of side OP and the area A of the triangle.

The following program, Listing 1, when compiled and linked with the library solves the above problem:

```
(1)   #include "headers.h"
(2)   #include "clifford.h"
(3)
(4)   int main()
(5)   {
(6)     clifford OP(2, 0, "4b1 + 3b2"),
(7)              OQ(2, 0, "2b1 + 3b2");
(8)
(9)     double r= OP.dabs();
(10)    double A= dabs( (OP ^ OQ)/2 );
(11)
(12)    cout << endl
(13)        << "OP = " << OP << endl
(14)        << "OQ = " << OQ << endl
(15)        << "length of side OP = " << r << endl
(16)        << "area of triangle OPQ = " << A << endl;
(17)    return 0;
(18) }
```

Listing 1

The explanation is as follows. Lines 1 and 2 specify inclusion of header files required by all programs using the library. The file **headers.h** is a header file of standard headers such as **iostream.h** (for input/output) and **global.h** (containing definitions of global data such as **const double EPS = 1e-6**). In lines 6 and 7, clifford elements **OP** and **OQ** are created using the three-argument *constructor,*
clifford::clifford(int p_sig, int q_sig, char* str), where the 2 and 0 refer to the signature $(2, 0)$, i. e. **p_sig = 2** and **q_sig = 0**. The value of the clifford element—a linear combination of products of the basis elements $e_1, e_2, \ldots, e_n$ —is given as a string. The symbol ' b ' is used in place of the usual ' e ' in mathematical writing. This is to avoid typing errors since ' e ' and ' E ' are used in the scientific notation for floating point numbers in C++. The coefficients may be string representations of numbers of type **double**; the following are examples of valid sub-strings that can appear to represent coefficients:

$$4, \ 2.1, \ 0.8976, \ 4.3e2, \ 5.4E-3, \ 1.003,$$

which represent the numbers $4, 2.1, 0.8976, 4.3 \times 10^2, 5.4 \times 10^{-3}$ and 1.003 respectively. The functions **clifford::dabs()** and **dabs(const clifford& cf)** appearing in lines 9 and 10 respectively, returns the absolute value of a clifford element as a **double** (real number). In line 10 the outer product of OP and OQ is computed (**OP^OQ**), the result is divided by 2 and assigned to **A**.
clifford::operator ^ (const clifford& cf) has been defined to compute the outer product. Lines 12–16 perform output: The global friend function
operator << (ostream& os, const clifford& cf) is invoked in lines 13 and 14 for displaying the values of **OP** and **OQ** respectively.

The output of the program in Listing 1 is given below:

```
OP = +4b1   +3b2
OQ = +2b1   +3b2
length of side OP = 5
area of triangle OPQ = 3
```

Example 2 Let a and b be elements in $Cl_{3,0}$ where $a = 2e_1 + 3e_2 + 3e_3$ and $b = 3e_1 + e_2 + 4e_3$. Determine the algebra product ab, the dot product $a \cdot b$ and the outer product $a \wedge b$; Verify that $ab = a \cdot b + a \wedge b$.

Here is the program, Listing 2, for Example 2.

```
(1) #include "headers.h"
(2) #include "clifford.h"
(3)
(4) int main()
(5) {
(6)    clifford a(3, 0, "2b1 + 3b2 + 3b3"),
(7)             b(3, 0, "3b1 + 2b2 + 4b3");
(8)    clifford ans;
(9)
(10)   a.write("a = ");
(11)   b.write("b = ");
(12)
(13)   ans= a*b;
(14)   ans.write("ab = ");
(15)
(16)   ans= a | b;
(17)   ans.write("a.b = ");
(18)
(19)   ans= a ^ b;
(20)   ans.write("a^b = ");
(21)
(22)   clifford lhs = a*b;
(23)   clifford rhs( (a | b) + (a ^ b) );
(24)   lhs.write("ab = ");
(25)   rhs.write("a.b + a^b = ");
(26)
(27)   return 0;
(28)}
```

Listing 2
   Explanation:
Lines 1, 2: see example 1.
Lines 6, 7: create elements **a** and **b** in $Cl_{3,0}$ and initialize their values given by the string arguments in the constructor calls.

Line 8: create Clifford element **ans** using the default *constructor* (no-argument constructor); **ans** is used as a variable for storing answers.

Lines 10, 11: write to **stdout** (the standard output) the message given by the string argument in the call to the member function
clifford::write(char* msg) followed by the value of the clifford object for which the function was invoked.

Lines 13, 14: compute the algebra product **ab** and display the answer;
clifford::operator * (const clifford& cf) computes the algebra product.

Lines 16, 17: compute the dot product **a.b** and display the answer;
clifford::operator | (const clifford& cf) computes the dot product.

Lines 19, 20: compute the outer product **a^b** and display the answer;
clifford::operator ^ (const clifford& cf) computes the outer product.

Line 22: compute the left hand side, **ab**.

Line 23: compute the right hand side, **a.b + a^b**.

Lines 24, 25: display the left and right hand sides.

Output of the program in Listing 2:

```
a =
+2b1   +3b2   +3b3
b =
+3b1   +2b2   +4b3
ab =
+24  -5b12   -1b13   +6b23
a.b =
+24
a^b =
-5b12   -1b13   +6b23
ab =
+24  -5b12   -1b13   +6b23
a.b + a^b =
+24  -5b12   -1b13   +6b23
```

Example 3 In the Clifford algebra $C\ell_{2,1}$ let $x = 2e_1 + 6.2e_2 - 4.1e_{23}$ and $y = 7e_1 - 2.1e_2 + 9.6e_{12} + 6e_{23}$. Compute $2x + 3y$, x^5 and the third outer power of y. Find the even part of xy and write the result to a file called **eg3.res**.

Here is the program, Listing 3, for Example 3.

```
(1) #include "headers.h"
(2) #include "clifford.h"
(3)
(4) int main()
(5) {
(6)    clifford x(2, 1), y(2, 1), ans;
(7)    ofstream out_file;
(8)
(9)    x = "2b1 + 6.2b2 - 4.1b23";
```

```
(10)   y = "7b1 - 2.1b2 + 9.6b12 + 6b23";
(11)
(12)   x.write("x = ");   y.write("y = ");
(13)
(14)   ans= 2*x + 3*y;
(15)   ans.write("2x + 3y = ");
(16)
(17)   ans= x.pow(5);
(18)   ans.write("x to the power 5 (algebra product) = ");
(19)
(20)   ans= y.outer_pow(3);
(21)   ans.write("y to the outer power 3 = ");
(22)
(23)   clifford xy = x*y;
(24)   ans= xy.even();
(25)   ans.write("even part of xy = ");
(26)
(27)   out_file.open("eg3.res");
(28)   ans.write(out_file, "even part of xy = ");
(29)   out_file.close();
(30)
(31) return 0;
(32)}
```

Listing 3

Explanation:

Lines 1, 2: see example 1.

Line 6: create Clifford elements **x** , **y** and **ans** in $C\ell_{1,2},$.

Line 7: create the file variable out_file, an instance of **class ofstream**.

Lines 9, 10: assign values to **x** and **y**, using the assignment function
    clifford::operator = (const char* str).

Line12: display the values of **x** and **y** (to stdout).

Line 14: compute $2x + 3y$; uses
    friend operator * (double r, const clifford& cf) which computes the
    product of a real number and a clifford element.

Line 15: display the answer.

Line 17: compute x^5; clifford::pow(int n) computes the integral power.

Line 18: display the answer.

Line 20: compute the third outer power of y;
    clifford::outer_pow(int n) computes the integral outer power.

Line 21: display the answer.

Line 23: create element **xy** with the algebra product of **x** and **y** assigned to it.

Line 24: find the even part of **xy**.

Line 25: display the answer (to stdout).

Line 27: open a connection which associates the file **eg3.res** on disk to the file
    variable out_file in the program.

Line 28: write the answer to the output file.

Line 29: close the file for writing; disconnect the file variable from the file on disk.
    Output of the program in Listing 3:

```
x =
+2b1   +6.2b2   -4.1b23
y =
+7b1   -2.1b2   +9.6b12   +6b23
2x + 3y =
+25b1   +6.1b2   +28.8b12   +9.8b23
x to the power 5 (algebra product) =
+15526.985b1   +23433.0395b2   +12049.08b13   -19382.84225b23
y to the outer power 3 =
0
even part of xy =
-23.62 -47.6b12   +39.36b13
```

Contents of the output file **eg3.res** (created and written to by the program):

```
even part of xy = -23.62 -47.6b12   +39.36b13
```

Example 4 Compute the Lorentz invariants, energy density and Poynting vector of the electromagnetic field $\mathbf{F} = \mathbf{E} - \mathbf{jB}$ with the electric component $\mathbf{E} = \mathbf{e}_1 + 2\mathbf{e}_2 + 4\mathbf{e}_3$ and the magnetic component $\mathbf{B} = 3\mathbf{e}_1 + 5\mathbf{e}_2 + 7\mathbf{e}_3$. The element $\mathbf{j}$ is the unit volume element of $\mathbb{R}^3$.

Here is the program, Listing 4, for Example 4.

```
(1) #include "headers.h"
(2) #include "clifford.h"
(3)
(4) const clifford j(3, 0, "b123");
(5)
(6) int main()
(7) {
(8)    clifford E(3, 0), B(3, 0), F, ans;
(9)
(10)   E= "b1 + 2b2 + 4b3";
(11)   B= "3b1 + 5b2 + 7b3";
(12)   F= E - j*B;
(13)
(14)   F.write("electromagnetic field F = ");
(15)
(16)   ans= F*F / 2;
(17)   ans.write("ans (Lorentz invariants): ");
(18)
(19)   ans= -F.involute() * F/2;
(20)   ans.write("ans (energy density and Poynting vector): ");
(21)   return 0;
(22)}
```

Listing 4

Explanation:

Lines 1, 2: see example 1.

Line 4: create the unit director $j = e_{123}$ of $\mathbb{R}^{3,0}$, a constant in the program (j).

Line 8: create Clifford elements E and B in $C\ell_{3,0}$ with their values initialized to zero. The three-argument constructor is used with an empty string as the default third argument. Create Clifford elements F and **ans** using the default constructor.

Lines 10, 11: assign values to E and B, calling the assignment operator, `clifford::operator = (const char* str)`.

Line 12: compute F = E - jB.

Line 14: display F with the message shown.

Lines 16, 17: compute F^2 / 2 and display the answer; Lorentz invariants: $1/2(\mathbf{E}^2 - \mathbf{B}^2) = -31$ and $\mathbf{E} \cdot \mathbf{B} = 41$.

Lines 19, 20: compute -F'F/2 and display the answer; Energy density and Poynting vector:

$$1/2(\mathbf{E}^2 + \mathbf{B}^2) = 52, \mathbf{E} \times \mathbf{B} = -6e_1 + 5e_2 - e_3.$$

Output of the program in Listing 4:

```
electromagnetic field F =
+1b1    +2b2    +4b3    -7b12    +5b13    -3b23
ans (Lorentz invariants) :
-31 -41b123
ans (energy density and Poynting vector) :
+52 -6b1    +5b2    -1b3
```

Example 5 In the Cayley algebra $C\ell_{0,7}$ compute the octonion product of $a = 3 + e_1 + 4e_2$ and $b = 2 + 3e_2 + 5e_3$.

Here is the program, Listing 5, for Example 5.

```
(1) #include "headers.h"
(2) #include "clifford.h"
(3)
(4) const clifford b124(0, 7, "b124"),
(5)                 b235(0, 7, "b235"),
(6)                 b346(0, 7, "b346"),
(7)                 b457(0, 7, "b457");
(8) const clifford e=(1 - b124)*(1 - b235)*(1 - b346)*(1-b457);
(9)
(10) int main()
(11) {
(12)    clifford a(0, 7, "3 + b1 + 4b2"),
(13)    clifford b(0, 7, "2 + 3b2 + 5b3");
(14)
(15)    clifford ans= real(a*b*e) + pure(1, a*b*e);
(16)
(17)    a.write("a = ");
```

```
(18)    b.write("b = ");
(19)    ans.write("octonion product of a and b = ");
(20)
(21)    return 0;
(22) }
```

Listing 5
Explanation:
Lines 1, 2: see example 1.
Lines 4-5: create Clifford elements b124, ... , b457 in $Cl_{0,7}$ as symbolic names
for the tri-vectors $e_{124}, \ldots, e_{457}$ respectively.
Line 8: create element e as shown.
Lines 12, 13: create Clifford elements a and b in $Cl_{0,7}$, using the string arguments
(see example 1 for details).
Line 15: compute the octonion product of a and b using the formula,

$$oct(a, b) = Re(abe) + Pu(1, abe).$$

The function `clifford real(const clifford& cf)` returns the real part of
`cf`. The function `clifford pure(int kvec, const clifford& cf)` returns
the pure `kvec`-vector part of `cf`. Line 15 could also have been written as
`clifford ans= (a*b*e).real() + (a*b*e).pure(1)`, using the member functions `clifford::real()` and `clifford::pure(int kvec)` respectively.
Lines 17-18: write a, b and the answer to stdout.
Output of the program in Listing 5:

```
a =
+3 +1b1   +4b2
b =
+2 +3b2   +5b3
octonion product of a and b =
-6 +2b1   +17b2   +15b3   +3b4   +20b5   +5b7
```

Example 6 In $Cl_{0,7}$ compute the outer exponential of the bivector $\mathbf{B} = e_{12} + e_{34} + e_{56}$.

Here is the program, Listing 6, for Example 6.

```
(1) #include "headers.h"
(2) #include "clifford.h"
(3)
(4) clifford expout(const clifford& f)
(5) {
(6) return( 1 + f + (f^f)/2 + (f^f^f)/6 );  // finite series
(7) }
(8)
(9) int main()
```

```
(10) {
(11)    clifford B(0, 7, "b12 + b34 + b56");
(12)
(13)    clifford ans = expout(B);
(14)
(15)    B.write("B = ");
(16)    ans.write("outer exponential of B = ");
(17)
(18)    return 0;
(19) }
```

Listing 6
Explanation:
Lines 1, 2: see example 1
Lines 4-7: declare and define the function **expout** which returns the outer exponential of its argument. The outer exponential is evaluated by the finite series (i. e. a series with finitely many non-zero coefficients), which converges to a rational value.
Line 11: create the bivector **B** in $Cl_{0,7}$.
Line 13: compute the outer exponential of **B**.
Lines 15, 16: display the answer.

Output of the program in Listing 6:

```
B =
+1b12   +1b34   +1b56
outer exponential of B =
+1 +1b12   +1b34   +1b56   +1b1234   +1b1256   +1b3456   +1b123456
```

Example 7 The example given in Listing 7 may be of interest to readers who are familiar with the *Triality* concept described in the second edition of Porteous' book, *Topological Geometry*. See (Porteous, 1981).

Here is the program, Listing 7, for Example 7.

```
(1) #include "headers.h"
(2) #include "clifford.h"
(3)
(4)  const clifford
(5)        v   (8, 0, "-b124 -b235 -b346 -b457 -b561 -b672 -b713"),
(6)        b1_7(8, 0, "b1234567"),
(7)        b1_8(8, 0, "b12345678"),
(8)        b8  (8, 0, "b8");
(9)  const clifford w = v * b1_7;
(10)
(11) clifford oct(const clifford& x, const clifford& y);
(12) clifford expo(const clifford& f);
(13)
```

```
(14) int main()
(15) {
(16)    clifford F(8, 0, "3b12 + 4b23 + 4b26 + 5b37 + b45 + 2b67");
(17)    F= F/10;
(18)    clifford u = expo(F);
(19)    u= u/(dabs(u));
(20)    clifford x(8, 0, "3b1 + 4b3 + 5b5");
(21)    clifford y(8, 0, "2b2 + 3b4 + 7b7");
(22)
(23)    clifford ans= oct( u*x*u.reverse(), u*y*u.reverse() );
(24)    ans.write("ans = oct( u*x*u~, u*y*u~ ) = ");
(25)    return 0;
(26) }
(27)
(28) clifford oct(const clifford& x, const clifford& y)
(29) {  return( pure(1, x*b8*y*(1-b1_8)*(1+w)) );  }
(30)
(31) clifford expo(const clifford& f)
(32) {  return(1 + f + (f^f)/2 + (f^f^f)/6 + (f^f^f^f)/24);  }
```

Listing 7

Explanation:

Lines 1, 2: see example 1

Lines 4-8: create constant Clifford elements in $C\ell_{8,0}$ in the usual way (see example 1 for details)

Line 9: create element w using the copy constructor,

    `clifford::clifford(const clifford& cf)`, w is created as a copy of v*b1_7.

Lines 11, 12: declare function prototypes— oct computes the octonion product of its two arguments (as in example 4) and expo computes the outer exponential of its argument (as in example 5).

Line 16: create F in $C\ell_{8,0}$ in the usual way (see example 1 for details).

Line 17: divide F by 10.

Line 18: create u, initializing it to expo(F).

Line 19: normalize u.

Lines 20, 21: create elements x and y in $C\ell_{8,0}$ in the usual way (see example 1 for details)

Line 23: compute the octonion product of $u x \tilde{u}$ and $u y \tilde{u}$.

Line 24: display the above answer which is shown below—the coefficient of the last term is "effectively" zero. Numerical range/domain/overflow/underflow error handling has been intentionally kept "open" to allow experimentations before implementing robust and conceptually correct algorithms.

Lines 28-29: definition of the function oct.

Lines 31-32: definition of the function expo.

    Output of the program in Listing 7:

```
ans = oct( u*x*u~, u*y*u~ ) =
+22.458192b1  -31.005692b2  -1.719467b3  +28.359217b4
```

-21.931417b5   -18.188871b6   -3.905487b7   +4.278559e-15b8

Note: The last term is "effectively" zero.

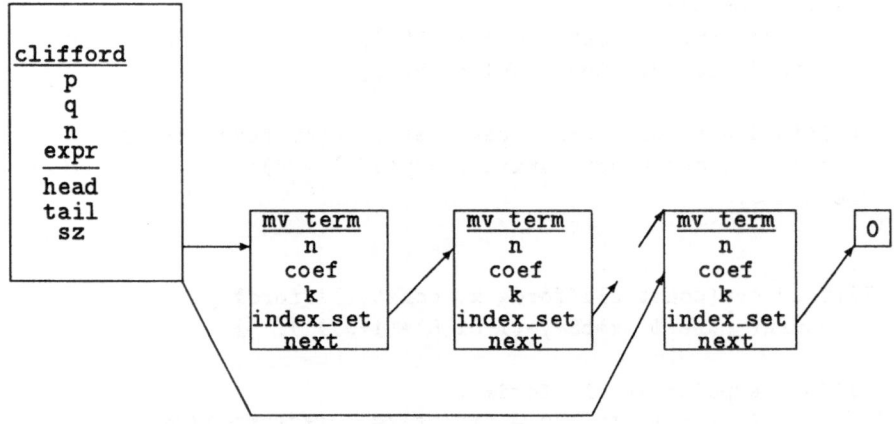

Fig. 1.   Data structures for the clifford class

3. Implementation details

In this section, I describe principal features of the implementation. Figure 1 illustrates the data structures used. The basic ideas depicted in Figure 1 were drawn from (Vierros, 1987).

class mv_term implements a multi-vector term—a node in singly linked list. For example the term **2b123** in $Cl_{2,1}$ is represented as an instance of **mv_term** with *private* data fields:
n=3 (= 2+1 = dimension of the underlying quadratic space $\mathbb{R}^{2,1}$) , **coef=2.0** (the coefficient), **k=3** (3-vector) and **index_set** = { 1, 2, 3 } (set of indices). **index_set** is an instance of **class int_set**, which implements a subset of integers within a specified range. The basic ideas for **class int_set** were drawn from (Näher, 1994) and (Mehlhorn *et al.*, 1994) —*LEDA, A library of Efficient Data Types and Algorithms.* In the author's opinion, *LEDA* is an excellent C++ class library, which is also very well documented. The **next** field of **mv_term**, a pointer to **mv_term** (**mv_term***), points to the next term as is usual with linked lists.

class mv_list is a singly linked list of **mv_terms**, representing the expression for a Clifford element. For example **2b1 + 3b12 + 8b234** in $Cl_{3,1}$ say, is realized by an instance of **mv_list**, containing three **mv_terms**. Sentinel (dummy) nodes **head** and **tail** are used in order to simplify (and make more efficient) the code for list operations such as **insert**. Basic ideas for list operations were drawn from the very readable book by Sedgewick (Sedgewick, 1991).

class clifford implements a Clifford element of $C\ell_{p,q}$ in general. The clifford class constitutes a comfortable high-level data type. It enables one to write programs for calculations in a form close to the usual text book presentations.

class clifford and **class mv_list** are *friends* of **class mv_term**. Functions for input/output such as

```
ostream::operator << (ostream& os, const clifford& cf)
```

are also *friends* of **mv_term**. This arrangement is not permanent and will change as the data protection and data hiding features of the library are further improved.

Appearing below is a collection of remarks and comments pertaining to various details.

3.1. REMARKS AND COMMENTS

– The function **clifford::clifford(int p_sig, int q_sig, char* str)** is an important one. It creates the data objects for a clifford element by parsing a string representation. The argument **str** is passed by reference to the **insert** operation of **expr**
(**mv_list::insert(int dim, char*& s)**) which in turn passes it by reference to
mv_term::mv_term(int dim, char*& str,mv_term* nx). Parsing for a multi-vector term is done by the **mv_term** constructor and the argument string is modified. The clifford constructor does this repeatedly until the string has been scanned. Further the indices for each multi-vector term are kept sorted and the sign of the coefficient is appropriately adjusted. Thus the string may have indices in any order, for example "**2b32**" forms valid input—the value corresponding to "**-2b23**" is stored. Repeated indices however are not allowed, e.g. "**2b33**" is not valid. (The insert operation does "too much" work for a critical "inner loop" function. This slows down computations considerably. It will be modified in future versions).

– At present, Clifford algebras for quadratic spaces of dimension greater than 9 can not be realized via the constructor mentioned above. This is only because the syntax for a string containing multi-digit indices hasn't been finalized. Here is an example of a plausible syntax: the string "**2b_3_10_11**" could represent a term in say $C\ell_{10,2}$ where the coefficient is 2 and the index set is $\{3, 10, 11\}$. However the present form of the library *does* allow users to implement higher dimensional Clifford algebras. The procedure is outlined as follows: Create the required dimension values, coefficients, k values and index sets (instances of **int_sets**); Insert them into an object of type **mv_list** using
mv_list::insert(int dim, double c, int kk, const int_set& iset).
Use the constructor
    clifford::clifford(int p_sig, int q_sig, const mv_list& mvl)
to create the required Clifford element. Section 4 documents the functions needed.

– The insert operations for **mv_list**,
mv_list::insert(int dim, double c, int kk, const int_set& iset)
and **mv_list::insert(int dim, char*& str)** maintain a sorted list as well

as performing "in-situ" addition and subtraction. The terms are kept sorted according the k field of **mv_term**; then the k-vectors for each **k** are sorted according to an ordering defined for the **index_sets**. For instance the index set { 2, 3, 5 } is "less than" { 2, 4, 6 }. Collection of like terms, i. e. addition of terms with the same index set (hence the same k value), is also done by the insert operations. The list is appropriately modified. Terms with zero coefficients are not inserted into the list.

— The argument for **k** does not need to be passed in the **mv_term** constructor, since **k** is always equal to the size (cardinality) of the index set. However it remains there, due to the author's personal preference (and laziness !)

— Operations and functions for computing the sum, algebra product, outer product, involute, conjugate etc. are based on standard results given in (Riesz, 1958), (Porteous, 1969) and (Vierros, 1987). The Pascal and FORTRAN source code for CLICAL was also referred to.

4. Function reference

Given below are some of the *global, member* and *friend* functions of the library.

4.1. Class clifford: public member functions

The inclusion of some of the member functions violates the principle of keeping the class minimal and complete (see (Meyers, 1992)), but are there at present for various reasons, such as reducing function call overheads.

```
clifford();
```
no-argument (default) constructor: sets p, q and n to 0, sets expr to an empty list, i.e. with head->next set to tail; k data field of tail is set to MAXDIM+1 (MAXDIM= 9 at present).
```
clifford(int p_sig, int q_sig, const mv_list& mvl);
```
constructs a clifford element with signature (p_sig, q_sig), data of mvl is copied into expr.
```
clifford(int p_sig, int q_sig, char* str="");
```
constructs a clifford element with signature (p_sig, q_sig); parses str for mv_term s and inserts them into expr.
```
clifford(const clifford& cf);
```
copy constructor, constructs a clifford element and makes if a copy of cf.
```
~clifford();
```
destructor, automatically called when a clifford element goes out of scope, destructors for mv_list, mv_term and int_set are called automatically.
```
clifford& operator = (const clifford& cf);
```
*this is made into a copy of cf.
```
clifford& operator = (char* s);
```
assigns clifford(p, q, s) to *this, returns *this
```
int dim_p() const;
```
returns p, "positive part" of signature.

```
int dim_q() const;
returns q, "negative part" of signature.
int dim_n() const;
returns n, n= p+q.
mv_list expression() const;
returns expr.
void set_sig(int p_, int q_);
resets p to p_ and q to q_.
clifford  involute()  const;
returns the involute (main involution) of *this.
clifford  reverse()   const;
returns the reverse (reversion) of *this
clifford  conjugate() const;
returns the conjugate of *this.
double   dabs()   const;
returns the absolute value as a double, for definite quadratic
forms only.
clifford abs()     const;
returns the absolute value as a clifford element - a 0-vector
with coefficient equal to dabs().
double  dreal() const;
returns real part, as double.
clifford  real() const;
returns real part, clifford element - a 0-vector with
coefficient equal to dreal().
double   dquad()  const;
returns quadratic form, as double.
clifford       quad()  const;
returns quadratic form, clifford element - 0-vector with
coefficient equal to dquad().
clifford       operator + (const clifford& cf) const;
ordinary componentwise sum, returns *this + cf.
clifford       operator - (const clifford& cf) const;
ordinary componentwise difference, returns *this - cf.
clifford       operator - () const;
unary minus, returns  -(*this).
clifford       operator * (const clifford& cf) const;
algebra product, returns *this * cf.
clifford       operator + (double r) const;
returns *this + r.
friend  clifford operator - (double r, const clifford& cf);
returns r - cf.
clifford operator * (double r) const;
scalar multiplication, returns *this * r.
clifford operator / (double r) const;
division by a scalar, returns *this / r.
```

```
clifford&        operator += (const clifford& cf);
returns *this = *this + cf.
clifford&        operator -= (const clifford& cf);
returns *this = *this + cf.
clifford&        operator *= (const clifford& cf);
returns *this = *this * cf.
clifford         pow(int m) const;
integral power, returns *this to the m.
clifford         outer_pow(int m) const;
outer product power, returns *this ^ *this ^ .... ^*this,
m factors.
clifford         operator ^ (const clifford& cf) const;
outer product, returns *this ^ cf.
clifford&        operator ^= (const clifford& cf);
returns *this = *this ^ cf.
clifford         operator | (const clifford& cf) const;
inner product, returns (*this) . (cf).
clifford         pure(int kvec) const;
returns pure kvec-vector part.
double  pure() const;
returns coefficient of this->pure(n), i.e. pure(n) / j,
where j=b12...n, the unit director
clifford         even() const;
returns even part of *this, i.e. sum of pure(k) with even k
friend  ostream& operator << (ostream& os, const clifford& cf);
writes cf to output stream os.
void write(char* msg="") const;
writes message msg to stdout, then writes *this.
void write(ofstream& ofile, char* msg="") const;
writes message msg to output file stream ofile, then writes
*this.
void error(const char* msg) const;
writes message msg to stderr and terminates the program with
a call to exit(1).
```

4.2. GLOBAL AND FRIEND FUNCTIONS PERTAINING TO CLASS CLIFFORD

Most of the functions listed or mentioned here are unnecessary because their tasks can be achieved using the functions in Section 4.1 However they are included, at present, to provide alternative functional notations. This is to help accommodate potential users' preferences. For example CLICAL users may prefer to write **pure(2, x)** instead of **x.pure(2)** to extract the pure 2-vector part of **x**—both expressions achieve the same task.

```
inline void clifford::set_sig(int p_, int q_);
resets the signature to (p_, q_), precondition: p_ + q_ = n.
inline clifford invo(const clifford& cf);
returns cf.involute().
```

```
inline clifford rev(const clifford& cf);
returns cf.reverse().
inline clifford conj(const clifford& cf);
returns cf.conjugate().
inline double dabs(const clifford& cf);
returns cf.dabs().
inline clifford abs(const clifford& cf);
returns cf.abs().
inline double dreal(const clifford& cf);
returns cf.dreal().
inline clifford real(const clifford& cf);
returns cf.real().
inline double  dquad(const clifford& cf);
returns cf.dquad().
inline clifford quad(const clifford& cf);
returns cf.quad().
inline clifford pure(const clifford& cf, int kvec);
returns cf.pure(kvec).
inline clifford pure(int kvec, const clifford& cf);
returns cf.pure(kvec), same functional as CLICAL
inline double pure(const clifford& cf);
returns coefficient of pure n-vector part
inline clifford even(const clifford& cf);
returns even part of cf, i.e. cf.even()
inline clifford operator + (double r, const clifford& cf);
returns r + cf
inline clifford operator - (const clifford& cf, double r);
returns cf -r.
inline clifford operator * (double r, const clifford& cf);
scalar multiplication from the left, returns cf*r.
```

4.3. PRIVATE MEMBER FUNCTIONS OF CLASS MV_LIST USED BY CLIFFORD MEMBER
     FUNCTIONS

```
mv_list operator + (const mv_list& mvl) const;
returns "sum" of *this and mvl.
mv_list operator - (const mv_list& mvl) const;
returns " *this - mvl ".
mv_list operator * (double r) const;
returns list in which each term of *this is multiplied by r.
mv_list operator / (double r) const;
returns list in which each of term of *this is divided by r.
```

4.4. PUBLIC MEMBER FUNCTIONS OF CLASS MV_LIST

```
mv_list();
create an empty list, i.e. head and tail are created with
sentinel values and head->next is set to tail.
```

```
~mv_list();
```
destructor, all terms in the list are freed from the free
store.
```
mv_list(const mv_list& mvl);
```
copy constructor, *this is made a copy of mvl.
```
mv_list& operator = (const mv_list& mvl);
```
assignment, *this is made a copy of mvl.
```
void insert(int maxi, double c, int kk, const int_set& iset);
```
inserts an mv_term via the term's info fields supplied by the
arguments.
```
void insert(const mv_term& mvt);
```
inserts mvt into *this.
```
void insert(int maxi, char*& str);
```
inserts an mv_term by parsing str, str is set to point to a
possibly next term.
```
void clear();
```
make list empty, i.e. all terms except head and tail are freed
from the free store, head->next is set to tail.
```
int size()  const;
```
returns the size = number of terms in list.
```
int empty() const;
```
returns 1 if list is empty.
```
friend ostream& operator << (ostream& os, const mv_list& mvl);
```
writes terms in mvl to output stream os.
```
void error(const char* msg) const;
```
writes the error message msg to stderr and terminates the
program with a call to exit(1).

4.5. PRIVATE MEMBER FUNCTIONS OF CLASS MV_TERM USED BY MV_LIST AND CLIFFORD MEMBER FUNCTIONS

```
int square(int j, int p_sig, int q_sig) const;
int dim_ok(const mv_term& mvt) const;
mv_term multiply(const mv_term& mvt, int p_sig, int q_sig)
const;
```

4.6. PUBLIC MEMBER FUNCTIONS OF CLASS MV_TERM

```
mv_term(int maxi, double c, int kk, const int_set& iset,
mv_term* nx);
```
constructs an mv_term object, data fields n, coef, k,
index_set, next are set to maxi, c, kk, iset, nx respectively.
```
mv_term(int maxi, char*& str, mv_term* nx);
```
constructs an mv_term, parses str for one term, str is set to
point to a possible next term, next is set to nx.
```
~mv_term();
```
destructor, automatically called when an mv_term goes out of
scope, destructor for index_set is called automatically.

```
mv_term(const mv_term& mvt);
```
copy constructor, *this is made a copy of mvt, except that
next is set to 0 mv_term& operator = (const mv_term& mvt);
assignment: *this is made a copy of mvt except that next is
set to 0 int dim() const;
returns n.
```
double coeff()  const;
```
returns coef.
```
int   multi()  const;
```
returns k.
```
int_set  indices() const;
```
returns index_set.
```
int   operator == (const mv_term& mvt) const;
```
returns 1 if *this' data fields are equal to mvt's data fields,
next is ignored.
```
int   operator != (const mv_term& mvt) const;
```
negation of above.
```
friend ostream&  operator << (ostream& os, const mv_term& mvt);
```
writes mvt's data to the output stream os.
```
friend  ostream&  operator << (ostream& os, const mv_list& mvl);
```
writes terms of mvl to the output stream os.
```
void error(const char* msg) const;
```
writes error message msg to stderr, terminates the program
with a call to exit(1).

4.7. PUBLIC MEMBER FUNCTIONS OF CLASS INT_SET

```
mv_term(int maxi, double c, int kk, const int_set& iset,
mv_term* nx);
```
constructs an mv_term object, data fields n, coef, k,
index_set, next are set to maxi, c, kk, iset, nx respectively.
```
mv_term(int maxi, char*& str, mv_term* nx);
```
constructs an mv_term, parses str for one term, str is set to
point to a possible next term, next is set to nx.
```
~mv_term();
```
destructor, automatically called when an mv_term goes out of
scope, destructor for index_set is called automatically.
```
mv_term(const mv_term& mvt);
```
copy constructor, *this is made a copy of mvt, except that
next is set to 0 mv_term& operator = (const mv_term& mvt);
assignment: *this is made a copy of mvt except that next is
set to 0 int dim() const;
returns n.
```
double coeff()  const;
```
returns coef.
```
int   multi()  const;
```
returns k.
```
int_set  indices() const;
```

returns index_set.
int   operator == (const mv_term& mvt) const;
returns 1 if *this' data fields are equal to mvt's data fields,
next is ignored.
int   operator != (const mv_term& mvt) const;
negation of above.
friend ostream& operator << (ostream& os, const mv_term& mvt);
writes mvt's data to the output stream os.
friend ostream& operator << (ostream& os, const mv_list& mvl);
writes terms of mvl to the output stream os.
void error(const char* msg) const;
writes error message msg to stderr, terminates the program with
a call to exit(1).

int_set(int l=1, int h=1);
construct a subset of integers from l (low) to h (high),
initialize to empty set.
int_set(const int_set& b);
copy constructor, *this is made a copy of b.
~int_set();
destructor, called automatically when an int_set object goes
out of scope.

int_set& operator = (const int_set& b);
assignment, makes *this a copy of b, returns *this.
int operator [] (int pos) const;
returns element in "position" pos.
int max_size() const;
returns maximum size (cardinality) of *this.
int low()   const;
returns lower bound.
int high() const;
returns higher bound.
int empty() const;
returns 1 if the set is empty.
int size() const;
returns current size (cardinality) of set.
void clear();
makes the set empty.
int member(int x) const;
returns 1 if x is a member of *this.
void insert(int x);
inserts x into *this.
void remove(int x);
removes x from *this.
int min()   const;

```
returns minimum member of *this.
int_set  operator + (const int_set& b) const;
returns union of *this and b.
int_set  operator * (const int_set& b) const;
returns intersection of *this and b.
int_set  operator~();
returns complement of *this.
int_set  operator - (const int_set& b) const;
returns the difference *this - b (all elements in *this which
are not in b).
int operator == (const int_set& b) const;
returns 1 if *this== b.
int operator != (const int_set& b) const;
negation of above.
```

```
friend ostream& operator << (ostream& os, const int_set& S);
writes elements of S to the output stream os.
void error(const char* msg) const;
writes the error message msg to stderr and terminates the
program with a call to exit(1).
int compare(const int_set& a, const int_set& b);
global function, linear ordering on int_set s,  returns -1
if a < b, +1 if a > b and 0 if a==b, e.g. {2, 3, 5} is "less
than" {2, 4, 7}; analogous to strcmp().
```

References

R. Lafore: 1991, *Object-oriented Programming in Turbo C++*. Waite Group Press.

P. Lounesto, R. Mikkola, V. Vierros: 1987, *CLICAL User Manual*. Research report A248, Institute of Mathematics, Helsinki University of Technology.

P. Lounesto: 1992, 'Clifford algebra calculations with a microcomputer'. In the book: A. Micali *et al.*, *Clifford Algebras and their Applications in Mathematical Physics*. Kluwer Academic Publishers, pp. 39–55.

K. Mehlhorn and S. Näher: 1994, *LEDA — A Library of Efficient Data Types and Algorithms*. Max-Planck-Institut für Informatik.

S. Meyers: 1992, *Effective C++*. Addison-Wesley.

S. Näher: 1994, *LEDA User Manual, Version 3.0*. Max-Planck Institut für Informatik.

I. Porteous: 1969, *Topological Geometry*. Van Nostrand Reinhold.

I. Porteous: 1981, *Topological Geometry*. Second Edition. Van Nostrand Reinhold.

M. Riesz: 1958, *Clifford Numbers and Spinors*. Maryland 1958, University of Maryland, The Institute for Fluid Dynamics and Applied Mathematics, Lecture Series No. 38; reproduced in E. Folke Bolinder, P. Lounesto (eds.) , *Clifford Numbers and Spinors*. Kluwer Academic Publishers.

R. Sedgewick: 1992, *Algorithms in C++*. Addison-Wesley.

B. Stroustrup: 1991, *The C++ Programming Language*. 2nd edition. Addison-Wesley.

V. Vierros: 1987, title in Finnish: *Cliffordin algebroilla laskeva ohjelma mikrotietokoneelle*. Title in English: *A program for calculations in Clifford algebras with a microcomputer*. "Diplomityö" (M.Sc. thesis), Helsinki University of Technology.

INDEX

A.

AXIOM vii, viii
affine space 147
affinor 59
ambient, coframe 128
ambient, frame 141
ambient, space 146
analytic function theory on quaternions 219, 220
antisymmetric tensors 59
approximants, diagonal 116, 118
approximants, Padé xi, 111, 113, 117
approximants, vector-valued rational 111
atomic surfaces, transversality of 264

B.

Baker condition 118
baryon octet 95
Bianchi identities 47
bilinear covariants 13, 18, 151
bilinear form, arbitrary xii, 168, 172
bilinear form, symbolic xii, 179, 184
bilinear form, with antisymmetric part xii, 167, 177, 180
bilinear forms, non-symmetric 22, 23
bipolar coordinates 63
bivectors 169, 286, 287
bivectors, exponentiation of 7
bivectors, orthogonal 7
blades 35, 234, 238, 259
boost 77, 81, 82
Bravais' Law 259

C.

C++ viii, ix, xiii, xiv, 285, 286, 290, 295
Cartan structural equations 47
Cartan subalgebra 103
Casimir operator 101, 108
Cauchy-Riemann equations 213, 220, 226
CAYLEY vii
Cayley algebra 233

Cayley-Dickson, algebra 26
Cayley-Dickson, procedure 229
Cayley-Dickson, process 26
Cayley-Klein parameters 37
cells, primitive 252
cells, unit 252
center of perspectivity 248
central operators 94
characteristic 2 22
charge conjugation 13, 18, 135, 136, 137, 141
charge conjugation, constant matrix of 139
charge density 39
charge operator 94
CLICAL vii, viii, xi, xiii, xiv, 3, 9, 20, 27, 70, 83, 89, 91, 96, 177, 233, 234, 244, 245, 250, 285, 295, 310
cliffor 47, 59, 63, 71
cliffor, Pauli 80, 81
cliffor-valued functions 78, 80
Clifford, William Kingdon xiv
Clifford algebra, as subalgebra of endomorphism algebra 167, 177
Clifford algebras, anti-involutions in 10, 12, 35, 123, 137
Clifford algebras, basis monomials in 185
Clifford algebras, central elements in 6, 69
Clifford algebras, Clifford conjugation in 4, 15, 35, 61, 66, 73, 155, 160
Clifford algebras, Clifford product in 35, 36, 59, 74, 75, 86, 121, 122, 168, 175
Clifford algebras, commutative xii, 189
Clifford algebras, commutative, idempotent basis in 189
Clifford algebras, complex 61, 192
Clifford algebras, complexified 10, 11, 126
Clifford algebras, deformations in 26
Clifford algebras, derivation in 24
Clifford algebras, even multivectors in 36, 37, 46
Clifford algebras, even subalgebra of 6, 272
Clifford algebras, exponentials in 5, 190, 219
Clifford algebras, generalized xii, 101, 102, 167, 187, 229, 230
Clifford algebras, generalized, left ideals in 103, 106, 272
Clifford algebras, generalized, matrix repre-

318

sentations of 101, 103

Clifford algebras, generalized, nilpotent elements in 103, 193

Clifford algebras, generalized, reversion in 181

Clifford algebras, gradation in 167, 171

Clifford algebras, grade involution in 5, 35, 60, 66, 87, 123, 155, 168

Clifford algebras, idempotents in 78, 126, 128, 189, 190, 193

Clifford algebras, idempotents in, primitive 88

Clifford algebras, in characteristic 2 23

Clifford algebras, invertible vectors in 5

Clifford algebras, left contraction in xii, 24, 167, 168, 170, 171, 178

Clifford algebras, logarithms in 5, 190

Clifford algebras, matrix representations of 121, 129, 155, 173, 184, 269, 272

Clifford algebras, minimal left ideals in 10, 11, 18, 88, 126

Clifford algebras, multilinearity in 121

Clifford algebras, multivector groups in 270

Clifford algebras, multivector structure of 4, 18, 22, 23, 167, 168

Clifford algebras, multivectors in 35, 270

Clifford algebras, multivectors in, isotropic 273, 278

Clifford algebras, multivectors in, simple 254

Clifford algebras, non-degenerate 4

Clifford algebras, orthonormal basis sets for 269, 274

Clifford algebras, outermorphisms of 248

Clifford algebras, partial involution in 87

Clifford algebras, polivectors in 102

Clifford algebras, projectors in 270, 272

Clifford algebras, pseudoscalar in 70, 235, 248

Clifford algebras, real, classification of 269, 270

Clifford algebras, real, factorizations of 270

Clifford algebras, reversion in xii, 5, 6, 8, 13, 35, 59, 60, 66, 113, 123, 155, 168, 180, 234

Clifford algebras, unitary multivectors in 36

Clifford algebras, wedge product in 170, 171, 178

Clifford group 5, 83, 132, 156

Clifford map xiii, 172, 269

Clifford numbers 193, 194

Clifford polynomial 177, 183

cognitive growth 3

color operator 90, 99

column spinor 18

complex plane 288

complex spinor algebra 121

complex spinor fields 140

computer algebra vii, viii, ix, 33, 34, 51, 83, 146, 213, 225, 251, 278

conformal transformations xii, 37, 20, 155

conjectures 3

connexion 1-form 47

connexion coefficients 41

connexion, metric compatible 47

connexion, Riemannian 41, 42, 48

continued fraction expansion 113

cotangent space 124

Coulomb field 45, 46

counter-examples 3, 22, 28

covariant derivative 40, 49

crystallography xii, 251, 255

cubic equation 189, 191

cubic equation, reduced 191

current density 1-vector 39, 44

curvature 38, 48, 141

D.

d'Alembertian 80

Darboux 2-vector 38

DERIVE vii, viii, 226

differential forms 34, 42

differential forms, Clifford algebra valued 47

dihedral group 83, 87, 89, 272

dilations 20, 37, 81, 163

Dirac adjoint 20

Dirac algebra 83, 84, 92, 94

Dirac bi-spinor equations 65

Dirac current density 19

Dirac equation x, 13, 15, 65, 140

Dirac equation, massless 142

Dirac matrices viii, 92, 109, 134

Dirac operator 40, 41, 52, 121

Dirac spinor 9, 19, 37

Dirac spinor fields 145

Dirac-Hestenes spinors 45

discrete Fourier transform 103

division algebras 201, 202, 229
division rings 10
dual frame 124
dual vectors 122
duality transformations 20

E.
eigenfunctions 101
eigenspinors 13, 17
electric conjugation 15, 16
electromagnetic field 39, 45, 301
electromagnetic moment density 19, 40
electromagnetic potential 140
ellipton 230
Emden's equation 213, 224, 225
energy-momentum 1-vector 45
Euler angles 81, 146
exterior algebra x, 3, 22, 121, 251
exterior algebra fields 58
exterior algebra, endomorphism algebra of 167
exterior algebra, gradation in 167
exterior derivative 47
exterior exponential 7
exterior forms 47
exterior product xiv, 58, 61, 252
extra special p-groups 270

F.
falsifying strategy vii, x, 3
Faraday 2-vector 44, 45
fiducial derivative 41, 43 field equations 213
field strength 49
fields, anti self-dual 50
fields, electrically charged 135
fields, self-dual 50
fields, square class-structure of 8
Fierz identities 18, 20
finite reflection group 83, 88, 91
flats 253, 254
flexible algebras 26
FORTRAN viii, ix, xiii, 270, 274, 275, 295, 308
Frenet equations 38
Frenet frame 38
fullerenes xiii, 251

G.
gauge fields x, 34, 49
gauge potential 49
Gell-Mann matrices 83, 94
Gell-Mann-Nishijima relation xi, 93, 94, 95
generalized spheres 157
geometric algebra xi, xii, xiii, 67, 233, 248, 255, 285, 290
geometric product 289
Gibb's calculus 255
Grassmann product 58, 59
Grassmann-Cartan differential forms 58
gravitation 121
group rings 213, 217
Gröbner basis viii
gyromagnetic factor 40

H.
hadrons 95
Hearn, Anthony C. viii
Hodge dual 76
Hodge duality operator 50, 64
Hurwitz Theorem 229
hyperbolic rotations 39
hyperbolon 230
hypercharge 94, 99
hypercomplex algebra 213, 216, 217
hyperoctahedral group 83, 87

I.
icosahedral symmetry 256
idempotents 9, 97
idempotents, mutually annihilating primitive 80, 189, 193
idempotents, primitive 10, 15
instanton number 51
intrinsic angular momentum 85
isospin 94, 99
isotropic subspaces, maximal 273
isotropic vectors 273

K.
k-cubic forms 229, 230
Kerr-Newman solution 49

L.
Lamé coefficients 63
Laplace-Beltrami operator 42, 44
Laplacian 42, 58, 64, 65
Last Fermat Theorem 230
lattices, n-dimensional 251, 252
lattices, direction of 253
lattices, primitive cells in 262, 263
lattices, reciprocal 252
lepton 95
Levi-Civita connection 121, 126, 137, 138
Lie bracket 41
Lipschitz group xi, 5, 6, 7, 132, 155, 156, 160, 163, 272
Lipschitz group, characterization of 157, 158
LISP viii
Lorentz group 7
Lorentz invariants 301
Lorentz space-time 15, 16, 135
Lorentz transformations x, 20, 34, 39, 46, 69, 77, 81

M.
MACSYMA vii, viii, 226
magnetic dipole moment 40
magnetic moment 39
Majorana conjugation 15, 17, 18
Majorana spinors 13, 16, 17
MAPLE vii, viii, ix, xi, 33, 39, 57, 69, 70, 76, 121, 131, 145, 146, 167, 177, 184, 226, 251, 295
MATHEMATICA vii, viii, ix, xi, xii, xiii, 33, 57, 58, 59, 61, 67, 189, 190, 194, 226, 251, 252, 257, 261, 266
mathematical arguments 4
mathematical proofs 3
MATLAB vii, viii, xi, 57, 63, 66, 67, 251, 295
Maxwell, James Clerk vii, x, xiv
Maxwell equations 44, 45, 58, 65
Miller indices 253, 254, 255, 259
Miller indices, generalized 255
minimal polynomial 189, 192, 193, 194
Minkowski space-time 4, 16, 21, 37, 39, 63, 65, 84, 99
modified Euclidean algorithm 111, 116
Möbius group 157
Möbius group, coverings of 164

Möbius transformations xi, 20, 155, 156, 163
muMATH viii
muSIMP viii

N.
nabla operator 57, 58, 63, 65
null coframe 124
Nuttal notation 116

O.
octahedral group 83, 87, 98
octonion product xiv, 303
octonion multiplication tables 202, 203
octonion X-products xii, 201, 207, 208, 210
octonions xii, 201, 229
octonions, pure 202
operators, raising and lowering 101
orientation group 87, 91
orientation matrices 98
orientation numbers 83
orientation quantization 83
orientation symmetry 83
oriented volume 289
orthogonal groups xi, 6
orthogonal groups, two-fold coverings of 6
orthonormal coframe 47
outer exponential of a bivector xiv, 303
outer product xiv, 298

P.
Padé approximants xi, 111, 112, 113, 117
Padé condition 118
Padé table 115, 118
parallelizable manifolds 201
paravectors 69
paravectors, complexified xi, 69
PASCAL viii, xiv, 308
Pauli algebra 70
Pauli algebra, generalized 230
Pauli algebra, Hermitian conjugation in 73
Pauli algebra, paravector basis for 74
Pauli matrices xi, 83, 85, 88, 99, 148
Pauli spin matrices 13, 70, 74, 85
Pauli spinor fields 145, 151
period-doubling 99
perspective transformations 248

Pfaff derivative 41
phason space 256
Planck scale 85
Plücker condition 254
Plücker coordinates 254, 266
Pontrijagin character 51
Potts models 230
Poynting vector 301
projective geometry xii, xiii, 233
projective geometry, duality in 238
projective geometry, flats in 235, 236, 245
projective geometry, homogeneous coordinates in 244
projective geometry, join in 236, 237, 243
projective geometry, meet in 238, 240, 243
projective geometry, oriented 233, 244, 245
projective geometry, signature of a point 244
projective geometry, Separation Theorem in 247
projective spaces 233
projective transformations xiii, 234, 248, 249
pseudodeterminant 21, 156, 161, 164

Q.
quadratic algebras 270
quadratic forms 22, 23
quantum chromodynamics 93
quantum computer 108
quantum fields 84
quantum geometrodynamics 83
quantum gravitational effects 85
quantum mechanics 101, 145, 153
quantum mechanics, finite dimensional 229, 230
quantum numbers of orientation 95, 97
quantum operator $\mathbf{J}_3$ 107
quantum operator $\mathbf{L}_3$ 107
quantum operators $\mathbf{J}_+, \mathbf{J}_-$ 106, 107
quantum states 83, 230
quarks 83, 85, 95
quasi-number systems 229
quasicrystals xii, xiii, 251, 252, 256, 259, 265
quasilattice xiii, 256
quasiperiodic structures 257
quaternion analysis 213
quaternion conjugation 13

quaternion group 272
quaternionic charge conjugation 15
quaternionic index triples 204, 208
quaternions xi, 5, 13, 37, 67, 213, 216, 229, 251, 290
quaternions, Clifford algebra representation of 218
quaternions, matrix representation of 218

R.
rapidity 77
REDUCE vii, viii, x, xi, 33, 34, 36, 39, 43, 47, 51
Ricci 1-vector 44, 48
Ricci rotation 63
Rich, Albert viii
Riemannian curvature 43
rotations 79
rotations in Euclidean plane 36

S.
scalar fields 42
Schwarzschild metric 43
Schwarzschild solution 47, 48
Schönfließ period-2 rotation 89
Schönfließ rotation 99
Schönfließ symbols 83, 85, 86, 96
SCRATCHPAD viii, ix
second quantization 85
septagon 203
shifting operators 101
SMP viii
spacetime algebra 39, 44
special orthogonal groups 6
special orthogonal groups, two-fold coverings of 6
special relativity 213
spin groups 3
spin group **Pin** 5, 156, 164, 272
spin group **Pin**$_+$ 7, 8, 164
spin group **Pin**$_-$ 7
spin group **Spin** 5, 7, 9, 37, 50, 156, 164, 272
spin group **Spin**$_+$ 5, 6, 7, 8, 12, 21, 46, 132
spin group **Spin**$_-$ 7
spin matrix 86
spin representations 83, 89

spin transformations 77
spinning particle 39
spinor adjoint 132, 137, 138
spinor calculus xi, 59, 121
spinor conjugation 131, 135
spinor covariant derivative xi, 121, 137, 141
spinor fields 145
spinor fields, rotation of 148, 149
spinor inner products x, 10, 11, 12, 121, 132, 137, 141
spinor metric, constant matrix of 138
spinor modules 126
spinor operator 19, 37, 86
spinor, charge conjugated 13
spinors xi, 3, 8, 11, 37, 45, 59, 83, 101, 145
spinors, automorphism groups of scalar products of 10, 11, 13
spinors, complex 135
spinors, Dirac adjoint of 134
spinors, dual 127, 129, 131
stationary states 106
Stoutemyer, David viii
strong interactions 83, 96, 98
structure equation 192
superluminary tachyon 15

T.
tangent space 40
tetrahedral group 87, 96
tetrahedral period- 3 operators 91
torsion 38, 48
translations 22
transversions 20
triality 9, 304

U.
uniformly accelerating orthonormal coframe 141
unipodal algebra 189
unipodal equations
solving of 191
unipodal numbers xii, 189

V.
Vahlen matrices xii, 21, 155, 158
Vandermonde determinant 193

vector fields 145
versors 8
Viskovatov algorithm 111, 113, 115

W.
wave equation 145, 153
Weyl group 91
Weyl spinor 37
Witt basis 35, 37
Witt ring 27
Wolfram, Stephen ix

Y.
Yang-Mills potential 49